THE ULTIMATE
Church Sound
Operator's Handbook

2nd Edition

Bill Gibson

Hal Leonard Books
An Imprint of Hal Leonard Corporation

Second edition published in 2012 by
Hal Leonard Books
An Imprint of Hal Leonard Corporation
7777 West Bluemound Road
Milwaukee, WI 53213

Trade Book Division Editorial Offices
33 Plymouth St., Montclair, NJ 07042

First edition published in 2007 by Hal Leonard Books

Book design, illustrations, DVD audio and video production by Bill Gibson
Cover design by Steve Ramirez
Front cover photo by Samuel Vert

Printed in the United States of America

Library of Congress Cataloging-in-Publication Data is available upon request.

ISBN 978-1-61780-557-8

www.halleonardbooks.com

DEDICATION

- To God for saving a soul such as mine.
- To my wife, Lynn. You're my best friend and the love of my life.
- To my daughter and son-in-law, Kristi and Paul. I'm so proud of you both.
- To Noah and Emma Zimmerman—you are most wonderful in every way!
- To Bob Marion for introducing me to God.
- To Steve Schell for his excellent teaching about God.
- To Jon McIntosh for showing me how to be a regular guy and still love God.
- To Robbie, Mike, Jim, Jamie, and Jon—better friends cannot be found.

ACKNOWLEDGMENTS

To all the folks who have helped support the development and integrity of these books. Thank you for your continued support and interest in providing great tools for us all to use.

Cover photo: taken by Samuel Vert at Christian Musician Summit, www.christianmusiciansummit.com

- Acoustic Sciences, Inc.
- Antares
- Apple Computer
- Avid
- Ben O'Brien: Production Assistant
- Bruce Adolph and Matt Kees at Christian Musician Summit
- *Christian Musician* magazine
- Jamie Dieveney: Vocals, songwriting, friendship
- Doug Gould: Continued support and encouragement
- Faith Ecklund: Vocals, songwriting, inspiration
- Gibson Guitars
- Glimpse: Josh and Jason, you rock!
- Steve Hill: Drums
- Mike Kay: Ted Brown Music,
- Mackie
- Mark of the Unicorn (MOTU)
- Monster Cable
- John Morton: Guitar and stories
- Native Instruments
- PreSonus
- Primacoustic Studio Acoustics
- Radial Engineering
- Robbie Ott
- Roger Wood
- Sabian Cymbals
- Shure
- Spectrasonics
- T.C. Electronic
- Taye Drums
- Universal Audio
- *Worship Musician* magazine
- Yamaha

CONTENTS

Chapter 3: The Spiritual Importance of Worship and Music   25

Chapter 4: Relational Considerations   33

Chapter 5: To Get Paid or Not to Get Paid

Chapter 8: Sound Theory 83

Chapter 9: Interconnect Basics 99

Chapter 10: The Front-of-House Mixer 125

Chapter 11: Signal Processors   165

Chapter 12: Microphone Principles and Design   187

Chapter 13: Wireless Systems — 215

Chapter 14: Loudspeakers — 233

Chapter 15: The Racks   249

Chapter 16: Basic Equipment Needs 277

Chapter 17: Monitor Systems 301

Chapter 18: System Configuration and Layout 319

Chapter 19: Acoustic Consideration 339

Chapter 20: Miking the Group

Chapter 21: Sound Check

Chapter 22: Creating an Excellent Mix

PREFACE

In any situation, there is likely to be good and bad—as students of life it's our responsibility to learn from both. Sometimes the greater value is found in lessons learned from painful mistakes than in lessons learned from glorious successes. Therefore, rejoice in your trials and count them as blessings stored up in history and instantly available for future wisdom.

When I was developing the material for this book, I indeed found that there was nearly equal value in the things I had learned to do well and the things I had done not so well. As a musician and technician, I have always defaulted to what I thought sounded and felt good. In some arenas this is cause for great joy—in other arenas this is cause for strife.

This is a handbook that focuses on functionality and practical application. It is not a book about the mathematics of sound engineering or the fine details of permanent system design and installation. There is some technical information, but it is always presented in a way that leads to a better understanding of how to be a better sound operator.

The Value in History

In 1991, I became the music director at a large church close to Seattle, Washington. Coming from the secular recording and performing world, I was probably in over my head, but I loved God and knew an awful lot about live and recorded sound. I spent several years submitted to my pastor and trying my hardest to find the happy compromise that served the majority of a multigenerational congregation. I have experienced many of the frustrations of trying to please a pastor, a congregation, and myself—this is frequently the most challenging test for the church sound operator and the music team.

Around 1998, I became a licensed pastor in our denomination (Foursquare) and continued to learn the complex role of the musician/technician in church. I've experienced my share of sticking up for my pastor, my team, and my family at church, and for myself. The truth is that no matter where you go, dealing with humans is a touch-and-go proposition. It's in our nature to sin—to lie (or at least skew the truth in our favor), covet, cheat, gossip, steal, and on and on. It is only with the grace of God and a lot of patience and prayer that we are able to function together powerfully in a ministry. But it can be done—never doubt that fact.

In 2001, my family and I helped start a church plant from the church where I'd been on staff for 12 years. The new church was very exciting—we were teamed up with the high school pastor and had a young, energetic congregation that grew very quickly. Our worship team was full of excellent singers and instrumentalists. It was, in many ways, a worship dream team.

We were also very solid relationally, shared in each other's lives, had a common vision and goal, and, surprise of surprises, we had no ego problems. This team grew very close and couldn't have been more unified and spirit-filled. In fact, we were getting asked to lead worship at some of the denominational events. I say all that not to boast or brag, but to emphasize the fact that after two years of worship bliss, everything blew apart, many people left the church, and most everyone involved was emotionally and spiritually crushed—there's a lesson well-learned. I will not point fingers as to why things fell apart, but they did.

The blessing that comes from that painful event could possibly be in the advice I can bring to you about how to work together as a team with the other sound operators, the musicians, and the pastoral staff.

The church I'm at now is a wonderful blend of folks. It is relatively young and there is a unified vision and mindset regarding music style, approach, and attitude. Everything seems to be going very well, and I continue to rely on the Lord and experiences of the past to help ensure that it stays that way.

This is the second edition of *The Ultimate Church Sound Operator's Handbook*, so several things have unfolded in our church life. My wife and I are still at the same church and on the worship team. Our daughter has become the music pastor and is doing a fantastic job shepherding a group of musicians ranging from teenage to old. The church is thriving, and our pastor is doing an excellent job teaching us about Jesus and how to serve our community in a loving, godly way. I could go on and on, but it all comes down to the fact that God is good and that he makes great things come out of what we might see as bad. If you've had a relationship with him for long, You may have noticed that frequently the struggles you endure in one season end up being just what was necessary to prepare you for the blessings you enjoy in the next.

The Road Is Narrow

The path of a sound operator is complex and perilous—the musical camp often conflicts with the pastoral camp. The staff is rarely in agreement about the direction of worship—heck, for that matter, the musicians often disagree about the direction of the worship team. The congregation is typically multigenerational, with each generation having a well-entrenched opinion of how things ought to be done—for some it will never be loud enough, and for others it will never be quiet enough. Some folks just prefer hymns sung in the traditional way with traditional instrumentation; other folks can't relate to anything without a few electric guitars and a drum set providing the foundation for aggressive vocals, a B3, and some pretty impressive visual imagery.

Sounds like a lot of fun, doesn't it? Not really, but once the team gels and you find the perfect balance that serves your congregation while they seek the heart of God in worship and the mind of God in the Word, there is no better job. The sound operator controls the fundamental communication vehicle that delivers a potentially life- and destiny-changing message to a lost soul—it's really a very serious responsibility, and it should never be taken lightly.

In any given church service, the sound operator needs to take joy in doing a great job, but rarely will there be accolades. The best sound operator is transparent to the service—the sound system should also be transparent. When the congregation has been at the receiving end of a perfect service, they have seen, heard, and experienced the heart and mind of God.

I was very impressed by an interview I read with Steve Gadd, one of the most talented and sought-after drummers of all time. He succinctly made the point that if someone listened to a song he recorded and came away talking about how great the drum track was, he had failed at his job. He understands the value of the team—the greatness of a combined effort that is much more powerful than any one individual. That's the way each member of a powerful worship team should function.

In this book, you'll learn ways to create great sounds, but you'll also learn ways to develop and function as a strong and valued team member. Strife, controversy, and disunity deal a crushing blow to ministry. Harmony, unity, and like-mindedness result in a spirit-filled and supernaturally powerful church experience, where the sum of the individual efforts is magnified into an exciting and unbelievably amazing and blessed event.

Terminology

It seems that each church has its own indigenous terminology. What I refer to as a pastor, you might refer to as a *priest, teacher, preacher, rabbi,* or *most high and revered one*. Some folks prefer to say *music team* instead of *worship team*, or *orchestra* instead of *band*, or *council* instead of *elders*. I'll endeavor to be sensitive to all denominations throughout this text, and I hope you'll get my point regardless of my terminology. Please understand that, although I approach sound and music from a contemporary Christian perspective, I am in no way attempting to proselytize or sway anyone's thinking in anyway. The principals in this book are foundational and their helpfulness is not dependent on denomination or religion.

The Accompanying DVD-ROM and Online Media

All media examples can be found on the accompanying DVD-ROM and online at www .ucsoh.com. Stream the examples online for super-easy access or, if you prefer, import all of the full-resolution media from the DVD-ROM directly into iTunes or any other full-featured media player for easy playback. This is especially convenient when you're on the go with your laptop, tablet device, or smartphone and unable to connect to the Internet.

All Audio and Video Examples are produced specifically to support the concepts and principles presented in the printed text, with the examples clearly marked in each chapter. The Audio Examples demonstrate many of the concepts that are explained in the text and accompanying illustrations. The Video Examples show you specifically how to build your skills as a sound operator. They are very powerful and they're produced with your education in mind. You won't find a lot of rapid-motion, highly stylized shots; you will find easy-to-understand instructional video that is edited for optimal instruction and learning.

Audio Examples are indicated by a gray bar.

Video Examples are indicated by a black bar.

QR Codes

Throughout this book, you'll see an occasional QR code. These codes are actually digital links to supporting websites; they're designed to be read by a QR code reader on a smartphone, such as the iPhone or an Android or on a tablet, such as the iPad or Xoom. Simply download a free QR code reader app, show the code to the smart device's camera, and the supporting link will automatically show up in its browser.

There is also a list of reference books and URLs at the end of this book, but these QR codes are a convenient way to quickly delve deeper into the supporting material used to help create this work. Once you have your QR code reader, let your smart device scan this QR code to be taken directly to www.ucsoh.com, the support site for the *Ultimate Church Sound Operator's Handbook*!

CHARACTERISTICS OF A SOUND OPERATOR

Most churches require just a couple of prerequisites for the sound operator position: candidates must be breathing and willing. The truth is, in many small churches it's not all that easy to find a sound operator, even with those standards. However, it is worth the time and effort to be continually on the lookout for a willing servant who is destined to excel in the sound ministry.

In most denominations, as well as in non-denominational independent churches, good music, sound, and visual presentation are fundamental to growth. It is, of course, acknowledged that God could anoint a donkey with a stick to grow a mega-church if He so desired, but that type of sensationalism seems to be rare. In reality, we are given tools to communicate a message—our modern tools include music, instruments, voices, sound systems, and video presentations, along with anointed preaching and teaching about God's principles, guidelines, and commands.

Avoid placing too much importance on the equipment and tools at your church. Developing an excellent sound system is usually a long-term endeavor. Do the very best you can with what you have at hand, and do it cheerfully. An excellent guitar player will make a marginal guitar sound good, and the same theory applies to a sound system. As you develop your technique and refine your ear for a good mix, you'll be able to get the best possible sound from your system. Then, like the guitarist, once you are using a fine instrument (sound system), you'll quickly achieve the next level of brilliance.

There's not usually enough funding available for such things as high-quality sound systems until it becomes apparent that there is a team capable of using such a sound system effectively. Even then, the church hierarchy needs to realize the power of a technically well-communicated service.

If you are currently a church sound operator or if you're a pastor or staff member trying to understand the makings of an excellent candidate, it is very helpful to understand the essence of a great sound person—to understand what makes a person perfect for this important task.

With the following points covered, even the novice can turn into a great sound operator—I've seen it happen again and again.

List of Qualifications and Characteristics

Notice that this list of qualifications and characteristics does not contain a subheading for "Is Already an Excellent Sound Operator." In most churches, the sound operator is a volunteer. Volunteers need to be supported in their ministry calling and given a chance to grow their skills. I find it very rewarding when anyone in a volunteer ministry tries hard, takes direction well, and succeeds.

I agree with the argument stating that since the pastor won't invite someone to preach if they don't know how to preach, why would you invite a musician who can't play or a mixer who can't mix to participate in these important tasks? However, there isn't always a willing volunteer with a sufficiently developed skill level ready to sing, play, or mix sound. There is always room for someone who has a fundamental gift and desire accompanied by a willingness to learn. It is up to the church leadership to provide the necessary resources for training and the patience to endure the learning process.

As the church gets larger, it becomes more important that the sound operator is very proficient and consistent. This need typically provides justification for a staff sound operator position with an actual salary.

If you can find—or become—a person who matches the following qualifications, there's a good chance that a brilliant sound operator will emerge.

Listens to Music

The fact that this is first on the list of qualifications is not random. If the sound operator doesn't listen to a lot of great-sounding music, he or she won't have an accurate frame of reference upon which to base decisions regarding the creation of a great mix.

If you want to be able to provide excellent mixes in church, listen to music that is like what your team is playing. When the team is copying an arrangement from a recording, listen to the original recording and try to emulate the recorded sound in your live setting. It is very common for the studio or live sound engineer to play his favorite recordings over the

Qualifications for Being a Sound Operator

- Listens to Music
- Loves Music
- Is a Proven Advocate of the Church, Pastor, Council, Organizational Structure, and Leadership
- Agrees with Church Theology
- Is a Church Member, Committed to the Church Mission and Vision
- Understands the Style and Spirit of Worship Music Desired by Church Leadership
- Is Servant-Hearted
- Has a Positive Outlook
- Does Not Have an Ego Problem
- Has a Consistent Spiritual Life

- Is Ministry-Minded
- Is a Lifelong Learner
- Is a Hard Worker
- Has a Long Attention Span
- Has a Stable Family Life
- Pursues Excellence
- Is Technically Gifted
- Has Good People Skills
- Has a Mature Perspective
- Likes to Help Others Succeed
- Does Not Participate in Gossip
- Is Determined to Become an Excellent Sound Operator

sound system just to hear what a trusted mix sounds like in that setting. It's not cheating or copping out—it's smart.

Loves Music

A great sound operator loves music. Sound operators not only listen to great music, but they really love listening to great music. They often have a very good music playback system at home, and they enjoy music on an emotional level—they go deeper into the experience than simply acknowledging the existence of music.

Sometimes, you'll find a highly trained and skilled musician who loves the technical aspect of sound so much that he prefers to be the sound operator over being a musician. Many of your music team members would be very good sound operators; however, most churches are unwilling to give up the musician to the sound team even though the church might benefit more from the musician running sound.

Is a Proven Advocate of the Church, Pastor, Council, Organizational Structure, and Leadership

It is important that the sound team and music team are filled with folks who have proven themselves as loyal to the church, the pastoral staff, and lay leadership. This is an important point even though it is often overlooked in the interest of simply engaging participants.

It is difficult in a small church, where resources are limited, to set too many prerequisites for participation—that's understandable and often necessary. However, it is very important that your team members are solid church members who appreciate what's going on and those who are ultimately responsible.

Agrees with Church Theology

The candidate should have demonstrated a love for God and a fundamental understanding of the weekly messages and sermons.

I have personally been through tough decisions regarding most of these qualifications regarding sound and music team members. This one is very important while being easy to overlook.

When a new member shows up who has experience and obvious talent at sound ministry, the already-frustrated and downtrodden pastor will be tempted to jump too quickly to ask for help. The immediate help is nice, but if the sound operator disagrees with fundamental theology being taught, his or her days are numbered. At best, the operator will eventually fade into the sunset to find a church that he or she agrees with. At worst, he or she will have started an eruption of angry and disgruntled church members.

John 17:20–23

My prayer is not for them alone. I pray also for those who will believe in me through their message, that all of them may be one, Father, just as you are in me and I am in you. May they also be in us so that the world may believe that you have sent me. I have given them the glory that you gave me, that they may be one as we are one— I in them and you in me—so that they may be brought to complete unity. Then the world will know that you sent me and have loved them even as you have loved me.

Is a Church Member, Committed to the Church Mission and Vision

This is a must. If the candidate doesn't know, understand, and buy into the mission for the church and the vision for its future, he or she should be bypassed. It is the church leadership's responsibility to clearly state these important points, and each candidate for ministry should read, understand, and agree to them. If there are questions, they should be asked and answered in a straightforward and concise manner. As an expression of submission to church leadership and agreement with the expressed vision and theology, the sound operator should become a church member.

Understands the Style and Spirit of Worship Music Desired by Church Leadership

There are many different styles of church music. Each style brings with it an inherent attitude. Some churches live on the edge—their music is loud, edgy, and potentially offensive to all those who don't share the emotional mindset that created it. Other churches are very traditional, appreciating hymns and non-offensive choruses. The sound operator must understand what the music director and pastoral staff expect for their church.

Though this point seems obvious, the inability to communicate exactly what is expected of the sound operator leaves ample room for disagreement, conflict, and strife.

Is Servant-Hearted

In any ministry arena, the willingness to serve is revered. A good team member will do what it takes to get the job done—he or she is willing to serve the church body. The sound operator position is one of the most important positions when it comes to the creation of an effective church service, yet there are few accolades and many who require assistance through the process.

The sound operator candidate who is always willing to help set up and strike equipment, get a cable for the music team member, show up early to rehearsal, or just get water for the team should be highly considered. In any ministry, the humble servant will win the race. Humility brings honor. Pride brings shame.

In my experience, I've found that the most impressive candidates for ministry are the ones who are willing to just show up at rehearsal, even though they don't have to. They come to observe, learn, or quietly sit and pray for the team. A person who is humble and willing to serve the needs of the team will typically be a person who is secure and emotionally stable—he or she understands a need and appreciates an effort to fill it.

Throughout this book I will strive to focus on positive solutions rather than the avoidance of negatives, but in a few circumstances I will make exceptions. This is one of those times—avoid the needy prima donna.

Has a Positive Outlook

It is important that all team members maintain a positive attitude—that they believe in the power of God and His willingness to help. Negativity is infectious, and it must be addressed immediately. Some folks just tend to look at the glass as half empty. They tend to be fatalistic, judgmental, over-critical, and defeatist in their outlook. Their poor attitude can infect the team—they need someone to help them work through their negativism while they begin to look on the brighter side of life.

Leaders should strive to build a team of people who see solutions that enable the achievement of dreams and goals. They should model the heart of an idealist who sees a goal as an opportunity for achievement rather than as a chance for failure. They should demonstrate the faith of a true believer, and it is their responsibility to help the team realize that with help from God and a little effort, virtually anything is possible.

2 Chronicles 20:15, 17

...the LORD says to you: "Do not be afraid or discouraged because of this vast army. For the battle is not yours, but God's.
"You will not have to fight this battle. Take up your positions; stand firm and see the deliverance the LORD will give you, Judah and Jerusalem. Do not be afraid; do not be discouraged. Go out to face them tomorrow, and the LORD will be with you."

Does Not Have an Ego Problem

This is one of the most important concerns in any team. Large egos are easy to spot—they shine brightly. If you find a candidate for the sound team who seems overly confident, always be sure that this candidate begins his or her participation in the team with a substantial probationary period.

Any person who talks himself or herself up like he or she is the best thing since sliced bread usually isn't. I've found that the best of the best are pretty humble, yet confident. Avoid introducing the large-egoed member to the team—he or she will typically bring strife, conflict, and anger along.

As with all dealings that include people, there are exceptions to all stereotypes. Each person deserves a chance to prove that he or she can be a valuable team member— insecurities are often manifested as excess confidence. A probationary period provides a great way to get to know someone while that person gets some space to show who he or she really is.

Psalm 51:17

My sacrifice, O God, is a broken spirit; a broken and contrite heart you, God, will not despise.

Romans 12:3–5

For by the grace given me I say to every one of you: Do not think of yourself more highly than you ought, but rather think of yourself with sober judgment, in accordance with the faith God has distributed to each of you. For just as each of us has one body with many members, and these members do not all have the same function, so in Christ we, though many, form one body, and each member belongs to all the others.

Has a Consistent Spiritual Life

A prerequisite for leadership is an active and passionate spiritual life. Any member of the music or technical team is a participating leader in church life. It's important that the sound team members are in stride with the spiritual growth and life of the church. Each service takes on its own life—a church service is not a lifeless and boring event.

The ebb and flow of a spirit-led church service is important. The music and technical teams should understand and sense the natural changes and tendencies of each service. Some moments call for gentle treatment, and other moments cry out to build to a crescendo.

The sound operator is crucial to the flow of each service. When the pastor steps up to speak, his mic needs to be on and turned up to the perfect level. When the worship leader pauses to share an inspiring story, his or her mic level needs to be up to speech level instead of the level it was at the end of the previous song. A good sense of the spiritual aspect of each service will enable the sound operator to be ready for what's going to happen before it happens.

Church leaders should help the technical and musical teams to be actively involved in the spiritual aspects of the church. Accountability is important—we all need it in one way or another. Modeling an active spiritual life is important—inviting your team members to participate is essential.

2 Chronicles 7:1–3
When Solomon finished praying, fire came down from heaven and consumed the burnt offering and the sacrifices, and the glory of the LORD filled the temple. 2 The priests could not enter the temple of the LORD because the glory of the LORD filled it. 3 When all the Israelites saw the fire coming down and the glory of the LORD above the temple, they knelt on the pavement with their faces to the ground, and they worshiped and gave thanks to the LORD, saying, "He is good; his love endures forever."

Deuteronomy 4:29
But if from there you seek the LORD your God, you will find him if you seek him with all your heart and with all your soul.

Is Ministry-Minded

It seems obvious from a distance, but some music teams lose sight of the ministry aspects that are inherent in church music. Though there are valid opportunities for musical performances in church that highlight the amazingly creative-spirited Creator of All, most church music has more personal goals for the congregation: to prepare them to receive the blessings of God, to glorify and magnify God, and to provide an environment that is tender and yet powerful. In an effective worship service, it is likely that individuals will receive physical and emotional healing or illuminations about their lives and the purpose God has for them. Damaged relationships are likely to be reconciled, and divinely inspired ideas are likely to pop into congregants' heads.

A sound operator who has a ministry-minded outlook will strive to provide an atmosphere where God can do what He does without distraction. In this media-savvy era in which everything is compared to high-budget film, television, and music videos, technical

Distractions during the Church Service

Distractions come in many forms:

- Mics being off when they should be on
- Mics being on when they should be off
- Unintelligible vocal sounds
- Leader mic being too soft during key ministry times
- Any instrument being too loud
- Feedback
- The mix set too loud
- The mix set too soft
- Leader too soft
- Leader too loud

- Backing vocals out of tune or not phrased properly
- Sloppy acoustic guitar performance
- Poor drumming
- Drums too loud
- Bad electric guitar sounds
- Too many vocal effects
- Overdriving the signal path
- Blown speakers
- Uneven sound system coverage
- Poor acoustics

mistakes are very distracting. However, in church, technical flaws distract the congregation's attention from ministry while, at the same time, distracting the pastor or song leader from his or her divine task at hand.

The list goes on and on. To top it off, many of the good musical and technical aspects of a service can be distracting to some, while they're a blessing to others. If the alto is singing a lot of licks to support the leader, some congregants will be blessed and inspired by the gifts of God; others might just wonder what that wacko girl is trying to prove. The sound operator must know the tendencies of the congregation, discovering their likes and dislikes and, over time, learning what the body needs in order to experience a wonderful church service.

Is a Lifelong Learner

Technical ministry endeavors are in a constant state of change. The advent of new and affordable technology provides access to a constant flow of new tools for the sound operator. A good sound operator is good at learning new things, whether they're about equipment, science, math, music, God, or virtually anything else.

If you have a sound operator who is intimidated by technology and prefers to live in the "old school," your church will have a difficult time moving ahead into the future.

Is a Hard Worker

Weekly services at an active church require a lot of extra work from the sound crew. Most churches share space between weekend services, youth group meetings, weekday services, adult ministry events, children's events, small group meetings, and so on. Setting up Sunday mornings often involves the sound operator starting from scratch to prepare the stage and sound system for service-ready functionality.

An experienced sound operator understands this aspect of church and happily works to serve the needs of the congregation and the music team. A smart sound operator communicates and supports the rest of the sound team in an effort to develop a system that makes setup and striking equipment consistent throughout the church technical life. At the same time, he or she will help develop a set of guidelines that make life easier for the entire technical team. We'll cover more about this later.

Has a Long Attention Span

Operating sound during a church service is a job that demands 100 percent of the operator's attention for 100 percent of the service time—this is definitely not a passive task. Running the sound system during any live performance requires constant attention to detail. An experienced and excellent group of singers and instrumentalists makes the sound operator's job much easier; however, there are enough variables in any live setting that the sound operator must devote his or her full attention at all times.

Any sound operator who has a hard time concentrating for long periods of time is destined to cause several distractions during the church service. Simply not paying attention to what is going on will result in mics being off when they need to be on (and vice versa), awkward pauses in media presentations, distracting noises when mics are removed from and replaced on their stand clips, and myriad other potentially annoying incidents.

Has a Stable Family Life

The greater the ministry requirement, the more we have to cultivate and develop a stable family life—it's scriptural, and it just makes a lot of sense. If an adult or young person has a chaotic family life filled with strife, anger, jealousy, and uncertainty, their ministry effectiveness will be substantially decreased. That person is likely to be erratic in church attendance, unreliable, and constantly under spiritual and emotional attack. He or she is likely to call in sick more often and will require much patience and gentle understanding.

Although we should all be patient and understanding with one another, the most efficient use of time and ministry resource lies helping the potential sound operator bring his or her family life into order before that person becomes an active team member.

It is also important that the sound operator's family be completely supportive and understanding of the time commitment involved in serving the church in this important capacity.

Pursues Excellence

People who are excellent at one thing tend to have the ability to be excellent at whatever they choose to pursue. If your sound operator hasn't achieved excellence at something in his or her life, he or she will require support, training, and much patience to achieve excellence as a sound operator.

The church setting provides a perfect opportunity for someone to develop his or her skills in a safe environment where his or her spiritual development is held above achievements. Joining the sound team could be the event that allows the development of excellence—a characteristic that could carry over into other aspects of that person's life. Training your sound crew is a very worthwhile investment.

Is Technically Gifted

As worship director, I have typically had four or five paid sound staff and a list of volunteers. I've noticed that some people walk in and immediately understand everything about the system, the equipment, and the responsibilities of a sound operator, while for others it is more difficult. The Lord gives us certain gifts that just make some things easier for some people—they get it!

If the sound operator doesn't seem to understand quickly, the technical gift might not be theirs; however, avoid rushing to a judgment. There is a lot involved in running sound for a complex church service, and people learn at different rates—they deserve a chance to serve if they have the passion and burning desire to help.

A sound operator who is responding to a need might not be technically gifted, but he or she is definitely blessing the pastor and the church by his or her willingness to sacrifice time. Continue to look for someone who is gifted and willing. Invest in his or her training and provide the tools necessary to do a good job. That person will thrive, and the church will be blessed.

Has Good People Skills

If there is one person who is capable of stirring up a hornet's nest, it's a cranky sound operator. Every area of ministry requires good people skills, but there seems to be a special sensitivity in the creative arts that demands an even higher level of grace, kindness, and understanding.

The sound operator has the ability to create tension and aggravation in the music team and the congregation—a small problem can easily blow out of proportion if handled poorly. On the other hand, a servant-hearted, ministry-minded sound operator can single-handedly calm the waters at any church. If he or she genuinely values people and gives credence to their concerns, and then communicates in a way that is sincere, authentic, and kind, life is much easier for everyone.

Psalm 19:14
May the words of my mouth and the meditation of my heart be pleasing in your sight, O LORD, my Rock and my Redeemer.

Has a Mature Perspective

Maturity is not all about age. Some aspects of maturity are cultivated by time, along with tempering from the successes and failures brought on by years of living. Other aspects of maturity are learned at a young age through excellent parenting and mature role models. Sound operators are closely involved with the worship team, the church staff, and the congregation; therefore, immaturity in their actions, reactions, and comments tends to be magnified.

I've had many teenagers on sound crews who were amazingly mature, hard-working, dependable, and respectful of authority—along with all of that good stuff, they were instant learners and technical gifted in every way. Some of them have gone on to great things in the technical world. The only problem I've found with including these bright young folks in the team is that they eventually tend to go off to college and develop active lives that take them out of the ministry.

An excellent candidate for sound operator is an intelligent person who has already gone to college and/or established his or her personal and family life. If the person is mature, servant-hearted, excited about the direction of the worship ministry, and in line with church's theological beliefs, that person can be a true blessing to the church. These people should also understand and appreciate the role they play as sound operators, secure in their role and not striving to increase or decrease its importance to the church.

A mature person understands that we are all going to make a mistake or say the wrong thing from time to time. Repentance and forgiveness are fundamental tools in the mature person's relational tool kit. Holding grudges and hypercritical outlooks are causes for potential strife and dismay.

Luke 17:3
So watch yourselves. "If your brother sins, rebuke him, and if he repents, forgive him. If he sins against you seven times in a day, and seven times comes back to you and says, 'I repent,' forgive him."

1 Corinthians 12:15–20 (NIV)
Now if the foot should say, "Because I am not a hand, I do not belong to the body," it would not for that reason cease to be part of the body. 16 And if the ear should say, "Because I am not an eye, I do not belong to the body," it would not for that reason cease to be part of the body. 17 If the whole body were an eye, where would the sense of hearing be? If the whole body were an ear, where would the sense of smell be? 18 But in fact God has placed the parts in the body, every one of them, just as he wanted them to be. 19 If they were all one part, where would the body be? 20 As it is, there are many parts, but one body.

Likes to Help Others Succeed

Ministry is much more about helping others succeed than it is about succeeding. A sound operator who is easily threatened and somewhat selfish isn't doing the best service to the church or the team. On the other hand, a sound operator who likes to help others learn about the system and what it takes to run it effectively is a true blessing to all.

There is plenty of opportunity in a growing church. If there is only one sound operator, and every church event depends on him or her, there will eventually be a time when burnout is inevitable, and then there won't be any sound operator.

The pastor, the staff, the music director, and the lead sound operator must model an environment that is conducive to learning. With any training scenario comes the opportunity for mistakes—they will happen. However, as long as the sound operator is trying and willing to continue to learn, the mistakes will decrease in frequency.

Romans 12:14–16
 Bless those who persecute you; bless and do not curse. 15 Rejoice with those who rejoice; mourn with those who mourn. 16 Live in harmony with one another. Do not be proud, but be willing to associate with people of low position. Do not be conceited.

Does Not Participate in Gossip

Gossip is unscriptural, unfair, destructive, immature, and unacceptable in ministry. When we have a problem with any person, we are instructed to confront that person about it—not other team members or everyone except that person.

Self-control and discipline are necessary to curb this natural tendency; however, if the church leadership leads the way in their intolerance of gossip, the congregation and all ministry participants will quickly catch on.

Tolerance of gossip can only lead to problems and strife.

Is Determined to Become an Excellent Sound Operator

It isn't necessary that the new sound operator is already proficient or excellent. However, he or she must be determined to become an excellent sound operator. It is ideal if the sound operator has previous experience and is naturally gifted, but if he or she is not supportive of the church vision, doesn't agree with the church theology, and is not dependable, hardworking, mature, and all the rest of the qualifications, that person will cause problems in any area of ministry.

Summary

Finding a candidate who meets all the qualifications on this list might be difficult, especially in a small church; however, these considerations should be understood as a point of reference. Simply providing your sound team with this list and discussing it openly could help bring the team into more cohesive unity. It might also help weed out some potential problems in the team.

chapter 2

DUTIES OF A SOUND OPERATOR

Believe it or nor, it can be detrimental to expect too little of the team members. Most people are more likely to enjoy giving more time to something that is high quality. If the expectations in any organization are so low that the quality of the ministry is mediocre or embarrassing, most people will end up fading away.

Sound candidates should be expected to invest their time and talents in the worship team. The following list of responsibilities might seem to be a bit much, but consistent participation at this level will help the entire team thrive. Most of these activities are simple and relatively brief, but some of them require daily commitment.

It doesn't take long to notice that the commitment of a dedicated sound operator involves much more than simply showing up and playing with buttons, knobs, and faders. There must be a desire and commitment to see progress in ministry. Without steady—although often slow—progress, ministry tasks tend to grow tedious, draining the team member of his or her desire to participate.

Pray for the Worship Team

Praying for the worship team is good in so many ways. Prayer works. Any time spent talking to God about your team is time well spent. Divine interaction in your team is what keeps it interesting. Any team that makes a place for God to participate regularly will grow and become fruitful. Any team that gets lazy and ignores or marginalizes regular input from the Creator will digress.

You might wonder how I can be so confident in my statements regarding this matter. It's really rather simple: I know God, and I've seen and participated in groups who have tried each of these approaches. Decide on a regular daily time to pray for your team. If you make it easy to remember and convenient, you'll succeed. If you complicate matters by setting unrealistic times and lengths of time that you'll pray, you might quickly burn out.

Commit to Spiritual Development and Growth

As you become more active in ministry, it becomes imperative that you pursue your own personal spiritual development and growth. The more effective you become in your ministry specialty, the more likely it is that you'll experience some sort of spiritual attack. Be prepared.

It's important that the sound operator have a chance to participate in the worship service as a congregant and that he or she has an opportunity to sit and really listen to the sermons without being preoccupied with technical thoughts. The focused sound operator is often so tuned into the sound system and its components that he or she is tuned out when it comes to what is being said, sung, or preached. Leaders should be certain that there are breaks for sound operators once or twice a month, so they can take the opportunity to just attend church with no other commitment than their own personal spiritual growth.

Sound operators should commit to:
- Church service attendance whenever they're not running sound
- Daily Bible study
- Participation in small group ministry
- Involvement in church missions and other activities
- Financial support of the church (tithing)
- Helping other sound candidates learn the sound system
- Helping other sound ministry members increase their technical and spiritual depth

Get to Know the Team Leaders

How are the team leaders ever going to discover what a magnificent person you are if they don't have a chance to get to know you? Team candidates should make time to communicate in some way with the leaders. Always be considerate of the fact that church leaders are often stretched beyond their comfort zone. Make your communications efficient and straightforward—do your best to avoid wasting the leaders' time or engaging them in rambling conversations.

A sure way to get church leaders to run the other way is to monopolize their time once you gain their attention; however, be sure they're aware that you're interested in getting to know them. Short conversations before or after church or rehearsal can lead to a lunch meeting. Before you know it, you'll have a friendly relationship.

Make yourself available to help in any way, and don't presume that you'll get anything in return. Sacrificial giving is scriptural—it will be noticed.

Many leaders are much more likely to return an e-mail than they are to return a phone call. Express your positive comments and constructive remarks via e-mail if you'd like, but keep them brief. Make it easy to respond. If your e-mail is long and rambling, it might never be read.

These comments might sound like I'm suggesting you should baby church leaders—in reality, I'm just being realistic, having been actively involved in church leadership myself. Use whatever efficient avenue you find to begin a relationship: in person, by mail, by e-mail, over the telephone, at lunch, at dinner, and so on.

Find Out What the Music Director Expects

Ask those in the leadership structure (pastor, music director, or sound team leader) what they expect of a good sound operator. Be direct and serious about this. Have a pen and paper available to take notes. If the leaders feel that a candidate is intent on providing the type of service they desire, they will be duly impressed—your forthright question might actually help them clarify in their own mind what they expect from the sound operators.

Be precise in your questioning and probe for specific answers. You might ask:

- Do you have a specific volume threshold in place?
- Has volume control been an issue between the worship team and certain members of the congregation?
- Are the mics and cables stored after every service?
- Is the platform rearranged during the weekday services?
- Have you been happy with your current sound team?
- Who is immediately in charge of the sound team?
- What is the time commitment involved in participating as a sound operator?
- Is there one worship sound that you prefer, such as Integrity, Vineyard, Passion, Maranatha, and so on?
- Is there a particular musical style that describes the musical tastes of the congregation?

The answers to these questions and others that occur to you will help provide a picture of how you can best serve the needs of your church.

Attend Rehearsals

Many worship directors don't require sound operators to attend rehearsals, especially if they're held on weeknights. As a new sound operator, make it a point to attend rehearsals anyway. If the music team is seriously striving for excellence, the sound operator is as much a part of the team as the instrumentalists and vocalists.

An aspiring sound team member should attend worship team rehearsals and quietly observe and pray. This speaks volumes to the leaders about the candidate's desire to serve and participate. Attendance at rehearsal will also reveal to the sound candidate the amount of need that the church has for another sound team member. This is an excellent time to learn the music and to meet the team leaders and members.

As a church leader, I have always been most impressed by someone who is willing to quietly and unassumingly attend several rehearsals, praying and observing.

Attend Small Group Meetings

Any potential team member should become actively involved in the small-group life at their church. Small groups that meet in homes or classrooms at church provide a means to more intimate and supportive relationships. Once a church reaches more than about a hundred members, the weekend service is a very difficult place to build relationships and to make good friends.

Some large churches offer interest-specific small groups. If there is a worship-team small group, join it if at all possible. It is important to keep in mind that the music and sound teams provide a very large portion of the weekend church experience. Sound operators participate in the entire service and have the capacity to make or break the service flow and mood. The worship team needs to be spiritually and emotionally bonded, as well as musically bonded.

Weekend services are often a blessing, and they're also a perfect opportunity for the enemy to bring strife, anger, and confusion. A wise sound operator will get involved in the social and spiritual aspects of the team. This provides an excellent opportunity for communication about potential problems while building a network of people that are likely to end up as lifelong friends.

Listen to Service Recordings

Most churches have the capacity to record the service mix straight off the house mix. This typically sounds pretty bad, but it does reveal several things about the mix. It gives a fairly accurate picture of the vocal balance in the house. It reveals the blend of electronic instruments fairly well; however, without a little help, it does a poor job of representing any acoustic instruments, such as acoustic piano and drums. These instruments are often loud enough without any sound reinforcement, so they are quiet on the recording. They also might be very distant-sounding because they are just being picked up as reflections of the church walls into other mics on the platform.

Listen to the service recordings to learn the music and to discover a way to help provide a better rendition of each service through the use of various other techniques. In upcoming chapters, we'll cover a few different ways to help increase the quality of the service recordings of the music and the sermon.

Listen to Current Worship Music

Find out the music team's favorite worship musicians. Each church develops its own musical preference, but it's fairly likely that they'll be drawn to one particular style of worship and that, within that style, they'll settle on a favorite worship leader or team.

The sound operator should spend a lot of time listening to the original recordings of many of the songs in the team's repertoire. There is no way to learn what the ideal mix is like without listening to the original recordings that inspired the worship team.

Focus on the team favorites; however, listen to several styles of worship music. They all have value. Find styles that are similar in instrumentation and energy level. Most modern worship teams like the music from:

- Worship Together (www.worshiptogether.com)
- Integrity Hosanna (www.integritymusic.com/)
- Hillsong (www.hillsong.com/music)
- The Lakewood Church (www.lakewood.cc)
- Tommy Walker (www.tommywalker.net)

There is a constant stream of new worship artists and styles. The sound operator must be familiar with the sonic qualities of several styles in order to have a solid sonic impression on which to base his or her mixing choices. Listen, listen, listen!

Be Punctual

This should go without saying, but always be punctual, which really means early. The entire team waits around if the sound operator is late or just on time. In the sound tech world, ten minutes early is ten minutes late—leave plenty of time to set up and troubleshoot any system problems.

It is an act of love to the team when the sound operator shows up early to set up and stays later to strike the platform. On one hand, it might not seem fair that the sound crew is putting more time in at rehearsals or services, but it helps to keep in mind that many of the

Make Notes on the Music

Use the same music that the instrumentalists use—make notes about mix considerations. Be as specific as is practical. A great set of notes will help the sound operator catch all of the mix changes, avoiding missed vocal parts and instrumental solos.

Make a note of all possible changes, including the likely order of the song—use simple abbreviations for the intro (I), verse (V), chorus (C), bridge (B), and solo (S). A typical song might follow this standard order:

I – V – C – V – C – C – S – C – C – end.

music team members spend a respectable amount of time practicing their instruments or voices during the week.

Also, the answers to long and intense setups are organization and excellent procedures. If the sound team is well-structured and organized, setup and striking can require minimal amounts of extra time—but more on this later.

Learn the Songs

The sound operator should take notes regarding each song. It is very simple to use the same chart that the music team uses. Mark each verse, chorus, bridge, or solo with reminders about focal point changes. Note any specific changes that require radical level or tone changes.

It is common to spread the melody between two or more singers, in which case the sound operator must mix the vocals perfectly so there is a constant focal point for the congregation. Instrumental solos are typically louder than the supportive comp, so the sound operator must adjust accordingly. Some songs include specific effects, such as delays or reverbs, in specific places. The sound operator should learn what the effect is, build it within the effects device, and note where it should be applied.

Cable Labeling Systems

Often, cables are labeled in the heat of the moment. A piece of duct tape or masking tape might be wrapped around one end of the cable, near the connector, and marked with a Sharpie or an ink pen. THIS IS A MISTAKE! It should only be used in case of an emergency repair and should be replaced with a permanent label immediately. Duct tape, masking tape, and Scotch tape are the worst methods of labeling anything for the long term. They lose their adhesive and fall off the cable in the course of several months, and they typically leave a sticky, messy residue behind that will require as much time to clean off as it would have taken to apply a permanent label to start with.

There are numerous commercially produced labeling systems—some are better than others. Any system that relies on adhesive-backed tape should be suspect. However, there are a few manufacturers who provide laser-printable labels in which the clear adhesive material extends past the printed label, continuing to wrap around and over the top of the printed surface. These types of labels tend to utilize an adhesive that has minimal residue, is durable, and is long-lasting.

Most adhesive labels can be easily made permanent through the use of clear heat-shrink tubing. This system is very effective, especially when constructing custom cables. Before the connector is applied, apply a neatly printed adhesive label, then cut a piece of heat-shrink tubing that is about an inch longer than the label. Slide it over the cable so that it lays centered over the label. Now use a heat gun to shrink the tubing snugly over the label—this system will last forever, and it looks very professional.

Always be as specific as possible when choosing how to indicate where the cable belongs. Imagine that you've never seen the system before and that you find this random disconnected cable and are wondering where it should be connected. Terms like MIC 12, LINE 12, MAINS L, MAINS R, AUX 1, FX 1, and CD IN are easy to decipher anytime—keep it simple.

Snake Configurations

Snakes are manufactured in many different configurations. Whether the sound system is portable and set up and stricken a few times a week or permanently installed, the snake provides the primary link between the stage and the sound board.

Standard snake configurations vary from 8 to 32 XLR mic lines, along with 4 to 8 returns. The returns, which are usually 1/4-inch TRS female or XLR male, carry the signal from main mixer outputs and auxes back to the stage, where they are connected to the power amplifiers or powered enclosures.

When a snake is referred to as "24 X 4" it means that there are 24 mic lines and four returns. Likewise, "32 X 8" means that there are 32 mic lines and eight returns. The sound operator determines the function of each snake channel—it is common to use extra mic lines for returns or to use extra returns as line-level sends to the mixer; however, beware that some snakes contain unbalanced returns, which are prone to noises and radio interference. Unbalanced lines are identifiable because they use 1/4-inch male (tip-sleeve) phone connectors at the mixer end of the snake.

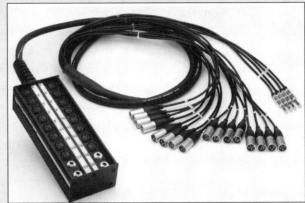

If the sound operator keeps his or her charts with mixing notes, each song will be more consistent, and there will be less stress than if the song must be relearned each time it's in the list. Spend the time to learn your team's music.

Set Up Gear

The sound operator should have the equipment set up and ready to go for each rehearsal, sound check, and service. An efficient, intelligent setup procedure is not that difficult to come up with—it should be implemented as soon as possible.

Strike Gear

Striking the gear (putting everything away) is really the first step in the setup procedure. An intelligent striking routine is designed to store gear safely and neatly while providing for the easiest possible setup.

The intelligent concept of striking equipment borrows from the life concept that acknowledges that a person's day starts with his or her bedtime. If you go to bed late and don't get enough rest, you will be continually playing catch-up the next day. You won't feel very well, you'll be inefficient, and it will be difficult for you to get into high gear. In the same way, a sloppy and inefficient striking procedure makes setup a difficult task. Scrounging for the correct cables and looking for direct boxes and mics is frustrating and inefficient.

Be sure that everyone on the sound crew knows and adheres to the same setup and striking procedures. If, during the stage strike, one crew Zmember puts something in the wrong place or carelessly stashes something just to save time after a gig, it could cost the entire crew several minutes, or even hours, during the next setup just to find a missing item or to re-think the system design to compensate for the missing ingredient.

Connect the AC Distribution System

If your church meets in a school, theater, warehouse, or any other location that requires setup and striking for each service, design an AC distribution system that makes sense and is easily set up from week to week. Find a reliable power source and set up the AC distribution system so that cables are hidden when possible and always well-secured and safely out of the way. This is a very important part of the setup—it is often overlooked, poorly planned, or performed in random response to immediate needs. Always keep safety paramount in your design and implementation. If a crowd will walk over it, tape it down or secure it well. Keep all the AC outlet boxes and connections out of the reach of children. Always ere on the side safety.

Determine where your AC outlets will be located around the stage. There should obviously be power available for all instrumentalists, the soundboard, outboard racks for the mixer, and onstage racks. It's also important to provide adequate AC outlets for stand lights and other onstage necessities.

In a small system, the inexperienced sound operator often considers a box of old extension cables as the AC distribution system. This approach typically results in a messy-looking stage and confusion during setup. There should be preplanning and organization before the gig.

A map of a typical stage layout should be drawn to scale, and the location of each power distribution center should be labeled. Determine the ideal maximum length of each

AC Distribution Map

Planning the distribution of AC around the stage helps the sound operator avoid wasting time. Random implementation and development of an AC delivery scheme typically results in a messy stage, an inefficient system, and wasted time.

AC Distribution System

This Whirlwind Power Link AC distribution system is rugged and extremely easy to route onstage. Even a complicated system is much easier with a distribution tool like the Power Link.

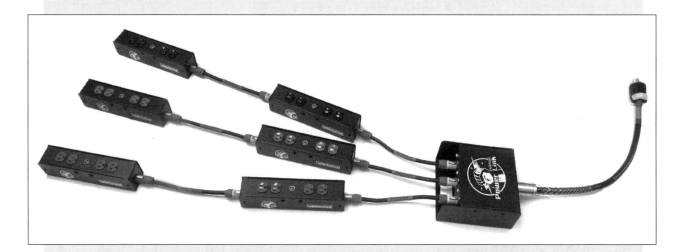

run to each location and construct (or hire someone to construct) cables using heavy-duty black wire with metal four-outlet boxes at one end and heavy-duty plugs on the other. A few of these AC boxes placed strategically around the stage are usually plenty for the instrumentalists. Each instrumentalist should have his or her gear racked together and plugged into a separate power distribution device, such as a simple rack-mounted six- or eight-way power conditioner. Each musician should require only one AC outlet.

In a small setup, the mixer and effects should ideally receive AC from the same circuit (the same outlet from the wall) as the instrumentalists' setups and sound system power amplifiers—this will help eliminate hums and noises caused by ground loops.

In a large setup in a professional venue, there should be designated circuits for all audio connections. Often, these AC outlets are a different color than generic AC outlets. If you find a set of orange outlets and a set of white outlets, it's likely that the orange outlets have been specially configured and verified to provide exact phase and conditioned power. They're often run through power conditioners to help eliminate noise and unacceptable variances in voltage. Ask the venue manager whether there are clean AC outlets specifically for audio connections. If there are designated outlets, there shouldn't be a problem connecting any device to any specified outlet. It also advisable in these venues to confirm and verify the location of all stage circuits on the circuit breaker panel.

Obviously, many road systems require far too much power for a single electrical circuit and, obviously, not every venue has specified clean audio circuits. If there is a noise problem with a system, disconnect components until the problem disappears. Do everything you can to isolate the source of the noise and then verify that the proper cables and connections have been made

Keep all lighting devices and fixtures connected to circuits that are isolated from the audio AC supply.

- First, lights require a lot of AC power. Each can typically require 5 to 10 amps of 110-volt current. Combining lighting and audio equipment on the same AC circuit is a surefire recipe for blown breakers.

- Second, light dimmers and controllers introduce noises into the power supply. When any dimmer system is connected to the same AC circuit with an audio system, there is an excellent chance that dimmer variations will be heard as changing noise levels in the sound system. Although some dimmer systems do a reasonable job of isolating noises from the circuit, it is still advisable to always connect lighting equipment to circuits that are separate from the AC supplying your audio and musical performance gear.

Develop Storage Systems

As a church grows, the sound system requirements become much more diverse. Therefore, there ends up being a lot of gear that's only used occasionally. Even though it might be used occasionally, when it's needed it becomes very important. If there isn't a well-thought-out storage system, there will be a constant energy drain as the sound team searches for gear and rethinks systems. You should always avoid wasting everyone's time while the sound team struggles through disorganization.

`If the church is portable, setting up for every service in a school, home, theater, or other public space, adequate and organized storage is fundamental to a decent church experience.

Labeling Systems

It is simplest if all cables are labeled and color-coded. A portable church that sets up the sound system before every service benefits incredibly from a streamlined system that is labeled well and color-coded. Develop a system that you could imagine someone with no experience or knowledge successfully connecting and striking. It should be obvious where the cables, mics, and components go at the end of the service, and it should be equally obvious where they go and what they're connected to before the service.

Cableorganizer.com and markertek.com carry several professional labeling systems, as well as other means to create an easy and repeatable setup procedure.

Makeshift labeling systems are almost always inadequate. Any system that includes tape labels must be designed for professional long-term use. Over the course of years, adhesive disintegrates, and there is nothing left but a nonstick label lying at the bottom of an equipment rack. Scotch tape is terrible. Masking tape is worse. Duct tape is disgusting. Seriously, invest in a professional labeling system.

The very best labeling system is sometimes difficult to use unless the cables are custom-made from bulk wire and connectors. Use a professional label system that is durable and easy to read. Apply it to the cable about six inches from the connector, then place a piece of clear heat shrink over the label. Apply heat until the shrink securely surrounds the label. This system is very easy to read, and it will last longer than even the most permanent install.

Snakes

For runs of several cables that always connect to the same points, use multi-pair snake cable to construct a custom-length multi-cable, use a manufactured multi-pair snake, or use a wire loom to combine single cables into a snake. Label each end of the cable and determine the best path to consistently run the cable.

It is always easier to run a snake than it is to run individual cables.

Storage Bins and Reels

Plastic storage boxes, such as the kind you would buy at a home-building supply store, work very well for storing certain pieces of a system. A box of XLR cables is very convenient as long as the cables are neatly wrapped and tied. Storing cables in a box is disastrous in the heat of battle if the cables are sloppily stored in a way that results in a rat's nest of audio cable.

Cable Storage

There is nothing quite as frustrating as a "rat's nest" of cables when the beginning of an event is impending and setup is going slowly. One of the most valuable skills a sound operator can develop is the ability to wind cables so that they are easily separated from the other cables. Some of the most valuable tools are those that help organize, store, and compact cables.

Plastic storage bins are commonly used to store cables. This method works well as long as the cables are neatly wound and securely tied. Loose, unwound cables in a bin are certain to create a messy, irritating problem during setup. Keep all mic cables of the same length and functionality together in separate bins. Keep all the 20-foot cables together and separated from all the 50- and 100-foot cables; likewise, keep all line cables separate from all mic cables.

When labeling the cables, it is convenient to color code them according to their lengths so they are instantly recognized for what they are. For example, a simple band of yellow heat-shrink tubing might indicate 20-foot cables, blue might indicate 30-foot cables, and so on.

Tools such as the snake reel can dramatically decrease striking and setup time, while guaranteeing that the snake is always safely stored and ready to go. Also, if your setup uses several mic cables of the same length, try storing them on a cable reel. Simply spool the first cable onto the reel, plug the next cable to the loose XLR connecter, and continue connecting a continuous string of mic cables until the reel is full or the cables are gone. During setup, just pull the cables off the reel as you need them.

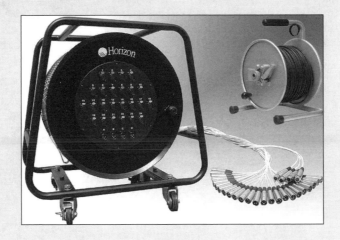

Avoid keeping multiple types of cables in the same box. Keep a separate place for mic cables, guitar cables, speaker cables, and digital interconnect cables. You'll be very glad you did when you're in a hurry to find a specific cable.

Try using a garden hose reel to store several mic cables. Wind the first cable on the reel just like you would a hose. Next, simply connect each cable to the loose end of the cable that's already wound. Continue until the reel is full or you're out of cables. This system keeps the cables very neat, and they can always be easily unspooled from the reel. When they pull off the reel, the cables are always straight, and they lay well across the floor. Be sure the cables are all the same length and that there aren't specifically labeled cables buried deep in the reel. This system is not very flexible, but it provides a very convenient way to store a lot of miscellaneous mic cables.

Miscellaneous parts, tools, and small devices should have a well-defined location. Avoid tossing a bunch of gear, unorganized, into a box or other type of container. When developing a good storage system, you should always plan for your needs during a catastrophic meltdown. If you suddenly must have a number 2 flathead screwdriver and an XLR male to ¼-inch female adapter in order to get the service started, your storage system must be clearly defined in a way that provides you with immediate success rather 20 minutes of plowing through a mass of messy mess.

Mic Cases and Racks

Use a professionally manufactured case for multiple microphones, especially in a portable church application. For a church in a permanent location, a honeycombed system of bins works very well for storing a lot of mics. Always be sure to store mics in their original storage boxes and either secure the area where the mics are stored or secure the cabinet containing the mics.

SKB manufactures several excellent and portable microphone cases with individual padded bins for each mic and a compartment for cables, DIs, or other miscellaneous gear.

Microphone Cases

Rather than tossing the mics in with cables or other gear, it is best to purchase a separate microphone case. These cases provide great protection for the mics and offer a a convenient and visual way to confirm that the mics are all safely stowed away—simply make sure that the correct number of mic pockets are filled up and do a quick scan to make sure all of the mics made it back in one piece.

Even if all of the microphones remain in their original containers, it's difficult to tell which mics are in the bin if all you see is a pile of black pouches. Many mic cases also provide an extra storage area for extra cables, DIs, adapters, and so on.

Storage Hooks and Racks

If there is space available in a permanent installation, a system of racks for mic storage and hanging brackets for cables works well. Be sure to develop an organizational structure and system before randomly hanging cables. Mark each area clearly—use durable plastic or laminated signs, securely fastened in place, to indicate each specific equipment type. Use labels such as "Mic Cables," "Guitar Cables," "Speaker Cables," "Direct Boxes," "XLR," "1/4-inch TS," "Phono," "1/4-inch TRS," and so on.

Plan for growth in your storage systems. Although signs should be securely in place, recognize that growth happens and things change. Leave extra room for additional cables and devices in each area.

If you are working with a group of people who are varied in their audio skills and understanding, it's a great idea to label the type and function of each cable. Line-level and instrument cables with quarter-inch phone connectors look a lot like many speaker cables with quarter-inch phone connectors. If an inexperienced helper uses a speaker cable for an instrument cable, you will have noise problems. Problems like this can be like a needle in a haystack when you're in the heat of the battle. If you hear a noise from the acoustic guitar and you're able to simply asks the acoustic guitarist if the cable says "Instrument" or "Speaker," what would otherwise take several minutes to determine might take a few seconds to solve.

Learn What a Good Mix Is

A diligent sound operator should spend time learning the essence of a good mix. He or she should listen to well-respected recorded music, other sound operators' mixes, music concerts, and music at church conferences. In order to craft an excellent mix, it is imperative to know what good music sounds like—that's not negotiable. Reading books such as this one helps the sound person understand what other experienced sound operators consider to be a good mix—that's a good start—but in order to be effective and excellent, a sound operator must at some time have a mountain-top musical experience that begins the process of imprinting the sound and emotional impact of a first-class musical mix.

Attend Seminars and Classes

Every year, around the globe, there are several seminars and classes that cater specifically to music team needs, from the worship leader to the sound operator. These events provide excellent instruction and examples of top-quality music and sound.

Many music and magazine publishers sponsor these types of events. If the music team frequently uses music from a particular publisher and reads certain church-related magazines, find out the events that they sponsor and attend them. Their style and philosophical approach should match the direction of your church, and the ideas and instructions you receive should be of great value.

Conferences and Seminars

Look for the conferences and seminars that match your church or denominational theology and style. The following list is a representative sample of the available conferences:

- The Christian Musician Summit. Exceptionally well run, these conferences feature the best of the contemporary musicians and technicians. They occur at Bayside Church in Sacramento, CA, Grace Chapel in Nashville, TN, Overlake Christian Church in Seattle, WA, and at Chapel at Crosspoint in Buffalo, NY (www.christianmusiciansummit.com).

- The Passion Conferences. Youth oriented and vibrant, the Passion Conferences are powerful and heartfelt. From the Passion website (268generation.com/passion2013):

  > Last year, our US gathering drew more than 40,000 students and leaders from around the world to the Georgia Dome in Atlanta. We were blown away by God's presence. Shattered and rebuilt by His Word. Challenged to the core. Repaired. Wrapped in love. Awakened to raise our voices for those who have no voice.

- The CCLI Conference List. This list is published by Christian Copyright Licensing International, represents many of the conferences that happen around the nation. (www.ccli.com /conferences/conferences.aspx).

THE SPIRITUAL IMPORTANCE OF WORSHIP AND MUSIC

When it comes to spiritual battles and warfare, we don't have anything to offer without God. When we rely on our own strength and power to live a Godly life, we typically fail. When we rely on God and his prescription for an abundant life, we succeed.

I have several personal stories that illustrate the power of God—in particular, regarding the importance He places on worship, music, and prayer. I'll share a few stories as a point of testimony and encouragement from my understanding and experience of how God works. However, the Bible is the source for all truth. Scripture refers often to the power and importance of music in life and worship—it is undeniably fundamental to a truly spiritual, Christ-centered life.

What Is Worship?

Worship is, in its most basic form, adoration of God. We can find more complex definitions, but everything seems to boil down to that simple adoring and honoring relationship between people and God.

We were created for relationship with a living God. To make it more complex, we messed up and allowed sin into the picture. However, God, being the great and loving Creator that He is, figured out a way for us to live in relationship with Him forever. His Son, Jesus, became a man and essentially paid the bill. Like a good earthly father loves to do for his children, God blessed us, took the check, and paid it in full. Worship is our way of saying thank you, I love you, you're amazing, and I place you at the very top of the list of all things and people.

Worship is really a lifestyle. It's not just music and songs that express reverence and adoration; it's not just taking in an inspired sermon and understanding a new and amazing facet of the living God; it's not just prayer; it's not just reaching out and sharing the love of God with someone who has never experienced it before. It is all of that and more!

The music team is often called the worship team—that's just how it goes, and there's nothing harmful about that. However, the entire church service is worship, and the rest of life throughout the week, before and after the church service, should be worship also.

The reality of life is that our time each week at work and play often makes us feel further from God rather than closer. A seasoned and wise Christian structures his or her week to include the things of God, such as prayer, Bible reading and study, Godly relationships, and worshipful music. As much as we strive for the ideal, it doesn't always happen. We watch questionable movies; get angry with our friends, family, and co-workers; and possibly overindulge in things we shouldn't. By the time we get to church, we're dragging from our weekly activities. Hopefully, we slowly and intentionally alter our weekly pattern so that we're poised for spiritual success rather than failure, but, at best, there is still a little ebb and flow in our spiritual focus.

I mention all of this to highlight the spiritual importance of the weekly song service. Music that worships God through melody and lyrics brings everything back into perspective. It's a little like this: At every church service, the congregation needs to be reintroduced to God. It is in these powerful times of worship that things change for us. We dial into God and out of the world. We are always given the opportunity to turn to God, but some of us just need a little kick in the behind to follow through.

The way the worship team presents itself displays or betrays their spiritual attitude. If the musicians are sleepy and lackluster, the singers are off key and unexcited, and the sound system is squeaking and unappealing, the spiritual tone has been set—BORING! If the entire team is excited about what God is about to do in the service they are honored to be part of; the musicians are playing skillfully; the singers are singing from their hearts instead of their tonsils; and the sound in the room is authoritative, powerful, and non-offensive, a completely different spiritual tone has been set—EXCITING!

The Authority of Excellence

1 Corinthians 14:8 says "For if the trumpet gave an uncertain sound, who would prepare himself for war?"

In this section of Corinthians, Paul is expressing the importance of interpretation of tongues, but this line jumps out in its importance of providing confident, well-rehearsed, and excellent worship. With uncertain music and sound, our congregations will probably just stand around in a daze while they're bowled over by the enemy.

The music team leads us into the presence of God, and we are often reluctant for any number of reasons. We might be insecure, discouraged, angry, or just drained by our weekly trials. It is undeniable that we're involved in a spiritual battle, and the worship team is calling us in. I've been at and involved in worship services where the team's call was a little reminiscent of Barney Fife in his weakest moment of cowardice. I've also been at and involved in worship services where the team's call was so confident and inspiring that you'd think General Patton himself was commanding you into battle. You're ready to go—fully charged and fully armored.

Our excitement and commitment to excellence demonstrates the importance we place on the worship service. Because this is the time that the congregation is brought back to the realization that God is alive and that He loves each of us, it is unlikely that the pastor will see the effectiveness of his anointed message if the worship team has given a half-hearted and unexcited song service.

In God's economy, we don't need to be world-class musicians and technicians to be loved, appreciated, and effective; however, out of respect, adoration, and praise, we must give a world-class effort in His name with a heart that is pure and loving.

1 Corinthians 14:7–8
Even things without life, giving a voice, whether pipe or harp, if they didn't give a distinction in the sounds, how would it be known what is piped or harped? 8 For if the trumpet gave an uncertain sound, who would prepare himself for war?

The Musicians to Battle before the Soldiers

2 Chronicles gives the account of King Jehoshaphat and an impending attack on his people.

2 Chronicles 20:1–3
It came to pass after this also, that the children of Moab, and the children of Ammon, and with them other beside the Ammonites, came against Jehoshaphat to battle.
2 Then there came some that told Jehoshaphat, saying, There cometh a great multitude against thee from beyond the sea on this side Syria; and, behold, they be in Hazazontamar, which is Engedi.
3 And Jehoshaphat feared, and set himself to seek the LORD, and proclaimed a fast throughout all Judah.

King Jehoshaphat and those in his charge were in hot water. They didn't have the manpower or the arsenal to beat their enemies back, so they turned to God.

2 Chronicles 20:14-22
Then on Jahaziel the son of Zechariah, the son of Benaiah, the son of Jeiel, the son of Mattaniah, the Levite, of the sons of Asaph, came the Spirit of Yahweh in the midst of the assembly;
15 and he said, Listen, all Judah, and you inhabitants of Jerusalem, and you king Jehoshaphat: Thus says Yahweh to you, Don't you be afraid, neither be dismayed by reason of this great multitude; for the battle is not yours, but God's.
16 Tomorrow go down against them: behold, they come up by the ascent of Ziz; and you shall find them at the end of the valley, before the wilderness of Jeruel.
17 You shall not need to fight in this battle: set yourselves, stand still, and see the salvation of Yahweh with you, O Judah and Jerusalem; don't be afraid, nor be dismayed: tomorrow go out against them: for Yahweh is with you.
18 Jehoshaphat bowed his head with his face to the ground; and all Judah and the inhabitants of Jerusalem fell down before Yahweh, worshipping Yahweh.
19 The Levites, of the children of the Kohathites and of the children of the Korahites, stood up to praise Yahweh, the God of Israel, with an exceeding loud voice.
20 They rose early in the morning, and went forth into the wilderness of Tekoa: and as they went forth, Jehoshaphat stood and said, Hear me, Judah, and you inhabitants of Jerusalem: believe in Yahweh your God, so you shall be established; believe his prophets, so you shall prosper.
21 When he had taken counsel with the people, he appointed those who should sing to Yahweh, and give praise in holy array, as they went out before the army, and say, Give thanks to Yahweh; for his loving kindness endures forever.
22 When they began to sing and to praise, Yahweh set ambushers against the children of Ammon, Moab, and Mount Seir, who had come against Judah; and they were struck.

Our spiritual battle is like the battle faced by Jehoshaphat—without divine intervention, we don't stand a chance. The battle is too big for us to handle, and if we rely on our own power, we're sunk.

Excellent musicianship and skilled technical abilities can present a problem in music ministry if their origin and purpose are not clearly understood—it's likely that our skill level will give us the impression that we are powerful in the battle. As worship team members, we must understand that God provides our skills and talents for His adoration and glorification, but the spiritual battle that goes on behind the scenes is His alone. To realize the powerful walk of a true believer, we must rely on our Creator, His expertise, and His will for our lives. Once we begin to take back control, we marginalize the truth of God and His Word—and that's a dangerous way to live!

So, it is very important that we give God our very best effort simply because He deserves it—He is worthy. It is also paramount that we trust in Him to do the real work in people's hearts and lives, and that when something really great happens we understand that, though it might have been a personal blessing to be included in the process, the glory goes straight to God.

Ego problems are common in the music business, whether secular or sacred. They cause strife, anger, hurt feelings, and a substantial decrease in divine potential. However, the most amazing things happen when we act in true humility, repentant for our sins and misgivings, remaining completely reliant on the Holy Spirit for our care and spiritual protection. A team that understands and lives out these concepts can be on the frontlines of the battle without fear. What an honor it is to be first in line to see the invincible, omnipotent power of God at work!

We Are Instructed throughout the Bible to Worship

Throughout the Old and New Testaments, we are instructed to praise God with music. It is a powerful ministry that we're called to, and one that breathes throughout the Word of God. The following is just a sampling of the numerous scriptures that highlight the importance of worship and praise in the lives of believers.

Revelation 15:1–3 KJV
And I saw another sign in heaven, great and marvellous, seven angels having the seven last plagues; for in them is filled up the wrath of God. 2 And I saw as it were a sea of glass mingled with fire: and them that had gotten the victory over the beast, and over his image, and over his mark, and over the number of his name, stand on the sea of glass, having the harps of God. 3 And they sing the song of Moses the servant of God, and the song of the Lamb, saying, Great and marvellous are thy works, Lord God Almighty; just and true are thy ways, thou King of saints.

Ephesians 5:18–20
...be filled with the Spirit, speaking to one another in psalms, hymns, and spiritual songs; 19 singing, and making melody in your heart to the Lord; 20 giving thanks always concerning all things in the name of our Lord Jesus Christ, to God, even the Father;

Colossians 3:16
Let the word of Christ dwell in you richly; in all wisdom teaching and admonishing one another with psalms, hymns, and spiritual songs, singing with grace in your heart to the Lord.

1 Chronicles 23:5
Four thousand are to be gatekeepers and four thousand are to praise the LORD with the musical instruments I have provided for that purpose.

1 Samuel 16:23
Whenever the spirit from God came upon Saul, David would take his harp and play. Then relief would come to Saul; he would feel better, and the evil spirit would leave him.

1 Chronicles 15:16

   David told the leaders of the Levites to appoint their brothers as singers to sing joyful songs, accompanied by musical instruments: lyres, harps and cymbals.

Romans 12:1–2

   ...I urge you, brothers, in view of God's mercy, to offer your bodies as living sacrifices, holy and pleasing to God—this is your spiritual act of worship. 2 Do not conform any longer to the pattern of this world, but be transformed by the renewing of your mind. Then you will be able to test and approve what God's will is—his good, pleasing and perfect will.

Hebrews 10:19–25

   ...since we have confidence to enter the Most Holy Place by the blood of Jesus, 20 by a new and living way opened for us through the curtain, that is, his body, 21 and since we have a great priest over the house of God, 22 let us draw near to God with a sincere heart in full assurance of faith, having our hearts sprinkled to cleanse us from a guilty conscience and having our bodies washed with pure water. 23 Let us hold unswervingly to the hope we profess, for he who promised is faithful. 24 And let us consider how we may spur one another on toward love and good deeds. 25 Let us not give up meeting together, as some are in the habit of doing, but let us encourage one another—and all the more as you see the Day approaching.

John 4:23–24 (New International Version)

   Yet a time is coming and has now come when the true worshipers will worship the Father in spirit and truth, for they are the kind of worshipers the Father seeks. 24 God is spirit, and his worshipers must worship in spirit and in truth.

Psalm 66:1–8

   Shout with joy to God, all the earth! 2 Sing the glory of his name; make his praise glorious! 3 Say to God, "How awesome are your deeds! So great is your power that your enemies cringe before you. 4 All the earth bows down to you; they sing praise to you, they sing praise to your name." Selah 5 Come and see what God has done, how awesome his works in man's behalf! 6 He turned the sea into dry land, they passed through the waters on foot—come, let us rejoice in him. 7 He rules forever by his power, his eyes watch the nations—let not the rebellious rise up against him. Selah 8 Praise our God, O peoples, let the sound of his praise be heard;

Take Your Duty Seriously

It is truly a humbling honor to realize that the Lord lets us participate in a life-changing and divine spiritual event such as a church service. Too often, church feels like an obligation or just another chance to hang out with our friends. In reality, it is a very important connection to the power of the living God. It's an encounter with the Creator of all things—the inventor of life, humor, love, lyric, melody, sight, hearing, and trilobites.

Communicating the Heart and Word of God

The sound operator is given the awesome responsibility of making sure that the heart of God (in worship) and the Word of God (in the sermon) are heard in an understandable way. It is possible for the sound operator to distract from the heart of worship and the message of the sermon; in fact, it is much more common than it should be.

     Most of the obstacles to proper communication in church are a result of a simple lack of knowledge and understanding. There is a simple way to make a vocal understandable without making it louder and to make a mix sound full and confident without overpowering the congregation. Once you've finished this book and have practiced applying the prescribed techniques, you should be able to provide an excellent-sounding mix that is full, intelligible, and musical.

Prayer and the Sound Ministry

Pray for your sound team and ministry on a regular basis. People will be people, and even the most pure of heart can be tempted into sin, enticed into arrogance, or just tricked into stupidity. We all need constant prayer, but the music ministries seem to come under as much or more spiritual attack as anyone. After all, Satan is a fallen music director.

Also, I know that spiritual attacks can come through technology—I've experienced and overcome them. I was producing a Christian album in Seattle at a studio that had just taken delivery of the brand-new digitally controlled analog mixing console made by a British company called Euphonix. The board was one of the very first CS IIs manufactured. It was brand-new technology, including real-time automation of every parameter. For the early 1990s, this was cutting-edge technology. The digital controls offered groundbreaking automation potential, and the well-designed analog circuits provided silky-smooth analog sound.

I was the only Christian producer using the studio at the time. I usually had an album or two going at once, and I taught recording classes a few times a week. I got along well with the owner, the manager, the house engineers, and the other freelancers that I'd see from time to time. The studio did a lot of advertising work and also a lot of the early Seattle grunge rock stuff from Nirvana, Soundgarden, and so on.

When it came time to mix the album I was producing, I was the first one in the studio to mix a full-scale album that took advantage of the console's capacity. We had track with two 24-track analog recorders running in sync, and we also had several virtual MIDI tracks and a pile of outboard processing.

When we started mixing everything worked fine, but every night for a whole week, we would get four to six hours into a mix that should have taken about eight hours total, and the computer would corrupt the automation data (the mix). There were constant calls to the manufacturer, and every time we'd start a new mix with their assurance that we were operating the console correctly and that they had fixed the problem. Sure enough, four to six hours into the mix, everything would freeze up and the data would corrupt.

This happened far too many times in a row. I don't know if it bothered me more that it happened or that we were crazy enough to start a new mix every time they told us they had solved the problem.

Finally, we showed up and started the mix process, and it occurred to me that we hadn't been praying before the sessions. In retrospect it seems pretty silly that we hadn't been praying before each session, but as you might guess, we had absolutely no problems with that console from that time on. However, the rest of the engineers continued to have same problems we were having. After several more weeks, the studio owner got the rest of the engineers together to have me show them what I was doing to make the board work. Uhh, would you believe prayer?!

Time, Just in Time

I was music director at a fairly large church near Seattle. We had a good group of musicians and some depth, but one of the things we decided to do was help the youth group worship team develop their skills. They had some excellent spiritual role models and needed some musical help. Each of the guys on the adult worship team committed to give the players lessons and to help them through the learning process while their team grew.

This was an endeavor blessed by God—you could feel it. Strong bonds formed between the youth and the adults. Our house was usually full of a bunch of high school worship team members who just loved to hang out and talk about music and worship and all the other stuff that gets discussed in a group of excellent church kids. They had to endure me showing them a DVD of Steely Dan and trying to get them to appreciate the brilliance of their musicality and the tightness of their grooves. It was really a lot of fun.

One day I was working with the entire rhythm section and trying to get them to lock into the same feel together as a unit, and they didn't quite get it. I asked how many of them had a metronome. Of course, no one had anything resembling a metronome, and some didn't even know what one was.

I remember praying that we could come up metronomes for some of the key players, thinking that I'd probably just go out and buy a few for them—oh, me of little faith. The very next weekend after the last service on Sunday, a guy came up who I had never met and asked me if we had a need for any metronomes. I couldn't believe it! I said sure, and told him that we were just praying about getting the youth group kids some. He seemed very pleased and ran out to his car. He came back with about 30 brand-new Seiko metronomes that were just perfect for our needs. He said he worked at a surplus import place, gave me some story about why he had them and why this wasn't a shady deal, and reassured me that it was all legal and aboveboard. I don't really remember all those details, but I do remember that 20 or so youth group musicians and bunch of adults were very impressed by God's provision for our youth and were encouraged in their prayer lives. Oh yeah, I never saw the guy with the metronomes again.

Prayer works, and your team is very worth praying for.

RELATIONAL CONSIDERATIONS

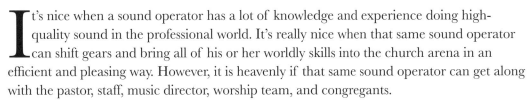

It's nice when a sound operator has a lot of knowledge and experience doing high-quality sound in the professional world. It's really nice when that same sound operator can shift gears and bring all of his or her worldly skills into the church arena in an efficient and pleasing way. However, it is heavenly if that same sound operator can get along with the pastor, staff, music director, worship team, and congregants.

In a church setting, relational considerations and the ability to put things into their proper perspective are fundamental to success. It is different working with volunteers than worldly professionals. It is sometimes difficult for the perfectionist mindset to adjust to the concept of allowing time for processes to develop and excellence to bloom. These challenges must be met and conquered. If they are, you'll discover years of personal growth and satisfaction in a worthwhile ministry. If they aren't, there's a good chance that your ministry efforts will crumble before your eyes.

The Worship Service from the Minister's Viewpoint

In any church setting, all of the teams must realize that they're being called alongside the senior pastor. It is the ministry that the Lord has called him or her to that everyone is joining to support. Therefore, when it comes time for a judgment call or a key decision, everyone must defer to the pastor. That's just how it works.

This doesn't mean that the team members must simply be robotic Kool-Aid drinkers, but it does mean that, after they've had their say and the matter has been discussed to the pastor's satisfaction, they'll need to support his final decision.

Time is the only tester that I've found for this one. When you find a room full of strong thinkers and there are disagreements, somebody gets to decide because everyone can't agree—and that's the pastor. Even if a team member still disagrees with the decision, he or she must put 100-percent effort in the quest of the vision. Trust in the fact that a good idea will always reveal itself over time—and so will a bad one.

Introduce the Congregation to God

One thing that is always at the forefront of a good pastor's mind is the provision of an encounter with God during the service. Many people feel God's presence for the first time during the music portion of the service. They might not know what the feeling was until later, when the pastor's teaching reveals a brand-new truth to them.

For the regular attendees, the song service should serve to bring them back to the reality that God is really in control after all, and that the life they just lived throughout the week, though it might have been tough, had purpose. The lyrics in most modern worship songs state simply, often in Scriptural form, the truths of God, providing encouragement and hope to an often discouraged and hopeless group. This important aspect of the song service should never be underestimated.

It is the sound operator's primary assignment to make sure that these introductions and explanations are presented in a way that is audible, intelligible, and not distracting. A bad-sounding lavalier, an unattractive sound, too little level, too much level, and feedback all offer the potential to distract congregants from the intent of the presenters. Do whatever it takes to get this right. We'll discuss several techniques and tools later in this book that will help the sound operator get the most out of these important times.

Prepare the Congregation to Hear the Sermon

The ideal scenario for the new attendee is that the song service will make him or her ask, "Oooh, what's that warm feeling of peace and contentment I have?" Then, the pastor's sermon comes along behind and says, "That was God. Would you like more?"

As the sound operator's skills become more refined, you'll find that the pastor will visibly relax more each week as a bond of trust forms with that operator.

Pastors want to begin preaching to a congregation that has had an excellent time of worship and who already feel close to God. The Holy Spirit will begin to work on the hearts in the church from the beginning of the service if we're well prepared—distraction will make things more difficult. If the sound or music is poor quality, many of the church members, much of the staff, and probably the pastor will be distracted. They'll tend to let their minds focus on the flaws, rather than being free to let their hearts focus on the things of God. Granted, in an ideal world, everyone would look beyond the technical or musical flaws to the heart of God, but in reality, most people don't have that level of self-control. It's best to not put anyone through that in the first place.

Create a Unified Spiritual Tone among the Congregation

Not everyone expresses his or her feelings and emotions in the same way, but I believe that, in a worship service, most people get it when it's right—they all check in together. Conversely, they also get it when it's wrong—they check out.

The very best worship services that I've seen and been a part of have invited and realized the participation of the church. Everyone joined in, worshipping out of true desire for a relationship with God. They literally worship Him because He's worthy to be worshipped.

This is a goal that can only be met when the entire team, submitted to the Holy Spirit and led by God, functions together in an efficient and inspiring way. The authority of excellence builds throughout the service, and everyone in the church building steps into the same boat.

When the worship team hands the pastor a congregation that has worshipped in heartfelt way and who are all together in their awe of God, great things can result from a divinely inspired message.

Getting Along with Everyone

The sound operator should strive to get along with everyone. It will help the senior pastor, and therefore it will help the church. Believe it or not, the church is often full of people who are less gracious, kind, and forgiving than the non-Christian workers that the congregation spends the rest of the week with.

If the sound operator is difficult to get along with, it will get back to the pastor in the form of an angry confrontation, letter, e-mail, or phone call. Each of those encounters adds up until the pastor overloads, burns out, or releases the sound operator from his or her duties.

In a church environment, kindness, grace, and forgiveness must rule the day. Take comfort that good and bad ideas reveal themselves quickly, and right and wrong opinions generally work themselves out in the long run. If someone is making it difficult for you to like him or her, be understanding and sacrificial—that's the Biblical model we've been given about how to interact with each other.

Bending Over Backward

Go the extra mile to make your pastor's life better. When disputes arise, try ridiculously hard to work them out. If the pastor asks you for something, even if it's at the last minute, and even if it stretches your limits of comfort, give it your best shot to see that it happens.

In modern secular times, we become so conscious of being taken advantage of that it is hard to shift gears to consider others' needs as more important than our own. In God's economy, we need to make that paradigm shift. He is our provider. He is our advocate. He is our teacher. We really don't need to defend ourselves because the Lord will do it for us.

Certain reactions become natural because we've learned that they're correct, but some just don't come easy. The Lord will provide the self-control to enable the willing team member to lower self and to exalt others. The truth is that a person who lives in this way—considering others' needs before his or her own—is almost always honored and blessed in the long run.

Grace Versus Judgment

When bad stuff starts happening around you, default to graciousness and kindness. Avoid judging the actions of others—it is typically fruitless. Judgment of people and their actions should be reserved for God.

In Matthew 7:3 and Luke 6:41, we are told that we should inspect the beam in our own eye before we condemn the speck in our brother's eye.

The longer we strive to live by Biblical standards, the more we see how true and wise they are. There have been too many times over the years when I have gotten worked up about something that someone else did that I thought was disgraceful, only to find out that there was a good explanation for that person's actions. I, on the other hand, had been a complete buffoon.

At the first sign of a problem, be gracious. Pray and seek the Lord's help in dealing with the problem. These verses from Proverbs 14 put it pretty clearly.

Proverbs 14:16–17

A wise man fears the LORD and shuns evil, but a fool is hotheaded and reckless.
A quick-tempered man does foolish things, and a crafty man is hated.

Keep Everything in Perspective

Church is a place where volunteers do most of the work. It's not the most efficient scenario, but it can be very effective. Some volunteers work at the pace of a skilled professional, whereas others are so beat up by the time they get to church that they're doing well to function at all.

Progress at church usually comes slowly and steadily. Church leaders must be sensitive to the needs and circumstances of the volunteers. Organization and follow-through are fundamental traits of a successful church program—and so are grace and understanding.

When programs and goals are kept in their proper places, it is easy to remember that the big goal is introducing people to God and helping them grow in their understanding of Him and His ways. The big goal is not a special program, a fundraiser, or a new sound system. Those are tools to help build effective ministry, but they aren't the end goal.

An Atmosphere of Peace

Pastors, ministry leaders, and staff are constantly under pressure. They are trying to be gracious and still get some powerful ministry done while working with a volunteer labor force. Spiritual attacks are commonplace, angry church members are on the rampage, and insecurities and frustrations abound.

The music and tech teams can either add to the chaos or provide an oasis of peace and tranquility for all. Most people can understand and participate in an atmosphere of peace if they understand a few things.

1. There is a forum for expressing concerns that provides a real avenue for discovering the truth without a mere lip-service Band-Aid.
2. Leadership really hears them when they have a concern.
3. Valid concerns receive immediate action.
4. When their concerns aren't valid or if they're a result of simple misunderstandings, church leadership will be honest and direct in their response.

Solid, adequate communication, along with intentionally kind and gracious attitudes, produces success in any endeavor—especially church.

The Sound Operator/Pastor Relationship

Often, the sound operator feels like the pastor sees him or her as an anonymous face behind the board. That's not true. Whether or not you realize it, the pastor, the musicians, and anyone who uses the sound system during the service is very aware of the person at the board—they're dependent on that person to help them communicate their message.

The pastor, especially, relies on the sound operator's support. Glitches in sound equate to distractions to ministry. Although a seasoned pastor will be gracious and kind to a volunteer sound operator, he will always be aware of the operators and their track records.

Be a Friend to the Pastor

Be friendly to your pastor. Sometimes the pastoral figure is intimidating, but it's helpful to realize that the pastor is as human as anyone. Pastors like to laugh and have fun, they make mistakes and sometimes say the wrong things, they have tempers and fears, and they are often insecure and need affirmation, just like anyone.

A friendly sound operator who genuinely cares about the pastor's opinion and who obviously tries to provide a high-quality service to the church can give a struggling pastor just enough of a boost to make it through the service.

Ask Your Pastor's Opinion

Many pastors claim ignorance about sound, music, and anything technical. They prefer to defer when it comes to any of these things. However, asking the pastor's opinion creates a point of contact and communication. A simple, "How's everything sounding today?" says to the pastor, "I appreciate you and care about your opinion."

Be Understanding

If you approach your pastor for small talk, a simple hello, or a question about the sound, you might feel as if you're getting the cold shoulder. There is a great likelihood that you've simply encountered a distracted pastor.

Having been part of a pastoral staff, I can attest to the fact that there are a lot of times when pastors are dealing with some outlandish things. Of course, we all should submit everything to prayer and place our trust in God, but some of the stuff people come up with in life is just plain baffling. Do your best to go with the flow and trust that if your pastor is acting a little weird, he or she is coping with something very unsettling.

Be Open to Criticism

I don't suppose there is anyone else out there besides me who has an inherently poor response to criticism. But just in case there is, we should talk about it.

Most sound operators can be in a church full of a thousand people and still see the angry person coming from the back of the room who's about to share his or her opinion.

For me (and, I believe, many others), it takes a conscious effort to flip the switch from "defensive-aggressive" to "ministry-minded." When someone offers his or her opinion, whether it's the pastor or a brand-new congregant, be gracious, kind, and understanding. Do your best to accommodate the person's request. This approach gives you time to regroup and reflect on his or her comments. If you disagree strongly with the input you've gotten, express it after the service or on another day. As a sound operator, you really should have someone who can help you work through these types of situations.

A good leader will determine the chain of command for the sound operator before any problems arise. In an ideal world, the sound operator won't be alone when it comes to a few key factors, such as determining the target volume, handling complaints, and making subjective mix decisions. If these types of decisions are predetermined and agreed to by the pastor, the sound team will have a much more enjoyable time.

If your pastor asks you to change virtually anything during a service, you should accommodate this immediately. If there is an inherent disagreement about artistic and creative aspects of the worship team, they can be addressed at any other time than during a service. Someone is ultimately in charge of church operations—that person should be given immediate results from any request. There are many factors involved in a decision made during a church service. Much like the military, respect and honor must be given to those in charge. If one has a fundamental problem with those given charge, then that problem must be addressed and resolved, because it is very unlikely to disappear on its own.

Don't Offend Quickly

Keep in mind that there is a lot involved in the ministry life at church. If a pastor or staff member throws a stinging comment your way, avoid an immediate response. Be gracious and kind and consider that, in the heat of battle, that person might be stretched beyond his or her comfort zone and might not quite be himself or herself at that particular moment.

It is our human tendency to bristle up when someone says something out of line. We might jump to the conclusion that there is no excuse for rudeness, anger, or just being inconsiderate. Although that is probably correct, we'll all be better off if we bend over backward to be gracious and loving.

Consciously deny offense. Give reality and truth a chance to sink in. If you are justified in taking offense, there will be plenty of time to settle the disagreement later. Above all, don't let an offense settle into bitterness. Communicate your feelings in an honest and loving way. Most of the time, offenses are a result of misunderstandings. Once the facts are on the table and both sides have presented their case, everyone involved typically has good intentions and at least a partial justification for his or her actions.

Take comfort in the fact that the process of working through disagreements almost always results in a greater understanding of, and a stronger bond with, those in your life circle, especially when the process is guided by love and submits to Biblical principles.

Pastors Are People Too

Every once in a while, your pastor will treat you in an amazingly human way. We typically expect pastors' actions and reactions to be Godlike, and often they are. However, pastors and church leaders are just as subject to anger, pettiness, jealously, impatience, rudeness, and an entire list of other emotional responses as the rest of humanity.

I have noticed that, in most cases, our character is defined by our reactions to our human failures, not by the failures themselves. The best pastors I've encountered absolutely had moments of failure, but their responses to those failures set the standard in my life for Christian living.

You Gotta Know When to Hold 'Em, Know When to Fold 'Em

Every church is not a perfect fit for every person. If you find yourself in a constant battle with the church leadership and you've tried in earnest to find a Godly resolution, but to no avail, you might be best suited to a different church. However, be careful with this decision. Pray and fast, ask God, and seek Godly counsel from those you trust before you decide to leave your church. If you handle this choice in a casual manner, you'll end up with a string of churches that you've left in an angry huff.

Exercise your prayer life and your spiritual insight to determine whether you're at the right church—don't use your temper and emotional immaturity. If you are certain you're at the right church for you and your family, you are obligated to work through any difference you might have with the pastors, staff, team, or anyone else in the church—it's not optional.

While you're in the process of working through difficult church problems, you should consider stepping down from any ministry duties. These times typically provide a foothold for anger, deceit, fear, and any other spiritual attacks, so it's best to work through these problems before re-entering the ministry force.

If it is confirmed in your spirit that you're at the wrong church, and there is agreement with the pastors and trusted friends, you should look for a church that is more in line with your beliefs and standards. Do your very best to resolve any disagreements before you leave. If you have any choice in the matter, don't leave in anger.

Communicate, Communicate, Communicate

Efficient communication is the antithesis to strife and misunderstanding. Almost all church problems are a result of poor communication skills. You might need to be the person who initiates all the communication, but someone has to do it.

Don't leave unresolved issues lying all around. You'll be happier, and your ministry will be more effective once you've sufficiently communicated your concerns. An efficient network of communication throughout and between each layer of authority will create an environment that is nearly immune from gossip and rumors. Granted we're dealing with human tendencies here, so it's impossible to squelch all vitriolic talk, but proper communication provides an atmosphere conducive to peace, love, and effective ministry.

Ask Questions

Ideally, the church sound operator will be included in the worship team information loop. Often, though, day-to-day, week-to-week, and month-to-month operations are whirling so fast for the music director that he or she forgets to let the sound person in on the plan. The sound operator should always feel free to initiate communications, asking questions that will help increase the efficiency and effectiveness of the team.

Of course, be sensitive to presenting well-constructed questions at the proper time. Sensitivity to the creative flow is a must, so a little courtesy and common sense will go a long way in helping you decide when and how to ask a question.

Keep Track of the Leaders' Opinions

Every few weeks or so, ask the pastor or other ministry leaders if they're happy with the sound during the services. This doesn't need to be a deep conversation, although it might turn into one, but it should mostly be a link to their thoughts about the effectiveness of the sound ministry.

Often, the pastors and leaders will be so involved in other areas of concern that they might not have a solid opinion available, but at least they'll know you value their opinion. If you receive some constructive criticism, do your best to respond in some way to rectify the concern.

It's important that you work your way through the layers of critique. If the pastor tells you the music is too loud, find out more specifics. Ask probing questions that will help you provide the best results. The answer to a volume problem is rarely found in the master level fader—it's usually found in a specific mix ingredient. So, you might respond to a volume complaint by asking:

- Is there one thing that seems too loud?
- Are the instruments too loud?
- Are the vocals too loud?
- Is the sound too boomy?
- Does the mix sound piercing or edgy?
- Is the sound too loud at a particular time during the service?
- Is the volume all right for the first few songs, but too loud for the rest of the songs?

Asking these types of questions can really help any sound operator respond to a concern in a way that provides a mix that better serves the congregation. Later in this book, we'll discover effective ways to respond to each of these concerns and more. If the master level controls are simply turned down, the mix will need to change to compensate. Because most teams combine acoustic instrumentation along with electronic reinforcement and vocals on mics, a simple reduction in the master volume control does not provide a linear reduction in level across the board.

There's a good chance that the pastor won't have the perfect answer to your questions because he or she hasn't listened for specifics; however, your probing questions will typically provide a reference point for future discussions. Once you find the factor that's triggering the volume concern, everyone will have won—you will have demonstrated that you care enough about the concern to find a solution, and the overall ministry will be just a little bit more efficient in its communication techniques.

Conflicting Opinions and Who Wins

There is a very good chance that the sound operator will receive conflicting criticisms. If the church leadership has done a good job in communicating the authority hierarchy, this question is a no-brainer—just follow the chain of command. It sounds so simple, but often something unhealthy gets in the way. Whether it's pride, insecurity, or just downright obstinacy, the human tendency typically leans toward providing results in response to opinions with which we agree.

A church provides a unique authority structure. In many ways the church functions like the military. Power comes from the top down, and most requests from the top are not optional. The pastor or church council is typically guiding the church in response to their discernment regarding spiritual, social, and financial concerns. The volunteer ministers must trust that their church leaders are actually receiving their directions from God, understanding that divinely provided guidance produces all things good for the church.

The true dichotomy in the Christian church is that the example set by Jesus Himself was melded between the absolute authority of the Word of God and the example He set as a humble servant to all. Pastors, at some level, understand the value of scriptural authority, but they have grown up understanding that a true pastor serves the needs of the church. The shepherd looks after the flock, feeding them, keeping them clean and groomed, and placing their needs above his or her own. So, there are often conflicting messages in pastoral responses to certain situations. A good leader and pastor knows how to provide great leadership and protection for the church while implementing a plan and realizing the culmination of a vision. Many pastors are in the process of developing such depth. What all our pastors and church leaders need from the congregation, staff, and volunteers is compassion, understanding, respect, and our sincere prayers that the Lord will guide them clearly through the process of fulfilling the ministry that He has called them to.

The bottom line is: Follow the church authority structure, respect it, and pray that the leaders are getting their marching orders from God.

Don't Avoid the Elephant in the Room

The elephant in the room is that invisible, yet all too present spirit of controversy and anger lurking in the corners of every room in which it is tolerated. Problems are almost always evident in a relationship. They might be easy to overlook or set aside, but they're always in the wings until they're addressed. We can make the choice to address issues intentionally in a setting that is civilized, loving, and controlled, or we can avoid them until they spring to center stage at the worst possible time.

An elephant in the room is really difficult to miss. Notably, it's interesting that our human spirit is so torn between addressing and avoiding conflict that we can avoid this obvious presence for years. We see the elephant and we even tend to understand that if the elephant stays in the room unattended for long, things will get very messy. However, most of the time fear gets in the way and procrastination takes control. For efficient communication and for effective ministry power, someone needs to shine the spotlight on the elephant and offer some suggestion about how to remove it.

It's not enough to simply enlighten the crowd about the elephant they already knew was there—a flexible plan for removal should be presented at the same time. The flexibility factor is usually fundamental in the successful removal of an elephant. There might be several ways to clear the room. Let your emotional flexibility help you avoid starting a completely separate problem by disagreeing on the removal technique in addition to the original problem that invited the elephant into the room.

I've avoided many elephants. I've also confronted many, and this is what I've learned.

- Be sure you have adequately submitted the issue for prayer so that your confrontation is presented with a kind and loving spirit.
- Timing is crucial to successful elephant removal.
- Sometimes an elephant can only be removed a little at a time, but if the effort is prayerful, consistent, and earnest, any room can be cleared of all unwanted elephants.
- Even senior pastors and lauded church leaders are capable of ignoring elephants.
- Every room has more room for happiness, love, and excitement when there are no elephants present.
- The byproducts of the elephant in the room are gossip, anger, bitterness, jealousy, deceit, and other sins of all shapes and sizes. Shall we just go ahead and name our elephant Satan?
- When the elephant is gone, the void it leaves is usually filled with love, affirmation, righteousness, wisdom, peace, kindness, and all the other characteristics and attributes of God.
- Fear keeps most people from initiating confrontation. Someone needs to step forward, and that someone might as well be you.

Tone of Communication

It's no secret that there is a difference between what you say, what you mean, and what the listener hears. It's important that we're honest and truthful in all of our communications. All conversations should be infused with kindness, understanding, and love.

Important conversations about controversial issues must also be covered in prayer. I'm convinced that statements that are backed by sincere prayer will reach past the listeners' ears and into their hearts—they will receive what was said in the spirit with which it was presented. I'm equally convinced that comments presented off the cuff with little thought and prayer support will probably be magnified in their condemnation—the listener will hear something much worse than was intended.

Avoid speaking around issues or inferring one meaning behind a barrage of verbal baloney. We've all experienced this communication technique, and most of us have dished it out on occasion. In the long run, it is far more effective to say what you mean and mean what you say. It is also of paramount importance that what you say is presented in a loving and caring way that values the individual.

Because sound operators are subject to a plethora of suggestions, it's very important that they remember to say what they'll do, and then do what they said. It is fundamental scriptural teaching that we all must be honest and follow through.

Mathew 5:37
Just let your "Yes" mean "Yes." Let your "No" mean "No." Anything more than this comes from the evil one.

As you abide by Matthew 5:37, you'll be guided through the process of avoiding conflict. If you are asked to turn the vocals up or the mix down, or whatever else you might be asked to do, be sure you can follow through on your response.

If the sound operator is in good communication with the pastor and authority structure, he or she should develop confidence in the mix. Although this confidence develops over the course of time, it does happen. If the sound operator has no line of communication with the church leadership, there is nothing solid on which to base a response when someone asks for a sound change. At that point, the operator, the congregant, and the leaders are left open to spiritual attack.

Reasons to Stand Up and Speak Your Mind

At all times, the sound operator should be humble, respectful, and accommodating. However, there are certain instances when the correct communication is the most difficult.

Countering Blasphemy

Blasphemy, being the unpardonable sin, is a serious issue, and when we hear it in church, we need to respond. Belief in God is fun and exciting—sometimes we take it for granted and treat our relationship with Him too casually. Sometimes, intentional or unintentional blasphemous comments slip into a conversation. These need to be addressed, preferably in a one-on-one setting with the offender. The sin must be acknowledged and recognized, and the offender must repent.

Mark 3:28–30 (New International Version)
I tell you the truth, all the sins and blasphemies of men will be forgiven them. 29 But whoever blasphemes against the Holy Spirit will never be forgiven; he is guilty of an eternal sin. 30 He said this because they were saying, "He has an evil spirit."

Defending the Pastor

When the conversation leads into areas in which the pastor is being criticized or spoken poorly of, speak up and bring the conversation to an end. This is gossip, it's unhealthy for the church, and it's unscriptural. If someone has an issue with the pastor, church leaders, worship team, or anyone else for that matter, that person should be instructed to approach the person with whom he or she has an issue directly to seek a resolution.

Proverbs 20:19
A person who talks about others tells secrets. So avoid anyone who talks too much.

When the Pastor Doesn't Understand the Facts

Sometimes the Pastor is unaware of the facts regarding a particular circumstance or situation. If the sound operator is certain of information that is unknown by the pastor or other leaders, he or she is responsibile for sharing that information in an effort to be helpful and supportive.

Don't Sweat the Small Stuff

Avoid petty grudges and conflicts. Many, if not most, of the issues that dominate our thoughts are inconsequential. It takes a conscious decision to think the best of someone, but that's what we should strive to do. Give people the benefit of the doubt. If a comment or action isn't a sin, and if it can be assessed from a positive viewpoint, it shouldn't typically be a source of much concern.

When emotions and feeling are involved, small casual comments can produce a disproportionate emotional response. Most of the time, the people who make brash and potentially controversial comments don't even know what they have done. Their comments were intended as casual nonsense, but these comments can escalate into a full-blown confrontation—this is the work of Satan himself.

Condemnation, controversy, strife, rage, insecurity, and confusion are just some of the traits of the devil. If you're in a situation that is infused with these feelings, you can rest assured that everyone is listening to Satan's whisper more than the Word of God.

A consistent prayer life, regular Bible reading, and a strong commitment to the ways of God will help each of us see and recognize the truth while avoiding Satan's lies—it will help us to not sweat the small stuff.

The Best Ways to Get Your Point Across

Having opinions and needing to express them is just part of being in any group or organization. It's easy to share an uplifting and supportive point, but sharing a critical comment is much more difficult. There are good and bad ways to get your point across to those in control.

Prayer

The more we trust in the power of prayer, the better off we'll be in every area of ministry. There are some fundamental disagreements that have no real absolute solution. These issues are best committed to prayer. If each of us would trust that the issues we bring before God are in good hands, a lot of controversy would be eliminated. We pray that God's will would be done in a situation, and this not only offers the potential to change the opinions and behavior of others, it also could possibly change us in the process.

It is a foolish person who prays that he will get his way in a battle over ideas, procedures, or anything for that matter. In prayer, we're drawn to trust that God has everything under control and that He'll provide the outcome that is best for us and all those who believe.

Follow Protocol

A good leader will implement a plan that is clear, concise, and accessible. The chain of command should be clearly defined. The sound operator should have a clear idea about to whom he or she answers. Always follow the direct authority path when dealing with a concern. The structure was designed with much thought and is meant to be followed. Going around someone to the top undercuts that person's authority and results in hard feelings and a potential foothold for spiritual attacks.

List Your Concerns or Joys

Schedule a personal meeting with your leader and make a list of your concerns. It's essential that you organize your thoughts and concerns in the order of their importance. There's a good chance that, when you make your list, you'll see at least a few concerns that are unfounded or so small that they don't create a problem for you or anyone else.

It's also important that you meet in person whenever you're presenting concerns. A one-on-one meeting lets each party see the intent of heart. Avoid e-mail communications regarding controversial issues. It's far too easy for the e-mailer to be misunderstood. Without visual cues and real-time human response, the real intent of the message is easily skewed. In addition, it is much easier to write a scathing condemnation than it is to deliver one in person. A personal meeting lets you see that you're dealing with a person with feelings, concerns, and vulnerabilities.

Not all meetings need to have a negative tone, but it seems that we meet with leaders when sharing concerns much more often than we meet to affirm their leadership and friendship. Schedule a meeting once in a while just to tell your team leader or pastor how much you appreciate him or her and why. Do your best to be thorough.

There's a good chance that the team leader or pastor will ask you whether you have any concerns or things that you think might make the ministry stronger. This is a good chance to share a vision and to help dream of what God could do with your team, but really resist any opportunity to turn the meeting into a downer. Be gracious and kind. Keep the environment upbeat and appreciative.

The drain on church staff and pastors can be dramatic. There are so many things that they do well. They pray regularly for the church, and they struggle to balance their home life with church. A minister can easily leave his or her family in the background, giving a total effort to the church in the name of God, but leaving behind a family in shambles. A positive meeting once in awhile with someone who genuinely cares is like a drink of cool, fresh water to a parched traveler. Your pastor will appreciate your time and will be grateful for your prayers and support.

The Worst Ways to Get Your Point Across

A lot of really bad communication techniques are born out of good intentions. You could be as right as right gets, but if a situation is handled poorly, your righteousness could result in damaged hearts and lives.

Gossip

The fruit of gossip is pain, hurt feelings, embarrassment, and humiliation, just to name a few things. Anytime you're involved a conversation in which you or someone else is putting someone down or telling incriminating stories, you're right in the middle of the gossip. You should immediately stop and apologize or ask that it be stopped. Talking poorly about people behind their backs is unfair to them and to the people who are hearing the information.

It is completely likely that just after someone has told an incriminating story, there will be reconciliation and everything will be patched up and better than ever. Meanwhile, the third party is still carrying anger or disgust for the incriminated person. It's really just not good, and it should stop.

I've been in churches in which gossip ran rampant, and I've been in churches in which it wasn't tolerated. There is a much better and completely different spirit in the church that doesn't allow talking behind people's backs. There are few things that I feel this strongly about, but I've seen its destructive power, and I advise against your participation.

Proverbs 26:20 (New International Reader's Version)
If you don't have wood, your fire goes out. If you don't talk about others, arguing dies down.

Proverbs 16:28
A perverse man stirs up dissension, and a gossip separates close friends.

Stirring the Pot

Immature and unethical people might think that they could get their way by stirring the pot—plotting one person against another, starting rumors, and enticing people to choose sides. Although these are common worldly techniques for individual advancement, they have no place in church. The fruit of this type of action is rotten.

There is beauty, peace, and harmony in achieving an objective through honest, prayerful means. Stirring the pot causes strife, anger, and resentment to turn into a raging, out-of-control tornado—bad idea!

Setting Up Camp

Avoid creating an "us versus them" mentality. For any argument, it's likely that there will be some who agree with each side. When camps start to evolve, a clear line of communication needs to be used to put this to an end. The only camp any of us needs to be in is God's—we should all be in it together. Any other division in forces only weakens the force. In this day and age, we need all the unity and strength we can get.

Well...I'll Just Do It Anyway

Not a good plan! Most of these really bad ways to get your point across seem obvious, but I'm mentioning them because I've personally seen them all happen—and some of them I've actually done. I know that if they were such obviously bad ideas, rational people wouldn't participate in them; however, they do participate—often!

Just in case there was ever any question, simply rebelling and doing something your own way, against the request or recommendation of your pastor and church leaders, is the wrong way to proceed. If you vehemently disagree with a plan, concept, or opinion, try some of the best ways to get your point across that I mentioned previously.

The E-Mail Bomb and Why You Should Avoid It

If you've ever written an e-mail bomb, you already know the feeling. It feels good to really give 'em what for—to give the knuckleheads a piece of your mind. While you're writing, you just might feel self-righteous and justified in every way. You're able to say things in an e-mail message that you've been too intimidated to say in person. These people need to see the light as you see it. Without your insight, they're just being really stupid, and if they would only adopt your ways and implement your solutions, everything would be just fine.

So, there it is—your masterpiece, ready to send. You do it. Send! Now you can hear the hissing fizzle sound of the fuse, PHZZZZZZZZZZZ.… Feels kind of good, doesn't it?

BOOM! It has landed in the recipient's e-mail box, and he or she has read it and is responding. If you're lucky, the recipient will e-mail his or her response filled with equally venomous tripe. However, the recipient will probably be mature enough to leave a little time for the fire to settle, and then he or she will call and ask to meet in person. Again, if you're lucky, the recipient might just rip you apart on the phone and say a lot of nasty stuff that he or she will be sorry about later.

Ah, but here's what will probably happen. You'll meet with the recipient in person. (It's much more difficult to be mad at someone in person.) The recipient will be very repentant about his or her actions, and then he or she will explain and—nightmare of nightmares—there will be a circumstance you knew nothing about that makes his or her actions completely sensible and, in fact, even downright brilliant. Ouch!

Yeah, the e-mail bomb is a bad idea—an idea that is archived in cyberspace for all time. Bummer!

Negative, Harsh, and Judgmental Attitude

If you find yourself constantly making negative comments about your pastor, staff, and church—and you'll know when you're doing it—evaluate the last couple years of your life. There's a good chance that someone really hurt your feelings, abused you emotionally, or was just mean to you.

From my experience, I've seen these actions as a reaction to pain—they're a defense mechanism that is holding everyone at a distance so you won't get hurt again.

I'm no psychologist, but I've seen this happen a few times. It is a tragedy to spend your whole life angry, keeping everyone from getting close by your poor attitude. God can easily heal these types of pain, and I know He wants to—He might just want to do it through a good Christian counselor. Think about it.

chapter 5

TO GET PAID OR NOT TO GET PAID

There are some really great paid sound operator positions around the country. They pay pretty well, have great benefits, and support world-class musicians. The house systems are amazing, money doesn't seem to be an issue when it comes to getting new gear, and they usually have really nice recording studios. The only problem is that there are only a handful of those gigs in each major marketplace, and the line forming behind the current operator is long and has been forming for years.

The Volunteer Sound Operator

Volunteers fulfill the vast majority of church sound operator positions. I've been teaching sound classes for a number of years and I've noticed that most of my classes are full of church sound people. You are on a quest for knowledge that is shared by virtually every other church sound operator.

It Doesn't Matter if You're Paid or Not

What we do, we do for eternal reasons. It really doesn't make any difference whether the sound operator is paid—the purposes of the church service far supersede the need for money. Therefore, our quest for excellence equals or surpasses that of the paid sound operator. The ability to participate in a service that promises life-changing joy for those in attendance is a privilege and an honor that can only be valued in eternal gratification.

Do Your Best

Always give your best effort. Let others on the team see your commitment in your diligence and desire to do a great job. Sometimes these things just need to be said. It seems obvious to many that we each should give our best effort to everything we do. Why, then, is it so common to see people slacking off or being lazy and acting bored?

Some people were raised in an environment in which excellence and a positive mindset were expected in everything. Other people were raised in a totally different environment in which laziness and sullenness were not only accepted, but modeled.

Just in case you need to hear this: Do your best, give your best effort, and expect great things from yourself and those around you.

Enjoy Being Part of Ministry

Part of doing a great job in ministry is enjoying what you do. At work, at home, or in church, it's pretty easy to spot the folks who don't enjoy what they're doing. Their negative attitudes are infectious. A positive attitude from someone who loves what he or she is doing is more infectious, so through simply enjoying what you do and sharing your excitement with others, you can change the tone of the team.

Certain people are simply charismatic. They are exciting to be around, they tend to lift the spirits of those around them, they're usually well-liked, and typically they're good leaders. People, in general, are drawn to those with charismatic personalities, largely because they're fun to be around and they tend to raise the excitement level of the activities in which they participate.

If you're excited about being part of the sound ministry, be encouraged to express and share your excitement. Don't worry if you aren't inherently charismatic. Your heartfelt and open expression of joy and excitement will do the work for you. You'll probably notice that other team members seem to like being around you, and the mood of the entire team might rise to a new high.

Be Sensitive to Others

We can only control our own attitude, outlook, and excitement. If other team members are sullen and negative, keep in mind that we are all in the process of working our way through life. They just might be at an emotional low point due to any number of factors that just happen throughout life.

Right now, I'm encouraging you to be excited about life and what you're doing, but you might be in the middle of a near-paralyzing life event. I'm still going to encourage you to be excited about being part of ministry because music and sound ministry are both exciting activities. However, we all need to be sensitive—we never know exactly where other people are in the emotional ebb and flow of life.

If I'm riding a crest today, you might be in a valley; tomorrow it might be vice versa. If I try to enforce or impose my excitement on someone, it might irritate that person into a deeper funk. On the other hand, if I'm genuinely enjoying the tasks I've been given and I'm very caring, understanding, and gracious, I'll be presenting the heart that pleases God.

It is God who heals brokenness. Consistent prayer for our friends, family, and ministry team members is essential. As we pray for the Lord's will and then as we align with Him, there is an amazing realization of power and effectiveness. Be sensitive to your team members, encourage them, and help provide opportunities for spiritual growth through prayer and purposeful fellowship.

Matthew 9:36
> *When he saw the crowds, he felt deep concern for them. They were beaten down and helpless, like sheep without a shepherd.*

Isaiah 42:3
> *He will not break a bent twig. He will not put out a dimly burning flame. He will be faithful and make everything right.*

Luke 24:45
> *Then he opened their minds so they could understand the Scriptures.*

Be Punctual

Here's another point that seems obvious—be punctual. Why, then, do so many of us show up late for rehearsals, services, and so on? Tardiness can easily become a lifestyle—some folks are just late for everything they do. Ministry is primarily a volunteer activity, so many people consider it optional to be punctual, if they even attend.

The fundamental issue in punctuality is the demonstration of the value we place on our fellow team members' time. If we're constantly late, we're essentially telling everyone that our time is more valuable than theirs—it's all right if they waste their time sitting around waiting for us to arrive. If we're always on time or early, we're demonstrating humility and a willingness to serve—we value the ministry and everyone involved enough to arrive on time, prepared, and raring to go.

The church leadership sets the tone for punctuality. If the team shows up just to sit around and wait for their leaders, they will soon catch on—they'll start showing up closer to the time things really start rolling. This trend can quickly spiral downward. Before you know it, the leader will start showing up later because the team isn't showing up, and then the team will show up later, and so on. Eventually, everyone will be upset about the trend to tardiness, a meeting will be called, tempers will flare and fingers will be pointed—it's an ugly scene. Sounds like I've been there, huh? Actually, I've been involved in this phenomenon from all sides. Let's just agree that it's best to be on time or early. Leaders should arrive soon enough to be well-prepared at least 15 minutes before the team arrives.

The Quest for Excellence Shouldn't Change with Pay

The sound operator donates his or her time and efforts in the name of God. We are actively involved in a process that introduces people to the creator of the universe. If we consider the full impact of our task—communicating the message of salvation and eternal life—we are left with no other option than to give 100 percent of our very best.

In ministry, volunteers often operate under the premise that says, "If they were paying me to do this job, I'd care a lot more about it." If that is your mindset, it might be time for a tune up. I can empathize with those who have gone through uncomfortable church or life experiences and who are involved in a rebuilding phase. There can be times when you're just helping fill a need while distracted by disappointment or a spiritual battle. However, if you'll continue to fight the good fight and to draw closer to God, those times will pass. In the meantime, value the ministry work you're involved in and give it your best effort—the dawn of an even better time is ahead.

1 Timothy 6:12

Fight the good fight of the faith. Take hold of the eternal life to which you were called when you made your good confession in the presence of many witnesses.

Philippians 4:8

Finally, my brothers and sisters, always think about what is true. Think about what is noble, right and pure. Think about what is lovely and worthy of respect. If anything is excellent or worthy of praise, think about those kinds of things.

Titus 3:3–8

At one time we too were foolish, disobedient, deceived and enslaved by all kinds of passions and pleasures. We lived in malice and envy, being hated and hating one another. But when the kindness and love of God our Savior appeared, he saved us, not because of righteous things we had done, but because of his mercy. He saved us through the washing of rebirth and renewal by the Holy Spirit, whom he poured out on us generously through Jesus Christ our Savior, so that, having been justified by his grace, we might become heirs having the hope of eternal life. This is a trustworthy saying. And I want you to stress these things, so that those who have trusted in God may be careful to devote themselves to doing what is good. These things are excellent and profitable for everyone.

The Paid-Staff Sound Operator

All right! You've studied, excelled, implemented your studies, and been a beacon of positive excitement. You've been faithful with your time and generous, caring, and sensitive to your fellow team members. And all of a sudden, you've become a paid-staff sound operator. Congratulations! A life devoted to ministry can be the most rewarding and worthwhile investment you'll ever make. Valuing eternal things more than earthly things has great valor.

The Part-Time Gig

There's a good chance that the staff sound operator position is part-time because most churches don't have the funding for a well-paid sound tech. If your family is supportive of your ministry efforts and willing to share you with the church, the part-time status might work really well. If your family becomes jealous of the time you spend at church, you might be heading down a difficult road.

Be forewarned that a part-time position at church is often only part-time on paper. It can easily demand your full-time effort, in addition to your other work and life responsibilities. It is up to you and your pastor to see that all things are kept in proper balance. A good leader will keep track of the staff's time expenditures and help keep them in line with realistic expectations, understanding that high achievers often place the realization of goals above relationships. It will do no one any good if the new hire burns out because of family problems, frustration, or just plain fatigue.

Accountability

Because it's already established that we should do our best, whether we're being paid or not—and that the cause, in ministry, is well-deserving of the highest degree of excellence we can provide—the accountability to do a great job as a staff member shouldn't matter.

Responsibilities of a Paid Sound Operator

Once a salary is involved, your responsibilities are very likely to increase. You were probably hired for a list of reasons:

- You are capable of running the system.
- You are capable of maintaining the system.
- Various ministry leaders wanted A/V support for their ministries.
- Special events require better A/V support.
- The equipment needs to be stored properly, and storage areas need to be constructed or designed.

- Someone needs to train the volunteer sound operators.
- There are far too many events and services to expect a volunteer to cover them.
- The pastor wants more consistently excellent sound at each service and event.
- . . . and the list goes on and on.

Once you're paid for your services, there is an increase in accountability. When you accept the paid position, you should receive a formal job description that outlines your specific responsibilities so you'll know just what you're committing to.

Once you make the commitment, you're obliged to perform each task with excellence. There's a good chance that you'll like some tasks on your list of responsibilities better than others, but you still need to do your best to perform them all well. Battle against procrastination! Your least favorite task can easily live at the bottom of a list that never ends—it'll never get done. If you make a commitment to perform the tasks, they need to be performed. Your "yes" needs to mean "yes," not "yes, if it ever makes it to the top of my list."

In the modern world, everyone is busy, overworked, and behind the eight ball. You'll receive sympathy when you neglect your responsibilities and, especially in church, you'll receive compassion, caring concern, and support. However, when you fulfill your obligations and produce results it will be noticed—you'll be representing yourself as faithful and trustworthy while you make the statement that God and your church are worthy of your best effort.

If you're being hired by your church and there is not a job description, ask whether you can write one. Leaving open-ended expectations and assumptions is a bad idea. Often, the sound tech is hired out of a desire for consistency of quality. Many pastoral-minded folks might not think of putting together a job description—all they want is release from a constant stream of sound issues. You'll be doing yourself and them a favor if you set down with them and specify what it is that you will be doing to help. This simple procedure will help avoid confrontation and misunderstanding once you settle into your position and realize that your new employer's expectations are completely different than what you thought you were signing up for.

Serving Multiple Ministries

As the staff sound operator, you'll probably be expected to serve the needs of all the ministries in the church. If you're highly organized and very administrative in your approach to life, this shouldn't be a problem. If you operate in a random fashion, responding to the immediate needs at hand, you could be headed for a growth phase.

It's unrealistic to expect even the most super of superheroes to be everywhere to serve the needs of everyone, all the time—that kind of activity is reserved only for God. Considering that fact, you'll need to put together a team, a schedule of events that need

Clearly Label All Cables and Components

If your team is inexperienced, don't take for granted that they'll be able to get a quality mix together quickly. Label everything that can be labeled so that most of the system settings can be roughed in without hearing a thing. Whereas we never want to end up depending on these markings, they guarantee—for the volunteer—the smoothest transition from rookie status to accomplished sound operator.

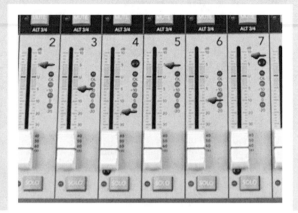

Stick-on markings are very convenient because they can be easily removed and repositioned. The mixer (right) has each fader marked. If each of the faders has been set to its mark, and if the input trims have been marked and set, there's a good chance that the mix will be close to an acceptable status. Considering all the potential variables in a sound system, the sound operator will still need to proceed carefully even if all the controls are on their previously-positioned marks.

A/V support, and a system that matches sound operators with events. You'll also need to train the sound operators so that they are fully capable to function independently—you should only receive occasional panicked calls from struggling sound techs about a problem that requires your depth of knowledge.

Constant communication with the ministry leaders is mandatory, and keeping a current schedule of needs is a must. Most churches hold regular staff meetings, which you'll probably be required to attend; these meetings provide the perfect connecting point. You should be able to do almost all of your personal interaction with ministry department heads in these meetings. Be sure to document all communications and events and follow up with confirming e-mails. E-mail is one of the most valuable ministry tools. With one e-mail all involved parties can quickly be informed of any conflicts, questions, or needs—technology is indeed our friend.

Training Ministry Leaders

The worst possible scenario for a staff sound operator sets him or her up to be the "Sound Savior." Sometimes, when a problem arises, it seems as if the easiest way to fix an A/V problem is just to fix it yourself. Establishing this type of pattern is very shortsighted and counter-productive.

In the heat of the moment, with disaster imminent, the staff sound tech might represent the only immediate solution to a problem. However, this is where a little pre-planning comes in very handy. A wise sound operator will set up a series of training sessions for all ministry leaders. This should include paid and volunteer staff and lay leaders—virtually anyone who has sound system requirements. Although it will require a substantial time investment to develop the training materials, the dividends will be glorious.

Most people are intimidated by sound equipment. It baffles most and irritates many, so you will need to provide small, bite-sized chunks of information—information that assumes absolutely no knowledge of the workings of music, sound, lighting, electronics, or fundamental logic. As a rule, ministry leaders require very functional instructions. They

Setup Diagrams

Create detailed diagrams of your sound system. Start with drawings that depict the basic connections (right) and then create detailed explanations with drawings and pictures (below) that illustrate each specific connection in the system.

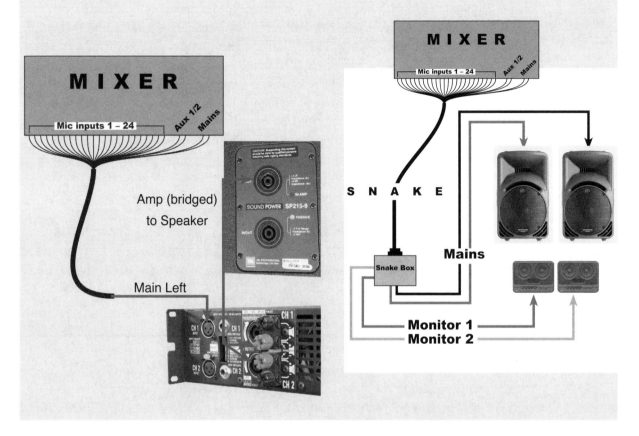

don't care how things work; they simply need to know how to connect the basic components and how to quickly achieve functionality.

Whether you're training the ministry leaders to set up portable systems or to power up and operate the main system in the sanctuary, simplify the process by eliminating unnecessary cables and components and by labeling everything extremely well. Use a professional labeling system and create large easy-to-read labels for:

- All components
- The primary controls on the mixer, such as auxes, master faders, subgroup assignments, connected effects returns, CD inputs, computer inputs, and so on
- Microphones
- Cables at the source end
- Cables at the destination end
- Snake boxes and pigtails (the fanned end of the snake with separate connectors for each channel)
- Direct boxes

- Key settings, such as suggested fader levels, EQ settings, effects settings, mute settings, wireless settings, and aux send and return levels
- Effects
- Any other important system components and settings

It is very important that you design each system so that there is little or no technical thought required to provide a successful setup experience for anyone.

Provide an accurate setup diagram with every connection, component, and setting labeled identically with the equipment labels. Take digital photos of the key connection points, as well as any suggested settings. Sometimes these diagrams end up several pages long, depending on the complexity of the system, but they provide an excellent training aid for leaders and staff, as well as providing fundamental information to new sound operators.

Once you've developed all of the necessary support materials, be sure that they are readily available for all who need them. Each system, whether portable or permanently installed, should have these materials close at hand. Portable systems should have a special place for setup documentation in a mic box, mixer case, or some other accessible road case or rack. Permanent systems should have a special place for all pertinent setup diagrams, equipment manuals, and other system documentation.

This type of attention to detail and support for the ministry leaders and staff is the mark of an excellent staff sound operator. Your efforts will be greatly appreciated and you will have formed the basis of the training materials you'll need to begin developing a sound ministry staff.

Building a Team

As a staff sound operator, your fundamental charge is to make sure the best possible audio is available for every church function—that seems fairly obvious. However, unlike most of the rest of the world, churches thrive because of the efforts of volunteers. With help from God, they achieve great things and do good works.

In the workforce, an employee commits to fill a job position, understanding that there will be certain demands that are not negotiable. If the employee expects to get paid, he or she will perform the tasks involved in the job. Volunteers, on the other hand, require care and support. Even though they should fulfill their commitments, their motivation must common from above—from sources that promise eternal gratification and quench a desire to serve a worthwhile cause.

Almost every church staff position requires the recruitment, support, and nurturing of volunteers. The sound ministry is intimidating for many; however, with a little grassroots networking at church services and functions, it is likely that you'll find a person who has some experience with sound equipment or, possibly even better, a person who loves music, listens to music all the time, and really wants to run sound.

Rehearsals are an excellent place to break in potential sound ministry members. Invite several people to join the worship team at rehearsal. Schedule them so they only show up one at a time, so you can get to know them and so they can show you some of what they know about audio.

Developing a Volunteer Crew

A team won't develop its synergistic potential unless it participates in group activities. A sound team is a little bit unique. Although the team might include several members, the task is almost always a solo activity. If there aren't intentional activities that build camaraderie, each team member will operate essentially as an island, surrounded by team life but never developing any connection to the other islands.

Prayer and Share Time

One of the quickest ways to build connections between team members is to implement group prayer times. Schedule regular meetings of the sound team and start each meeting with prayer time. Ask the team members whether they have any prayer requests or praise reports. There are all sorts of ways to cover everyone's prayer needs, but it needs to be done.

The obvious spiritual benefit is in the submission of prayers to God and the faith-building process of seeing them answered. There is no stronger bonding experience than sincerely praying for someone's heartfelt need. We learn to trust God to answer the prayers in His way, and we learn to comfort one another during the process.

The benefit of prayer and share time is amplified as individuals begin to learn more about their fellow team members—the islands connect! I've seen this process completely change the team dynamics in multiple churches. Once the members get to know each other, their ministry potential increases dramatically.

As the team leader and staff member, it is your responsibility to guide and direct the team in the direction of a common vision. Once the group is praying together and participating in each other's lives, potential begins developing. In order for the relationships to realize their full potential, the leader must provide an atmosphere in which dreams and desires unite—in which the team shares a common vision. Once everyone catches the vision, the ministry is truly birthed. The collective power of shared prayer provides spiritual muscle and intellectual clarity. The collective power of a shared vision multiplies the potential impact for the kingdom of God.

Special Functions

Team-building activities are an excellent way to solidify the unity that builds as the team grows closer. When the team begins to get together outside of the regular church services or team meetings, several things happen at once.

- The team grows closer.
- Greater capacity for accomplishment is a byproduct of increased unity.
- Your team members can realize the life-enriching process of developing true friends—friends who care enough to get to know them and who know them enough to care what happens to them.

Special functions can include training sessions, guest speakers who are experts in the audio community, or even a visit from the senior pastor as a point of encouragement and vision sharing. These are all great activities, but be sure to blend in some functions that are just plain fun. The collective personality of your team will determine the most appropriate activities; however, here are a few ideas of activities that I've seen work well:

- Go-kart racing and dessert

- Paintball and breakfast
- Weekend retreats to a mountain chalet
- Christmas parties
- Summer barbecues
- Movie nights at the theater
- Movie nights at a member's home
- Thematic dinners at a member's home
- Potlucks for various purposes, such as birthdays, anniversaries, births, and so on
- Concert nights (Christian or secular)
- Team-member appreciation functions with gifts, food, and acknowledgements of achievement and faithful service

Your imagination and the likes and dislikes of your team should easily direct you to fun team-building activities. The most important thing you need to do, as a leader, is schedule the events. Put them on a calendar. Plan for the entire year. Schedule all of your events so that you can easily see spaces where an event doesn't conflict with other events, holidays, or special services. In my experience, if you don't put these valuable group activities on the calendar and commit to following through, you probably won't get around to them.

Motivating and Assisting

It is important that the leader strive for a strong spiritual and emotional tone in the team. The leader typically sets the mark for the rest of the group. If the leader is sullen and introverted, the team will probably lean toward being sullen and introverted. If the leader makes the effort to be upbeat and excited, the team will probably be upbeat and excited.

A motivating atmosphere is full of positive reinforcement. Ideally, comments and critiques should be presented in a friendly, supportive, and uplifting manner. The way you communicate with your team is somewhat dependent on their skill level and individual personalities. You'll probably discover that the higher the team member's skill level, the greater the likelihood that he or she will respond well to constructive criticism.

Most people are fairly insecure—they might respond poorly to any criticism simply because they're uncomfortable and tense in their situation. A good leader will assess each individual and adjust instructions to help him or her succeed.

If you are extremely knowledgeable, you could easily overwhelm the newcomer. Dole out your instructions a little at a time. Learn to read your team members. It's fairly easy to tell when someone is becoming overwhelmed; however, in your excitement to share knowledge with that person, you might overload his or her brains. It is better to hand out bite-sized pieces at an even yet manageable pace than it is to spew forth a barrage of knowledge, expecting that the person will assimilate it over the course of time.

One of the best ways to establish the proper learning pace for your team is to develop a curriculum. Organize all of the tasks involved for sound operators during each service. Some tasks will be very simple to explain; others will be simple to say but a little less simple—possibly even difficult—to explain.

"Set worship leader's mic on a stand at front, center stage," is obviously a very straightforward instruction—it simply is what it says it is. "Mic and EQ the kick," sounds like a simple command, but it is a very involved command for most new sound operators.

Develop a Curriculum

A written curriculum provides a road map for the efficient instruction of new team members. Without a tool like this, instruction easily become an aimless and rambling exercise in futility. The time spent developing a logical and well-thought-out outline of your instructional plan is an excellent investment in the development of your team—it helps your new team members succeed, and it shows that you value their participation and respect their efforts enough to give them the tools to do the job correctly.

When developing a curriculum, break your expectations for sound operators into small bite-sized pieces. Check them off as you cover each small module. Keeping the modules small makes it easy to slip an instructional time into the regular flow of setup and striking—you can always cover more modules if you have time.

Be sure to cover:
- Setup procedures
- Storage
- On-stage connections
- Mixer connections
- Outboard processors
- Monitor systems
- FOH systems
- Playback and recording devices
- Mix techniques and common settings
- Acoustic considerations

To properly mic and EQ the kick drum, you must explain the detail involved in microphone placement—the way to find a good sound before you adjust equalization—and then you must explain some basics about signal path management. Finally, you need to explain something about the theory of equalization.

You must decide whether you want to train the sound operators in a way that they'll eventually grow into excellent sound techs who understand what they're doing—and can apply the concepts they know independently to each new scenario. Or, do you want to achieve more immediate results by simply telling them where to set the controls? Practically speaking, a blend of these two approaches is advisable.

Sunday church services happen every Sunday like clockwork. Sound operators must immediately function as well as possible. The success of the service is intertwined with the successful orchestration of several practical and spiritual aspects. However, a poorly executed technical service distracts immensely from the power and impact of the other ingredients. It is important that the sound operators are given the information they need in order to provide transparent technical support—support that doesn't call attention to itself while providing excellent communication of music, media, and ministry.

On the other hand, the team leader should train the team so that the members are self-sufficient. If the leader has to do all the thinking for the sound operators at a large church, he or she might look very intelligent and feel very important, but, in the long run, he or she will burn out quickly. The energy involved in training a crew to be independent is very front-loaded. You'll have to spend considerable time to develop materials that show people why things are done a certain way, rather than simply how they're done. It's worth your time and effort to produce this type of training. It is useful in many ways:
- It produces a team that can work their way through problems without calling the leader.

- The collective knowledge accumulates quickly as team members build on the knowledge they receive.
- The church has documentation of the leader's responsibilities and how they're accomplished.
- The leader can spend more time building more efficient systems and procedures, instead of spending hours and hours handholding unknowledgeable team members.
- It is much simpler to include new sound operators because the system is already in effect.
- The team members become much more intellectually involved in the ministry—the knowledge they get intensifies their desire for even more depth.
- Everything runs efficiently, and the entire team lives a more stress-free life.

Training others well is sometimes uncomfortable for the leader. There is an inherent fear that the leader will lose importance and influence if the team members get too good at their jobs—the church could realize that they don't really need the leader after all. In reality, a good leader is highly valued by any intelligent organization. Developing a strong, highly skilled team is the mark of an excellent executive. A strong leader with excellent people skills, the emotional security required to help others succeed, good administrative skills, and a heart for ministry is powerful for the kingdom of God.

When it's time to train new sound operators, follow a very simple and time-tested prescription for success. Work through the process following this simple order:
- I do. You watch.
- You do. I watch.
- You do.

If you'll simply let them watch you do sound, they'll pick up many things before you even need to tell them. Be focused and attentive when you run sound for them. Avoid any passive actions at all. Demonstrate the actions of a person who is completely devoted to creating the best-run service of the decade. They should be able to see your attention to detail, and they should feel your intense desire for perfection. We obviously don't need to hold everyone to the quest for a perfectly run service, but everyone must know that we're always trying for that very goal.

When the person does sound and you watch, take notes and make constructive comments about what they could do to provide a better service for the church. Most people will want to know the truth about their performance of the sound operator duties. Tell the truth in a positive way. Avoid degrading and cutting comments.

If someone did a really bad job of riding the leader's vocal mic level, instead of saying, "You did a really bad job of riding the worship leader's mic," you could say, "Try to focus on the leader. Keep your attention on the mic level and tone, adjusting it so it is always audible and pleasing to the ear." This suggestion is full of doorways to further instruction, it doesn't degrade the person or their skills, and it contains keywords (focus, audible, and pleasing) that will help produce positive results.

Training volunteer sound operators is an exercise in motivational procedures. On one hand, the leader wants them to do an excellent job soon. On the other hand, they are volunteers, and this is an optional activity—if the leader is negative, harsh, and critical, the volunteer might just stop volunteering.

The best leaders show people what excellence is—they either model it themselves, or they bring in outside help to give the trainees an idea of what it looks like to do a great job. Next, they'll break the job up into bite-sized pieces—tasks that are not so daunting that they present an insurmountable challenge. As the trainee builds confidence through the successful performance of each portion of the overall job, he or she will soon be functioning far beyond even his or her own expectations. In addition, a good leader demonstrates that he or she cares about the person doing the job.

We've all had teachers, leaders, or bosses that we would do anything within our power for. Though we might have been impressed with them because of their excellence or charisma, the ingredient that leads to loyalty is care—they genuinely seem to care about us as people. Yes, they're interested in us as workers or volunteers, but they take an interest in us reaching our full potential for our sake more than theirs. They take joy in our professional and personal development, achievements, and successes. However, they often share in our pain. They show compassion when we hurt, they help us up when we fall, and they have genuine expectations that we will excel—that we'll succeed at what we do, and that we have value and worth whether we're at the top or bottom of our game.

If you care about your team, if you're willing to help them achieve, and if you provide sufficient resources for them to learn, you will develop a strong team over the course of time.

Job Descriptions

The sound team leader should develop job descriptions for each position. Most churches have a wide range of sound support needs, and each task is unique. Youth groups, various adult ministries, children's groups, dramatic presentations, and regular Sunday services could all utilize different sound systems.

You will help yourself and your team if you prepare a detailed description of setup and striking procedures, as well as operational requirements, insights, and tips. Again, this is a front-loaded activity that requires substantial energy to initiate, but offers a huge payoff. The team members will be happier because they'll know what they're getting in for, the ministry leaders will be ecstatic, and all the sound techs will be better at what they do.

Be sure that each job description includes:

*   An explanation of the ministry
*   The potential range of the time commitment involved, including typical event times, sound team staff meetings, setup and strike times, and the overall length of the commitment to serve
*   A list of required equipment and cables
*   The location of the equipment
*   A detailed setup map showing the physical location of everything and everyone on stage
*   A diagram showing how every piece of equipment is connected
*   Suggestions for an efficient setup procedure
*   A list of instruments and voices on the music or drama team
*   Suggested equipment settings
*   The location of all circuit panels and a digital photo of the circuit breakers that control each power outlet in the room

- The striking procedure and special instructions regarding storage locations after the event
- Contact information for the sound team leader, music director, ministry leader, facilities administrator, and so on

Over the course of time, develop these job descriptions so that they are complete, accurate, and helpful. This type of job description can serve as a handbook for each team member, helping to eliminate confusion and increase efficiency.

Scheduling

Sometimes, the paid sound operator just shows up for church and runs sound, possibly maintaining the gear and making sure everything gets set up before the service and put away after. However, typically the staff sound operator is given the primary responsibility of building and administrating a team of volunteers. In a ministry application, there is a high level of importance placed on providing an opportunity for the congregants to participate in each ministry—the sound ministry happens to be an area in which volunteers can be trained to function well, as long as they have a sufficient aptitude and a strong desire to learn.

Once you build a team of several sound operators and as your church ministry needs expand, scheduling will become a concern. Consider that most thriving modern churches have enough different ministries to fill up several evenings each week, along with several weekend activities, so scheduling audio support for all of them is an administrative extravaganza. The need for organization is imperative in these situations.

There are many excellent scheduling software packages available on the Internet; plus, the calendar programs available in the Mac OS iCal, Outlook, Entourage, FileMaker Pro, Datebook Pro, and so on are becoming increasingly powerful and useful. Whatever software you decide to use to keep your team organized and dependable, use it regularly and keep it up to date.

There are several features that you should look for in your scheduling system, including:

- Wireless bidirectional synchronization with your PDA, phone, laptop, desktop, work, and home computers
- Automatically generated calendars in HTML format for Internet distribution
- Individual calendars for each team member
- A comprehensive reminder architecture that automatically generates e-mails for sound operators, ministry directors, staff, and pastors
- Easy-to-use functionality
- Color coding so that each ministry category displays in its own unique color for easy visual recognition
- The ability to generate calendars that filter by ministry category as well as by team member

Be sure to take full advantage of the Internet in your scheduling communications. Once all the team members understand that e-mail is an important form of communication, they'll check it often and be appreciative of this vital connection point.

Most modern churches have a website, and the system administrator should be able to supply you with a sound team page so that you can post the schedule online. Even in your e-mail connections, link to the online calendar so that any changes that are made are instantly available and accurate. If your team gets used to checking the website calendar instead of looking at an old e-mail with an expired version attached to it, you'll save yourself and them the heartache of missing a service or showing up for one that's been cancelled.

Whatever system you implement, try to find the routine that makes it convenient for you to update the information and easy for your team to access it. Automate as many features as possible. The Internet is full of interesting utilities, and they can be helpful, but in addition, try your hand at scripting languages. Scripts can automate almost any operation, function, or routine. The language isn't always intuitive, but it's also not that difficult to conquer. The beauty of writing your own scripts is that you can easily incorporate multiple applications, documents, and utilities. Scripts provide a convenient way to combine the functionality of your favorite software packages.

chapter 6

VOLUME ISSUES

V olume control is a controversial issue in many churches. There seems to be a constant battle between the music-minded folks on the worship team—who just want the music to sound and feel good—and the pastoral-minded folks who don't want to offend congregants.

As with most of the points I make in this book, I have lived through this dilemma and have learned some valuable lessons in the process. I offer these suggestions so that you can avoid some of the problems I've had, and so you can benefit from some of the things I've learned.

I'm a musician and an audio engineer/producer. I've also been a music pastor. I understand the pastoral side of the volume issue, and yet it is important to me that the mix sound full and musical.

Pastoral Considerations

The pastor is the boss in many churches. The council governs some churches. Each denomination has its own political nuances, but typically the pastor has the controlling vote in things that relate to music and sound. Even if the sound operator answers to the music minister, the music minister probably answers to the senior pastor.

The pastor typically wants to meet the needs of the congregation on a variety of levels, including:

- Introducing new people to God

- Providing an atmosphere that's conducive to spiritual growth

- Providing an environment that's conducive to spiritual depth

- Presenting divinely inspired teaching in a way that inspires the congregation to a closer walk with God

- Providing an atmosphere that supports and encourages worship

- Providing an opportunity for members to exercise their talents and spiritual gifts

- Helping members through life's trials and challenges

- Teaching the congregation to be financially faithful to the church because of the blessing they'll receive from God

- Building a church that is fully functional, effective, and happy

    And there are other very lofty and worthwhile considerations.

    The pastor also has to meet the needs of the church council or governing body. These are very real and practical concerns, and they weigh heavily on most pastors. They include:

- A steady increase in the number of members
- A steady increase in the church income
- Good stewardship of physical resources, such as the building, grounds, furnishings, and equipment
- Wise use of available funds
- Maintaining an effective and motivated paid staff
- Increasing the number of tithers in the congregation

    And there are other very practical and worthwhile considerations

    There is a long list of factors that tug at the pastor, and there are many different viewpoints that need to be considered in day-to-day, week-to-week, and month-to-month church life.

    Frequent complaints about sound system volume, even when they come from just a couple of members in a large church, add to the cumulative stress felt by the pastor. If a pastor consistently walks into the sanctuary knowing there will be an angry member complaining about the volume of the music, it will cause a drain on his or her enthusiasm. Most pastors are very gracious and thankful for help from their volunteers and staff, but there will probably be a point at which all he or she wants is to walk into the church and feel peace. In other words, the sound operator will probably have to overcompensate to help achieve a peaceful church environment.

    Most pastors want similar things from the worship team. They typically want:

- The music to support the congregation
- The team to model how to enter into worship, through their attitude and demeanor
- The congregation to be able to hear themselves sing
- The congregation's attention to be drawn from their troubles in the world to the love, provision, and great news of God by means of an uplifting music service
- The music director and sound operator to work in unity with them to provide an excellent church experience, week after week
- Regular participation in church life and activities
- Respect for authority and consideration of others
- A happy church

    The heart of a pastor cares about people and their spiritual growth. The heart of an administrator cares about organization and structure. The heart of a musician cares about providing great music that sounds good and inspires worship. When the pastor, administrator, and musician agree to work together in the interest of the spiritual growth of the congregation, church begins.

Sound Operator Considerations

The sound operator usually just wants to make the mix sound good. The more exposure the operator has had to great-sounding music, the more likely it is that he or she will be able to construct an excellent mix.

The sound operator typically wants to please everyone, including:

* The pastor
* The staff
* The music director
* The drummer
* The bassist
* The keyboardist
* The guitarists
* The vocalists
* The leader
* The young crowd who thinks the music is always too quiet
* The old crowd who thinks the music is always too loud
* The family members of all of the above
* The newcomers
* The charter members
* And on and on

The diverse demographic that comprises most churches brings inherent possibility for disagreement about the perfect volume. It's not easy to please everyone.

Often, the sound operator is so stretched in all directions that he or she gives up or burns out. Keep reading! It helps to understand more aspects of the volume issue. As we progress into the technical aspects of shaping sound and creating a good mix, we'll cover techniques to help you build a mix that is full, sounds good, and is non-offensive.

Congregational Considerations

The congregation just wants to come to church. They'd like to:

* Learn something new about God
* Be inspired to live a more righteous life
* Spend time in worship
* Feel closer to God when they leave

Sometimes, in the politics of church, we forget how simple church is for the majority of the congregation.

Veteran church members often become embroiled in the structure of the church and express their opinions frequently, but it is easy, even for them, to forget that most people who wake up, get ready, and drive to church care very little about behind-the-scenes politics. They're just looking for a little hope.

Why Volume Issues Are So Common

I'm addressing volume controversies here. If you are a sound operator and you never have volume or sound complaints, then you're very good, very well-liked, very diplomatic, or very scary-looking. I have taught hundreds of church sound operators, and most of them fight frustration about volume issues—they're trying their hardest, but can't find the perfect balance that pleases the musical, pastoral, and congregational crowds.

The volume knob, in general, is the problem. Everyone has one, and they get to set it exactly where they want, whether they're at home or in the car. They're in control of the volume almost everywhere, except at church.

We all have different tastes. For some folks the volume will never be quiet enough; for others it will never be loud enough. Some folks like the vocals louder than the band; others like the band louder than the vocals. Some people like guitars more than keyboards. Some like the drums and percussion to be very predominant; others prefer that they be understated.

As a church body grows and matures, members become a family—they take ownership in the church. Though this is healthy and is the sign of a good church, it also means that more and more members feel comfortable enough to share their opinions. Their opinions about sound will be shared with the sound operator, staff, and pastors.

Most of the time, good leadership will consider all aspects of the volume issue, communicate well to all parties, decide on the correct way to proceed, and everyone will live happily ever after. Sometimes, there is so much tugging from all sides that the pastor is fed up, the sound operator is very frustrated, and the congregants are angry. This is a difficult state to work through, but there's really no other option but to find a peaceful agreement between all concerned.

The Pastor Versus the Music Director

Often, the sound operator is caught between the pastor and the music director. Good musicians will always want the music to sound great. They'll want the mix to have a good balance and blend, they'll want the lows to be full and the highs to be clean, they'll want the vocals to blend, and, of course, they'll each want their own instrument to be the loudest.

The pastor and the music director might not agree on the perfect volume. If you find yourself in this situation, as the sound operator you need to ask everyone involved to meet as a group to help you determine how to please them all.

Agreement Is the Solution

Volume issues that become a problem must be addressed, and all parties must agree upon a resolution. Nothing good can come from ignoring the problem. It will get worse, people will get angry, feelings will be hurt, and people will leave the church.

At some point, there must be an agreement between the pastor, staff, and sound operators about how loud the music service and the sermon should be.

To accomplish an agreement, invite the pastor to spend at least the musical portion of the service at the soundboard. Set up the mix and ask him or her to help you decide on a proper volume—use a decibel meter to quantify your settings. Once you've been through the technical portions of this book, you should have plenty of techniques available to help you build a pleasing mix at all reasonable volume levels.

Invite both the pastor and the music director to sit at the mixer during the music service. If you can get them both to agree that the sound is excellent and representative of the church personality, you are well on your way to peace.

Often, the pastor and music director don't take the opportunity to actually hear what the team sounds like. Pastors usually sits in the first row, where the mix is rarely accurate, and when they are told that the sound is too loud, they tend to believe it. The music director is typically on the platform and is probably in the band, or leading, or singing. It's very difficult to get an accurate picture of what the congregation hears from that vantage point. Simply bringing both the pastor and the music director into the sound operator's area is the best way to resolve a potential volume problem. It is best to establish an agreed-upon target volume before there is a problem, but once there *is* a problem, this step is essential.

It is best to get everyone together at the soundboard during a service. It might seem easier for everyone to get together at a rehearsal, but that doesn't provide an accurate representation of the sound when the congregation is present. The congregants absorb reflection and reverberation, and their singing volume provides a baseline for the live mix level.

Once the pastor and music director have experienced the sound at the mixer, they will have a better idea of just how the team is perceived. Make changes while they're with you, and try to set up a mix that everyone agrees is acceptable. Once you get the stamp of approval from both the pastor and the music director, do your best to maintain a similar-sounding mix at all times.

Use a decibel meter when the pastor and music director are with you so there is a quantifiable measurement of acceptability to which you and the other sound operators can refer in the future. Although a decibel reading is somewhat a vague representation of the musicality of any mix, it does provide a point of reference.

EQ Versus Volume

Very often, someone will complain about the mix being too loud, but what they're really hearing is a particular frequency that is out of balance. Vocals that are piercing typically contain an abundance of frequencies between two and three kilohertz. To many people, especially the elderly, those frequencies are painful. For them, the mix might sound too loud even when the decibel meter reads in the acceptable range.

Once you understand the proper use and application of equalization, you'll be able to construct a mix that is full and pleasing to the majority of the congregation.

Whereas the highs are often too edgy and harsh for some people, low frequencies can be tolerated by most. Try to create a mix that is full in the low end and inoffensive in the highs. This type of mix will typically receive the most accolades from the largest segment of the congregation.

Questions for the Pastor and Music Director

Use a checklist to help establish clear communication. Think of the factors that should be discussed in advance of the service, then when the pastor and music director are listening, you can refer to your list.

- What is your impression of the overall sound of the mix?
- Is volume too quiet?
- Is the volume too loud?
- If it is too loud, what in particular seems too loud? (Leader vocals, instruments, backing vocals)
- If something seems too loud, is it really too loud, or is the sound just abrasive or irritating?
- Can you hear the leader well enough to understand what he or she is saying?
- Can the congregation be heard well enough?
- Is the level of any instrument too high or low?
- Does the mix provide a sound that supports the type of worship service everyone envisions?
- What adjustments should be made to the mix?
- Are the effects too dominant in the mix?
- Does the mix sound too thin?
- Does the mix sound too boomy?

Listen to the mixes of actual church services in Audio Example 6-1. They are all the same peak amplitude, but notice how different they sound in volume. Also, notice how they compare when played over your church sound system.

Audio Example 6-1

The Effect of Equalization on Perceived Loudness
Mix Peaked at 2 kHz, 3 kHz, 4 kHz, 100 Hz, 200 Hz, 50 Hz

Painful Frequencies Versus Comforting Frequencies

The previous examples reveal that even mixes created at identical signal levels (readings on the output meter) can sound louder or softer depending on the frequency balance across the audible spectrum. Typically, elevated frequency levels between 1 and 4 kHz seem louder than elevated frequency levels below 1 kHz or above 4 kHz.

The human ear does not have a flat frequency response across the frequency or volume spectrum. It is most sensitive to the frequency band between 1 and 4 kHz, especially at volumes below 80 dBSPL. We'll cover this topic in greater depth in Chapter 9.

Low frequencies, below about 150 Hz, tend to round out the sound—they help the sound operator create a full-sounding mix. A mix with clean highs above 7 or 8 kHz and warm lows below about 150 Hz typically sounds good at reasonable levels—it usually doesn't need to be loud to sound full. As we work our way through equalization and other sound-shaping techniques, we'll discover how to balance the frequency spectrum throughout the mix in a way that sounds good and is non-offensive.

Sound Reinforcement Versus Music Presentation

Sound reinforcement systems support the acoustic sound coming from the stage. A music presentation system is a playback system that amplifies a musical source, such as a CD, DVD, vinyl record, MP3, or other prerecorded music program.

The greater the amount of acoustical sound coming from the stage, the more difficult it is to create a mix with stunning clarity and intimacy. Many of the volume problems we face are caused by too much stage volume. It's not uncommon for a worship team to be about 90 dBSPL in the sanctuary before the main speakers are turned up.

Acoustic drums, guitar amplifiers, and vocal monitors are often so loud that there is no way even the best sound operator of all time could create a good mix at an acceptable volume. To achieve intimacy and intelligibility, the mains must be so loud that the volume would only be embraced by the youngest and most aggressive congregation.

If you want to eliminate volume issues in your church, the first step is to find ways to minimize the stage volume.

The Decibel Meter

The decibel meter is a valuable tool in any live sound environment; however, it is important that the sound operator understand the usefulness and uniqueness of all user-selectable parameters. To simply proclaim that the music team is at 90 dB is somewhat meaningless unless the settings are specified. The sound operator must set the meter to read peak or average levels, the weighting scale must be noted, and the details of how the meter is held and pointed must be considered. All of these variables affect the accuracy of the meter reading. In Chapter 9 we'll study the intricacies of the decibel meter (sound pressure level meter).

Who Is in Control?

In the typical chain of command, the sound operator is at or near the bottom of the chain of authority. If the operator is well respected, his or her opinion might carry more weight, but the person in control is the senior pastor in most churches. There are denominations in which the church council or governing body trumps the pastor, but in most cases the sound operator's direction comes from the pastor.

As we've already seen, a conscientious sound operator will invite the pastor and other key authority figures to sit at the sound board during a service so that the volume and sound-quality parameters can be specified. Once this has happened, the sound operator must honor the commitments he or she made to conform to the set standards.

As long as the standards are met, any volume or sound control complaints should be directed up the authority chain. There will be congregants who want the operator to exceed the limits of acceptability in either direction. In one service, there might be complaints that the sound is too loud, followed by comments that the mix is too quiet or too weak—that's just the nature of the live sound gig, especially in a church environment. Be aware that there are usually as many people who are bothered by a weak and quiet mix as there are people who are bothered by a loud and aggressive mix.

Learn the chain of command—it is your link to peace. The music director is usually the first point of contact for the sound operator, followed by an administrative staff member, and then typically the pastor. It is best to follow the organized chain of command and avoid directing any complainant directly to the pastor. The less often the pastor needs to handle sound complaints, the better it is for everyone.

Can the Congregation Hear Themselves?

Most modern churches want the congregation to act as the choir so that they are an active and important part of the worship experience. If your church leadership ascribes to this way of thinking, the volume of the congregation determines the volume of worship.

You must find a way to test the volume of the congregation when they are singing strongly. There are several opportunities to break the instrumentation down to a minimum in modern worship music. When the band thins out and the bulk of the sound is coming from the congregation, be sure to assess the decibel meter. Be sure it is weighted correctly and that the speed is set for an average level, and note the volume with the congregation singing alone.

Talk with the worship leader when you think there will be an a cappella section for the congregation—ask him or her to get the congregation started, and then to step back from the mic once they're on autopilot. Note the decibel reading.

Typical Organizational Structure

There should be an understanding between the pastor, staff, and ministry team members regarding the official chain of command. Knowing in advance who is in charge provides a pathway for complaints and concerns. Any sound operator who is earnestly doing his or her best to provide an excellent mix deserves support from the pastor and staff—they shouldn't be the punching bag for angry congregants.

In an ideal church environment, information flows in two directions in an effecient and structured manner:

- From God, through the senior pastor, to the congregation and staff

- From the congregation, through the proper channels, to the senior pastor

The senior pastor is responsible for remaining in the will of God and for constantly listening for God's desire for the church. The best way for anyone to influence the senior pastor is to pray directly to God that His will would be done in the situation—this is much more efficient than starting a war between rivaling church factions.

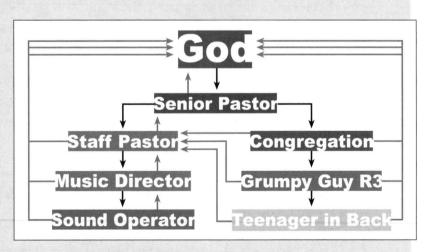

You might be surprised at the results from this test. When I was music director for a large church in the Seattle area, I conducted this test with the senior pastor joining me at the board. When the a cappella congregational section came up, we were both surprised that the congregation alone was consistently at 88 dB, and that was in the back of the room at the mixer position. We had a couple members pushing hard for us to never exceed 85 dB—this simple test quickly put those requests to rest.

Earning the Trust of the Team

If the instrumentalists and singers hear that the sound was good during the service, they will soon develop trust that they are in good hands. If they hear that the sound isn't very good or if they're told that certain instruments or voices aren't being heard, they'll quickly lose trust in the sound operator, especially if they're the ones being left out of the mix.

If the sound operator expects to gain the trust of the team, he or she must learn to do a good job. This book will help a lot, but the techniques must be practiced, and the principles must be applied.

Sometimes the factors leading to a negatively perceived mix are outside the sound operator's responsibility set. Sometimes, an instrument must be understated in the mix because the instrumentalist is musically undependable. As the team gets better at playing musically and as they provide a more consistent source for the operator to mix, everyone will appear to be better at what they do.

A great band and excellent singers provide a source that is easy to mix. On the other hand, a team that provides a jumbled mess of unstructured playing and out-of-tune singing is very difficult to mix. The sound operator ends up arranging during the mix by continually highlighting one instrument or voice while intentionally understating others.

MULTI-GENERATIONAL CONSIDERATIONS

In a typical secular concert setting, there might be a few complaints because the sound is incredibly loud or because the sound quality doesn't meet expectations, but generally the audience shows up because they love the act and they usually know what to expect. If an audience member doesn't like the sound, the music, the lights, the theatrics, or anything else, they simply won't come back. This is a fact of life in the concert scene. Although the act might lose a few ticket sales because they are what they are, the following they develop will become more and more loyal and avid over the course of time, especially if the act remains true to their original vision.

Churches operate differently. They're typically made up of a wide demographic and consist of multiple generations. It's true that there are a lot of churches that focus distinctly on one age group or demographic, ignoring or alienating other segments of the potential congregant pool.

It's healthy to realize exactly what the personalities of the pastor and staff are and to build a church around their comfort zone; however, most pastors prefer a multigenerational congregation for a number of reasons.

- It is stabilizing for the younger congregants to benefit from the maturity and life experience of the seasoned generations.
- It is vitalizing for the older generations to have relationships with the younger generations.
- Most pastors grew up in multigenerational churches—this is their comfort zone.
- The older members are typically more stable professionally and personally than the younger members are. Their lives are in order, and they understand the value of tithing.

Understanding Hearing Differences

It is very important for everyone involved in sound operation and music in church to understand that, aurally speaking, we are not all created equally. Some people are quantifiably more sensitive to certain sounds than others—it's just the way they are made.

Mixing sound for a group of people is challenging in many ways. A hundred people might all have slightly or radically different impressions of the mix you construct. Ear damage, age, and physiology combine throughout the church demographic into a potentially troublesome situation.

Many musicians and music lovers enjoy the sensation of loud music. As the sound gets a little louder (90 dB and above), the low frequencies can be felt as well as heard, and the highs are more clear and precise. Loud volume isn't perceived as anything other than a pleasurable listening experience. When someone complains about loud volumes or about loud sounds causing pain, it's difficult for these lovers-of-loud-music to relate to his or her complaint or empathize with his or her pain.

The wise sound operator will become educated about hearing issues in order to serve the needs of the church. When people express their opinions about volume or sound issues, they're often treated like complainers or their requests are trivialized. As a musician and sound operator, I have to admit that I've been through this. In the past, if a congregant complained about volume or sound issues, I might have been polite, but inside I was a little miffed and couldn't understand what the big deal was. To me, the sound was great. It felt good, and it had clean highs and a warm, full low end. What could be wrong with that?

As I've researched degenerative and physiological hearing conditions, I've grown more compassionate and sympathetic to the problems loud sounds present to many people. There are three basic types of hearing loss (presbycusis, tinnitus, and conductive) and two fundamental physiological conditions that cause hypersensitivity to loud sounds (hyperacusis and recruitment).

Types of Hearing Loss

Hearing loss can be caused by loud sounds, age, and obstructions, such as swelling and wax buildup. Each person seems to experience a unique set of symptoms in this area. When subjected to the same sound source, some experience little damage, while others experience catastrophic damage.

It is important for us to understand the basics of hearing issues in order to better serve a variety of church members we encounter weekly.

Presbycusis

Presbycusis is hearing loss that is primarily related to aging; however, illness, prescription drugs, circulation problems, loud noises, heredity, infection, or head injury can also cause it. This loss occurs gradually over time and is typically due to sensorineural damage in which there is damage to parts of the inner ear, the auditory nerve, or hearing pathways in the brain. The progression of presbycusis is often analogized to the body's transformation of one's original hair color to gray—it is gradual, steady, and relentless, and it happens at differing times and speeds for everyone.

People with presbycusis usually experience the regular discomfort of hearing loss— difficulty understanding speech and other sounds that contain abundant high-frequency information. In addition, this condition can cause intolerance of loud sounds.

Once we understand that there are multiple physiological causes for discrepancies in opinion regarding sound, it is easier for us to provide the type of compassion required to serve the audio needs of a multigenerational church body.

Tinnitus

Tinnitus manifests itself as a ringing, roaring, or other noise that's heard separately from acoustic sounds. Although tinnitus accompanies other forms of hearing impairment, it also adds to hearing loss in varying degrees, depending on the severity of the internal noise.

The noise of tinnitus is typically constant although it varies in intensity, dependent on several factors. It can be caused by loud noises, hearing loss, medication, circulatory problems, jaw misalignment, certain tumors, allergies, ear or sinus infections, wax buildup, and head or neck trauma.

People who attend loud concerts often experience a ringing in their ears after a show—this is tinnitus. Sometimes the ringing will disappear over time, but hearing damage is cumulative. Continued concert attendance could result in an increase in the length of time the ringing persists. Eventually, the ringing might remain permanently.

There are some touted aids that claim to decrease the symptoms of tinnitus, but most in the medical community consider this condition to be not yet curable. Some herbal remedies claim to help improve the condition, and vitamins C and B-12 are said, by some, to help decrease the ringing noise. Some say that aspirin, alcohol, and smoking worsen the condition.

Conductive Hearing Loss

The tympanic membrane, also called the *eardrum*, receives the sound waves—it vibrates sympathetically with the sounds to which it is subjected. Next, the sound information moves to the inner ear. Conductive hearing loss results when the sound information is blocked from the inner ear.

This type of hearing loss can be caused by abnormal bone growth, infection, earwax buildup, fluid in the middle ear, or a punctured eardrum.

Aural Hypersensitivity

Certain individuals are more sensitive to sounds than others. To the hypersensitive listener, even normal sounds can cause pain—loud sounds might be intolerable. There are two primary types of hypersensitivity to sound: hyperacusis and recruitment.

Hyperacusis

People with hyperacusis have essentially no hearing loss. In fact, they are abnormally sensitive to certain sounds—even common sounds cause pain or discomfort. Those who experience this condition often rely on earplugs and other noise reducers to participate in everyday activities.

Occasionally, tinnitus accompanies hyperacusis. The potential causes for this hypersensitivity to sounds are suspected to be noise exposure causing subtle damage to the ear, head injuries, and dysfunctions in brain chemistry.

Many people with hyperacusis rely on earplugs to shield their ears from common everyday sounds, even at low volumes. It has been discovered that this practice increases the problem rather than aiding in it. Patients are encouraged to use ear protection only when they are in the presence of potentially damaging sounds above 85 dB.

Recruitment

Recruitment refers to abnormal loudness sensitivity, which often accompanies sensorineural hearing damage. A person with recruitment differs from a person with hyperacusis in a couple primary ways:

- Someone with recruitment is primarily bothered by loud sounds, whereas the person with hyperacusis also experiences discomfort with moderately loud sounds.
- Recruitment is present in persons with hearing loss—those with hyperacusis do not have hearing loss.

A person with recruitment might simply feel bothered by loud sounds, but in acute cases this condition is very uncomfortable. In addition, the difference between acceptable and unacceptable volume might only be a matter of a few decibels—sound pressure levels don't necessarily need to increase dramatically to surpass the threshold of comfort.

When someone complains about volume during a church service, he or she might be suffering from recruitment and not know it. That person might sincerely believe your mix is way too loud—that it hurts—and you might sincerely believe that your mix is perfectly acceptable in volume, tone, and balance. Each of you is correct from your own perspective.

Earplugs are not always helpful in this instance—there is already hearing loss present, and it is probably most extreme in the high frequencies. Simply using foam earplugs might minimize the high frequencies that are necessary for speech discernment, while negligibly affecting the frequencies that cause discomfort.

A possible solution for the recruitment sufferer is the modern hearing aid. Digital units are capable of adjusting for many deficiencies in hearing, including overall damage, specific frequency deficiencies, and recruitment issues.

Education Is the Answer

Once you understand some of the causes behind comments that you receive during the course of operating sound, you're likely to find it easier to sympathize with your congregants. They're not crazy, and neither are you.

Try to find ways to help each demographic have the best possible church experience.

- Locate the quiet spots in the sanctuary for those who like less volume.
- Locate the sweet spots in the sanctuary where the mix is clean and loud for those who enjoy a punchy worship experience.
- Have earplugs available for those who seem hypersensitive.
- Have a discussion with those whom you suspect might have recruitment, and suggest they visit an audiologist for a professional opinion.
- Keep a copy of this chapter around to provide some basic insight into the considerations of sound in church.

Spend some time doing further research online. The more you know, the better you'll be at helping your congregation. There is probably no one else who will take the time to learn about this subject—it's up to you.

If you are educated about potential hearing differences and the effects of loud sounds and hearing damage, you'll be a much greater asset to your pastor. Also, if everyone up the chain of command agrees that the volume and sound you provide are appropriate, you'll know whether you should change what you do for one vocal congregant or whether you should help provide the tools necessary for that person to enjoy the service that's provided.

The worst way to handle the sound demands of a multigenerational church is to just ignore the comments with which you don't agree, while you embrace the ones with which you do agree. Ignoring a complaint with no further conversation devalues the complainant and typically results in an angry and bitter church member.

Make the effort to communicate all the way through the problem to a resolution. There is no gain in leaving these issues alone. Even if the resolution is that the member might not be right for the church or that you might not be right for the sound operator position, resolving the conflict is important. There is a good chance that the truth will reveal itself eventually, and it is much better to walk in the light—to find out the truth of the situation—in order to eliminate what could be weeks, months, or years of pain and anguish.

Senior Citizens (Born Before 1946)

There are churches that cater to the needs of senior citizens. They understand the needs and tastes of this valuable demographic. They provide music and teaching that is non-offensive and traditional. However, often the older generation doesn't feel all that old, and they want a church experience that is vibrant and exciting—which often means young and loud.

There is great value in making room for the mature generations in church. They bring insight and wisdom that only comes from living life—from succeeding and failing at enough things that their opinions are based on real experience and not simply blind speculation.

If your church is multigenerational, embrace the gift that God has provided.

- Keep a stock of earplugs handy at the sound board.
- Learn to create a mix that is less offensive to the older crowd while still pleasing to the younger folks.
- Keep an open line of communication.
- Show that you value all congregants
- Keep your pastor in the communication loop. He or she will appreciate your ability to handle a potentially uncomfortable situation.
- Reconfirm the target sound levels with your pastor.

If you use good communication skills, you'll find that your life will become simpler and people around you will be happier. With your pastor on your side, you can be sure you're supporting the ministry to which God called him or her, while building your own ministry and increasing your communication skills.

Boomers (Born Between 1946 and 1964)

The baby boomers, sometimes referred to as the "aging hipsters," are coming of age. They're beginning to retire and it still remains to be seen how their retirement will affect

Hearing Comparison between Men and Women

The charts below are the result of a study performed by Harry F. Olson and published in Modern Sound Reproduction. They clearly show the marked difference in the hearing loss tendencies between men and women. Notice that an average man at age 55 could expect a 30-dB hearing loss, while the average woman at age 55 could expect only a 17- to 18-dB loss.

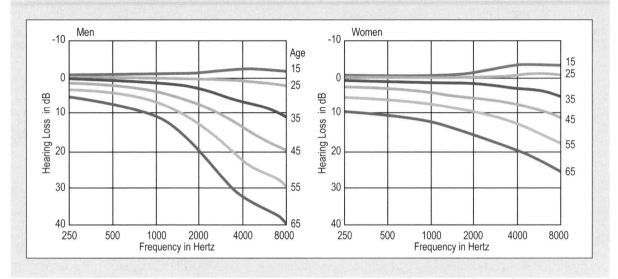

the healthcare industry and social security. In addition, they are the first generation to live through the amplification age, in which rock concerts, home hi-fi systems, and car stereos became really loud. It used to be the senior citizens who suffered the most from hearing loss—after all, they were the first generation to endure automated factories with loud mechanical equipment, the internal combustion engine, and jets. Now, the age at which hearing problems commonly occur is well into the boomer generation.

Sometimes, members of this demographic complain about volume, but they are still less likely than the senior citizens to experience extreme discomfort. Besides, they mostly grew up liking loud music, and certain habits are hard to break.

Hearing damage that comes as a result of age occurs gradually over time and primarily in the high frequencies. In addition, it is usually more extreme in men than in women. Whereas a 65-year-old man might experience a 30+-dB loss at 4 kHz, a 65-year-old woman might experience about a 20-dB loss at the same frequency.

Generation X (Born Between 1965 and 1980)

The X generation grew up during the 1980s and early 1990s and learned from the examples set by their boomer parents that selfishness isn't all it's advertised to be. They are often turned off by societal class, status, and the worshipping of money.

This is also the first generation to grow up with computers in the household. Video games became increasingly sophisticated during the growth of Generation X, and the advent of Walkman cassette and CD players brought sound and music into people's own personal space. Headphones became commonplace, and the music and recording industry provided increasingly hi-fi audio. The advent of the digital era in the early 1980s saw huge strides in technology and music production.

The people of Generation X grew up with a different set of values from previous generations. Their focus on nontraditional activities and a narrowing of their social contact created a generation that values individual friendships more than corporate popularity.

In a church setting, Xers are sometimes difficult to please. They've grown up with good sound and media presentations, and they've had the opportunity to develop their own musical tastes. Whereas previous generations collectively experienced most of their music in person, over the radio, or later on television, Generation Xers could isolate themselves, listening increasingly to their own personal blend of musical styles and influence.

It is sometimes more difficult to compel those from Generation X into a corporate worship experience in adult church. As with all generations, there is a gap that takes a while to bridge. Eventually, Xers begin to blend into the adult world and actually tend to thrive.

As the youngest of this group approaches the age of 30 and the oldest approaches 50, this demographic is coming into power. They hold a high standard for sound, video, and multimedia presentations. They know what they like musically, and they appreciate technological advancement.

Volume isn't usually a problem with this generation, although they have been subjected to loud bands and high-volume headphones for most of their lives. They are in danger of hearing damage soon, if they haven't already experienced it.

Youth Groups

Junior high, high school, and college groups are made up of the most technologically sophisticated young people the world has ever seen.

- They have grown up on computers.
- They understand technology.
- They appreciate great multi-channel audio.
- They have typically been exposed to a constant barrage of world-class productions, from Dolby digital theaters, to HDTV, to MTV, to iMovie.
- Many of them create their own audio and video recordings and then edit them using the software that came on their computers.

These technologically savvy young people know what a quality production looks like, and they also recognize a bad production. If your church expects to hold the attention and interest of this young crowd, they'll need to provide high-quality video—there goes the overhead projector into the trash—and high-quality audio. Adios, 1960s center cluster!

Some pastors and leadership councils resist technology. They're a little afraid, and they don't quite understand it. Often, they can't quite envision how high-quality presentation tools, such as video projectors, screens, and monitors, will be installed or how they'll be operated. Frequently, they consider the addition of an up-to-date sound system as frivolous—after all, the current system has worked fine for 30 or 40 years.

The truth is, once a church decides to step into the third millennium, they will discover a brand-new force: young people. Youth pastors are often on fire for God, and they typically do a really great job of inspiring their young church members to long for the things of

God. Relationships are very important. Many young people find their only role models of family life in church. With both parents overworked, the *Leave It to Beaver* generation has disappeared.

These intelligent young people sometimes look a little scary. In many cases, they seem to do everything they can to shock and scare all the rest of us, but inside they understand and long for relationships. If your worship team, tech crew, and pastors will invest in the lives of these young people, they'll find parched seeds, struggling to bloom. Many youths long for an adult to take the time to be with them, to help them through the mess that their family life might have become, and to tell them that they are valuable. Given a little water, these important church members will thrive!

To thrive, the modern church needs to trust in God in every way. They also need to communicate their message in a way that holds the attention of the church. If you need help with technological advances in church, tap into the resources you've already been given. The young people in your church are ready. Invest in some good training so they get off to a good start, and then watch them grow. They'll surprise you. They'll probably be the first ones at every event, setting up and then sticking around to help strike the gear. And you know what's even better? They'll be serving the purposes of God alongside you. Your investment in them will be good for the church, and your rewards—in terms of relationship and purpose—will be immense.

Children

So often, children's church gets the leftovers when it comes to technology and music. The adults that help are amazing. They give and give and do everything they can to help all of our kids to grow up knowing God. However, they could use a little help.

The children in all of our churches deserve the best we can offer. They don't usually require the highest technologies, but they deserve a system that communicates the message well and sounds good. They are also worthy of our time. If everyone on the adult team would invest a little time in helping the children's ministry, it would make a big difference. Willingness to help is important. Letting the children's ministry staff know you're willing to help is even more important. Following through and actually helping when asked is crucial.

All of us find it pretty easy to say that we would help anywhere in the church if asked. If we'll all just stand up and help, we'll make a difference for the children, the staff, and ourselves.

chapter 8
SOUND THEORY

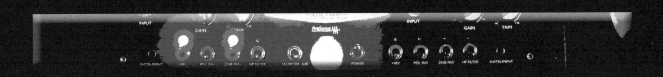

It's very important that we understand the principles of sound in order to accurately capture music, speech, sound effects, or even noise. The sound source helps define the tools we should use to record it. Most of the time, we strive to preconceive the sonic impact of the recording, imagining the sound that provides the desired musical or artistic result, and then go after it. Other times, we stumble across a sound that inspires a complete redirection of the artistic process. In either case, a thorough understanding of your recording tools is essential.

Anytime you're recording an acoustic instrument, listen to the instrument first. Stand beside the musician and hear what he or she hears. Listen to the sound of the instrument or voice decaying in the room. Stand close and move away. Assess the sonic differences in the acoustic space. If you really want to capture the true essence of the sound, you'll need to make excellent decisions about where the instrument is placed in the room, what microphone you'll use, and where you'll place it in relation to the instrument. If you want to capture something other than the true sound of the instrument or voice, you'll need to be fully aware of the options available to you, and you'll need to be able to use them in a creative and artistically supportive manner.

The information we're about to cover is fundamental to the understanding of sound. Carefully study this material. It will help you make great recordings of great music.

Characteristics of Sound

Sound is energy that travels through air. Air molecules move in relation to the sound that moves them. When something vibrates, such as a drum, string, vocal chord, and so on, it affects the air around it. The air responds to the vibrations directly, contracting and expanding as the vibrating material completes its cycle of vibration. These vibrations cause continuous variations in the existing air pressure.

Visualize sound in air like a wave in water. Any sound creates a disruption in the stillness of air, just like dropping a rock in a lake creates a disruption in the stillness of the water. In fact, sound is referred to as *sound waves* because of this simple concept. As with so many concepts

Sound Wave Reflections

Sound waves move in air like waves move in water. Interactions occur between waves and their reflections spherically in all directions from the sound source.

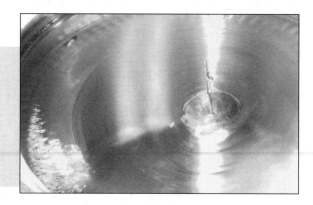

in art or science, the basic principles are easy to understand. Most complex theories and concepts can be stripped down to a fairly reasonable string of simple ingredients.

As you watch the waves in water travel, they are minimized until they disappear, unless they reach an outer boundary, in which case they reflect back toward the center. The amount of reflection depends on the energy at the source and the distance to the boundary. As the waves radiating from the source meet the waves rebounding from the boundary, it's easy to see the waves interacting and influencing each other's shape and size. That's exactly what happens to audio in an enclosed space. The reflections combine with the source audio, each influencing the other. This is the reason any given instrument or voice takes on a different character, or timbre, depending on the space it is in.

Our perception of sound is directly related to the waves in the air. Consider that a lack of sound is completely still air—mighty difficult to find, but let's just imagine it for the sake of understanding. As soon as there is vibration by anything in the still air, waves begin. In relation to still air, each wave contains a crest and a trough.

When the wave touches our eardrum, the membrane vibrates in sympathy with the source, being pushed in by the crest and pulled out by the trough. This is explained by the principle of sympathetic vibration. When a sound wave strikes a body, which will naturally produce the same wave, the vibration of the body is called *sympathetic vibration*.

The simplest of sound waves is called a *sine wave*. When we chart the rise and fall of the crest and trough, the sine wave is perfectly smooth and provides an excellent illustration of the basic aspects of sound. To illustrate a sound wave, we consider a straight line as still air. As the wave ascends above the line, creating a crest, our eardrum is pushed in—this is also called the *compression* portion of the cycle. As the wave descends below the line, creating a trough, the eardrum is pulled out—this is called *rarefaction*. Compression causes the air

Acoustic Reflections Combine with the Source

Sound emanates omnidirectionally from the source. However, when we place a microphone in front of a source, we also get the reflections off each surrounding surface, combined together at the mic.

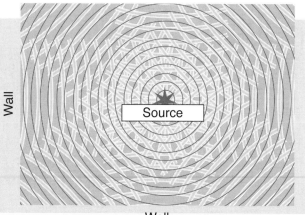

Wall

Source

Wall

Crest and Trough

Audio waveforms consist of a series of crests and troughs that push and pull on the eardrum. This simplest waveform, called a sine wave, has the smoothest sound and the smoothest curve.

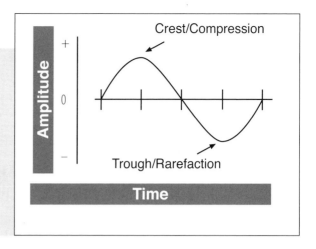

molecules to bunch together, causing an increase in air pressure. Rarefaction is the result of the air molecules filling in behind the compression, resulting in a decrease in air pressure.

Keep in mind that individual air molecules don't travel from the sound source to the listener's eardrum. Sound is merely causing a chain reaction, which moves air molecules back and forth, causing the air molecules they're touching to move back and forth, and so on, until the air molecules that touch your eardrum initiate its movement. At football games in Seattle, we all get a big kick out of doing "the wave," where a chain reaction flows all around the stadium. As soon as the person next to you stands up, you stand up, and as they're sitting down, you sit down. A huge wave moves all around the stadium, which looks really cool and for some strange reason makes you feel good about life. No one has to run around the stadium, but the wave makes it all the way around. That's how sound transfers through air.

Speed

In normal atmospheric conditions sound transfers through air relatively slowly, at the rate of about 1,126 feet per second (about 340 meters per second, 30 cm per millisecond, or just over one foot per millisecond.) Elevation, temperature and humidity affect the speed slightly.

Push and Pull on the Eardrum

The pinna focuses sound toward the eardrum—it is fundamental in localization of sound. Compression and rarefaction are channeled into the ear canal, where the changing air pressure vibrates the tympanic membrane, which begins the process of sending corresponding electrical impulses to the brain.

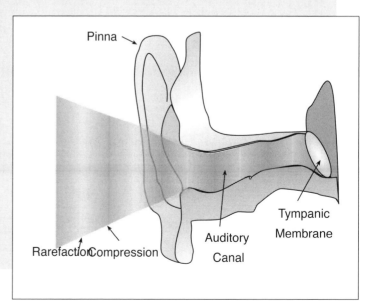

360-Degree Complete Wave

From beginning to end, a complete cycle of any waveform is quantified as 360 degrees. Halfway through the wave cycle is 180 degrees, one quarter of the way through is 90 degrees, etc.

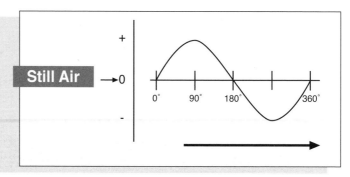

The speed of sound in air is determined by the conditions of the air, not by waveform characteristics like amplitude, frequency, or wavelength. There are plenty of formulas available on the Internet that take all factors into consideration for the speed of sound. For the purpose of our illustrations and calculations, 1,126 ft./sec. will usually suffice.

Cycles

When a source has completed one crest and one trough it has completed one cycle. This represents a push and pull on the eardrum—one compression and one rarefaction of air molecules. We quantify position during the cycle in degrees. There are 360 degrees in one complete cycle. The zero degree mark denotes the beginning of compression (the crest); 360 degrees denotes the end of rarefaction (the trough). 180 degrees marks the midpoint of the cycle, where the crest ends and the trough begins.

Frequency

The frequency of sound quantifies the number of times a wave completes its cycle in one second. In relation to pitch, higher frequencies complete more cycles each second. Frequency is expressed in Hertz (Hz) or cycles per second (cps). 100 Hertz, or 100 cps, represents a waveform that completes its cycle 100 times each second. 1,000 Hertz equals 1 kilohertz (kHz). As frequency increases, it's common to refer to v multiples of kilohertz. 2,500 Hertz is typically referred to as 2.5 kHz; 5,100 Hertz = 5.1 kHz. In common usage, we often refer to kilohertz simply as "k." For example, it's common to say, "I boosted the vocal track at 4 k."

The frequency response range of the human ear is roughly from 20 Hz to 20 kHz. Brand-new ears, like those in a baby, tend to be able to hear frequencies above 20 kHz, sometimes approaching 23 kHz. Old tired ears, like those found in many musicians, probably don't hear high frequencies as well as they used to.

Pitch

An octave on the piano is the distance from a note to the next note of the same name. From middle C to the C above middle C is one octave. Mathematically, an octave above any pitch is twice the frequency of the pitch. On octave below any pitch is half the frequency of the original pitch.

Each pitch (note on a piano, guitar, trumpet, etc.) has a specific fundamental frequency within our system of tonality. We use a 12-tone system (12 notes per octave), tuned in a

Frequencies Mathematically Related to Pitch

Each musical note is related to a specific frequency. In the illustration to the right, we see whole number frequencies that, though they've been rounded off, indicate the way the frequency range of most music relates mathematically to pitch. The specified number of Hertz indicates the fundamental frequency—the sine wave that defines the specific pitch and octave.

These frequencies are mathematically just. In our tonality, the piano is tuned to a tempered scale, which adjusts some pitches in the chromatic scale, especially in the high and low registers, to provide for consistent intonation when playing in a variation of keys. The tuned piano notes don't all match these exact frequencies in every case.

Note	Frequency
C	4186 Hz
B	3951 Hz
A	3520 Hz
G	3136 Hz
F	2794 Hz
E	2647 Hz
D	2349 Hz
C	2093 Hz
B	1976 Hz
A	1760 Hz
G	1568 Hz
F	1397 Hz
E	1319 Hz
D	1175 Hz
C	1046 Hz
B	988 Hz
A	880 Hz
G	784 Hz
F	698 Hz
E	659 Hz
D	587 Hz
C	523 Hz
B	494 Hz
A	440 Hz
G	392 Hz
F	349 Hz
E	330 Hz
D	294 Hz
C	262 Hz
B	247 Hz
A	220 Hz
G	196 Hz
F	175 Hz
E	165 Hz
D	147 Hz
C	131 Hz
B	123 Hz
A	110 Hz
G	98 Hz
F	87Hz
E	82 Hz
D	73 Hz
C	65 Hz
B	62 Hz
A	55 Hz
G	49 Hz
F	44 Hz
E	41 Hz
D	37 Hz
C	33 Hz
B	31 Hz
A	27 Hz

specific, tempered way. There are many other tonalities throughout the world, utilizing different numbers of notes per octave.

Middle C on the piano has a fundamental frequency of 262 Hz. The A above middle C is often used as a standard tuning reference, and is called A 440—indicating a fundamental frequency of 440 Hz.

The frequency that defines the pitch name is called the fundamental frequency. In reality there is much more to a note than its fundamental frequency. Aspects of the sound wave called harmonics, overtones, and partials determine the individual character of a sound. The fundamental only determines the name and octave of a note.

As a point of reference, the lowest note on a standard 88-note piano keyboard has a mathematically calculated fundamental frequency of 27 Hz. The highest note on the piano has a mathematically calculated fundamental frequency of 4,186 Hz. A modern piano is tuned to a tempered scale, in which certain pitches are not mathematically precise. Therefore, a piano tuning technician references a slightly different frequency grid for frequency-pitch calibration that moves the pitch slightly sharper as the notes progress from A 440 upward and slightly flatter as the notes progress from A 440 downward. For our purposes and understanding, it is most appropriate to reference the keyboard as a series of frequencies represented as whole-number multiples of a fundamental frequency moving up and down from A 440. For a precise list of frequency offsets in cents (hundredths of a half step) visit: www.mts.net/~smythe/st-6 .htm#tuning

Wavelength

Low frequencies have longer waveforms than high frequencies. The physical distance in air from the beginning of one cycle to the beginning of the next cycle is the length of the sound wave. The wavelength is often indicated by the Greek letter lambda. To calculate the wavelength (λ), we use a formula consisting of the frequency (f) specified in cycles/second, and the speed, or velocity, of sound specified in feet/second (v). Wavelength (λ)= Velocity (v) ÷ the frequency (f). $\lambda = v/f$

Instrument Ranges Compared to the Piano Keyboard

Piccolo

Violin

Flute

Female Voice

Male Voice

Guitar

Snare

Toms

Kick

Bass Guitar

To calculate the length of a 1,000 Hz tone, simply plug the variables and constants into the formula. The constant is the speed of sound (v), clocking in at 1,126 feet/second. In this case the wavelength (λ) = 1,126 (v) ÷ 1,000 Hz Therefore, λ=1,126/1,000, which equals 1.12 feet long.

The lowest note on the piano (27 Hz) is calculated λ=1,126/27. The result of this equation indicates a wavelength of about 41.5 feet. The highest note on the piano (4,186 Hz) is calculated λ=1,126/4,186. The result of this equation indicates a wavelength of just over three inches (.27 feet).

Our understanding of wavelength is crucial to our understanding of acoustics and how sound reacts to, and interacts with, its environment. There are some situations where we need to calculate the frequency of a specific wavelength. This is a simple task of cross-multiplication in which we find that f=v/λ. These equations will be important in our studies on basic acoustics.

Wavelength - Proportional Length Comparison

This illustration would be a much more impressive wavelength comparison if it were full scale. The lowest discernible pitch (20 Hz) is 56 feet long, just about the length of four Ford Explorers; the highest discernible pitch (20 kHz) is barely longer than half an inch.

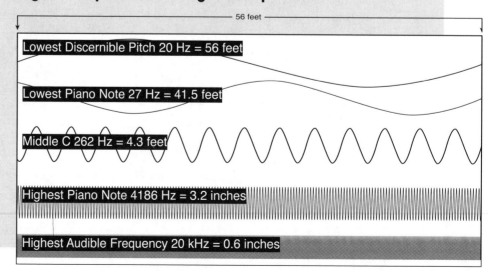

56 feet

Lowest Discernible Pitch 20 Hz = 56 feet

Lowest Piano Note 27 Hz = 41.5 feet

Middle C 262 Hz = 4.3 feet

Highest Piano Note 4186 Hz = 3.2 inches

Highest Audible Frequency 20 kHz = 0.6 inches

Amplitude

Amplitude expresses the amount of energy in a specified sound wave. When comparing two waveforms on a graph—one with twice the amplitude of the other—the waveform with twice the amplitude has a crest that rises twice as high and a trough that dips twice as low.

Amplitude only compares the energy of a sound wave. It's a simple comparison: A waveform with maximum amplitude that's 2.5 times higher than another waveform contains 2.5 times the energy. The unit commonly used to quantify amplitude is dB SPL (*Decibels Sound Pressure Level*). This is an objective scale based on mathematical logarithmic comparisons expressed as $decibel = 10 \log 10(P1/P2)$.

Any increase in amplitude indicates an increase in volume. However, the correlation is not always direct—twice the amplitude does not always indicate twice the volume. The relation is dependent on the frequency and loudness.

Loudness

Loudness is a sound characteristic that involves the listener—it is a perceived characteristic that can be charted and averaged, but it's not simply a mathematical calculation. The common unit, used to quantify loudness, is the *phon*. Loudness is a subjective, perceptual aspect of sound.

The human ear is not equally sensitive to all frequencies. In fact, as amplitude varies, so does the frequency response characteristic of the ear. The ear is most sensitive between 1 and 4 kHz. This frequency range just happens to contain the frequencies that give speech intelligibility, directional positioning, and understandability. Hmmm…it's almost like it was designed that way. In fact, as the amplitude decreases, our ears become dramatically more sensitive in this frequency range.

So, yes, there is a difference between amplitude and volume. They are very similar at a certain point, though. Two scientists at Bell Laboratories in the 1933 charted a survey of perceived volume. They compared actual amplitude to perceived volume throughout the audible frequency range (x-axis) and the accepted range of normal volume (y-axis).

The results of their survey involved generating pure tones through the audible frequency and volume spectrum at a specific amplitude, then asking numerous individuals to subjectively identify whether the sound was louder or softer than the reference. Their survey, referred to as the Fletcher-Munson Curve, is a very visual representation of why music sounds

---

The Loudness of Everyday Life

Examples of everyday noise levels in dB SPL

Weakest sound heard	0 dB
Normal conversation (3–5')	60–70 dB
Telephone dial tone	80 dB
City traffic (inside car)	85 dB
Train whistle at 500'	90 dB
Subway train at 200'	95 dB

Sustained exposure may result in hearing loss at these levels

Possible hearing loss	90–95 dB
Power mower	107 dB
Power saw	110 dB
Pain begins	125 dB
Pneumatic riveter at 4'	125 dB
Jet engine at 100'	140 dB
Death of hearing tissue	180 dB
Loudest sound possible	194 dB

Government Regulations

OSHA Daily Permissible Noise Level Exposure

The Occupational Safety and Health Administration (OSHA) is part of the U.S. Department of Labor. This organization has studied and prescribed maximum sound pressure levels in the workplace, in relation to the number of hours per day the worker is exposed. These guidelines are useful to help audio engineers guard against permanent hearing loss.

Hours per day	SPL
8	90 dB
6	92 dB
4	95 dB
3	97 dB
2	100 dB
1.5	102 dB
1	105 dB
.5	110 dB
.25 or less	115 dB

fuller at loud volumes and thinner at soft volumes.

Each curve on the graph represents perceived constant volume throughout the audible frequency range. This, for example, shows us that perceiving 70 phons of loudness at 1,000 Hz requires 70 dB SPL (amplitude). However, in order to perceive 70 phons at 50 Hertz, 80 dB SPL is required. At 10 kHz, to perceive 70 phons, a similar 10 dB SPL boost is required.

As dB SPL decreases, the contrast becomes even more extreme between loudness and the actual amount of dB SPL required. At 20 phons, 20 dB SPL is equal to 20 phons. In contrast, at 50 Hz almost 65 dB SPL is required to maintain the perceived 20 phons.

Analysis of the Fletcher-Munson Curve points us to the dB SPL range at which the human ear is most accurate throughout the audible frequency spectrum. Notice that

The Fletcher-Munson Curve of Equal Loudness

This graph plots results from a survey that relates amplitude (dB SPL) to perceived volume. This curve is valuable because it highlights the frequency response characteristic of the human ear. Since amplitude is a quantifiable energy level and loudness is a subjective characteristic, based on the listener's opinion, there's no better way to discover perceived volume than to ask human beings and then chart the results.

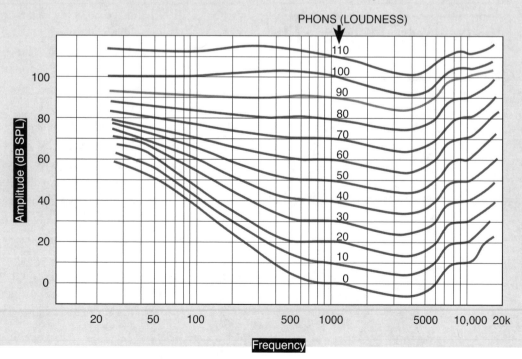

between roughly 700 Hz and 1.5 kHz, phons are essentially equal to dB SPL at all volumes. Also, notice that the center of the graph is where more often than not dB SPL is most similar to phons.

From this graph it is generally held that the most sensitive frequency range is from 1 to 4 kHz, although the graph might indicate an extension of that range from about 700 Hz to 6 kHz or so. Because this is a subjective study, some generalities apply, but it is obvious where the consistencies and trends are.

For our recording purposes, it is constructive to find the flattest curves on the graph. A curve with less variation indicates a volume at which the human ear's response most often matches loudness to dB SPL—the level where the most accurate assessments can be made regarding mix and tonal decisions.

The Loudness button on your stereo is an example of compensation for the fact that it takes more high and low frequencies at a low volume to perceive equal loudness throughout the audible spectrum.

The most consistent monitor volume for our recording purpose is between 85 and 90 dB SPL, according to the Fletcher-Munson Curve. Notice on the graph that the 80 and 90 phons curves are the flattest from 20 Hz to 20 kHz.

There are a few different devices available to help you quantify specifically how loud, in dB SPL, you have your system set. The simplest and least expensive way to assess dB SPL is with a handheld decibel meter. These are available at most home electronics stores and, depending on features and manufacturer, typically range in price from about $40 to $300. Most of these instruments offer A- and C-weighting, along with slow (average) and fast (peak) attack times.

C-weighting is optimized for full-bandwidth sources at levels exceeding 85 dB. A-weighting filters out the high and low frequencies and is optimized for lower volumes.

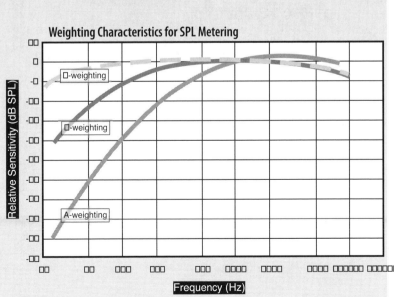

A-, B-, and C-Weighting

Any piece of gear that quantifies amplitude must specify whether it's sensitive to a full or limited bandwidth. Weighting is the qualifier for sound pressure level measurements.

C-weighting closely approximates full-bandwidth sensitivity. This is the scale that most accurately represents amplitude.

A-weighting closely approximates loudness, attenuating the lower frequencies to resemble the response of the human ear (which is most sensitive to frequencies between 1000 and 4000 Hz).

B-weighting includes more of the mid frequencies in its sensitivity than A-weighting. It's usually used in conjunction with A- and C-weighting in analysis of acoustical anomalies.

Weighting Characteristics for SPL Metering

Relative Sensitivity (dB SPL)

□-weighting

□-weighting

A-weighting

Frequency (Hz)

The A-weighted scale more closely reflects perceived volume, whereas the C-weighted scale measures amount of energy (amplitude).

Phase

We discovered previously that a sound wave is represented by one complete cycle—a crest and a trough—which is measured along the timeline in degrees. The beginning of the crest is at zero degrees, and the end of the trough is 360 degrees. The way multiple sound waves interact in the same acoustical or electrical space is called *phase*.

Because a sound wave has a crest, which pushes on your eardrum, and a trough, which pulls on your eardrum, it's fairly simple to visualize that if two identical waveforms happen simultaneously and follow the exact same path, their energy would increase as they worked together. In fact, they double in amplitude, meaning the peak is twice as high, and the trough is twice as deep. As experienced by your eardrum, the compression and rarefaction are doubled. Two identical waveforms that start at the exact same point in time and follow the identical path through the crest and trough are said to be in phase.

If two signals are out of phase, their waveforms are mirror images of each other. The electronic result of this combination is silence. When this happens electronically, the energies oppose each other completely—for each push, there is an equal pull throughout all 360 degrees. Because we refer to a complete cycle as 360 degrees, we mark the center point of the cycle at 180 degrees. By delaying one of two identical waveforms so that the beginning of the trough of one coincides with the beginning of the crest on the other (180 degrees into the cycle), we create a scenario of complete phase cancellation. When this happens, we say the two waveforms are 180 degrees out of phase.

It's easy to create a scenario, electronically, in which two waveforms combine 180 degrees out of phase. It rarely happens acoustically because of the predominance and complexity of reflections, along with the fact that we hear with two ears, which already

Phase Relationship

Wave B is 180° out of phase with Wave A. The result of opposing crests and troughs is no air movement. No air movement means no sound.

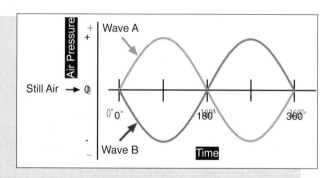

Conversely, two identical waveforms that start at exactly the same time (at right) are in phase. They combine, resulting in twice as much energy.

The height of the waveform (the distance above and below the center line) is referred to as the amplitude. Amplitude corresponds to the amount of energy in the waveform.

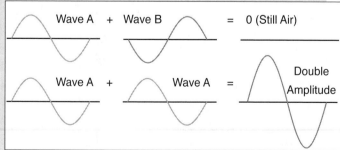

receive the same waveform at slightly different points in time. Interactions between acoustic sound waves are, however, still an important factor in understanding music and recording. For our study of acoustics, it's most enlightening to recognize the concept that multiple sounds work together to form the whole.

Waveforms can combine out of phase at any point in the cycle. If two waveforms are 90 degrees out of phase, they interact together to change the resultant sound. Even though there isn't complete phase cancellation, we still experience the result of the opposing and summing forces.

There are a few different ways in which negative phase interactions provide obstacles:

- When multiple microphones are used in the same room, sounds can reach the different mics at different times and probably at different points in the cycle of the wave. They combine at the mic out of phase. That's why it's always best to use as few mics as possible on an instrument or group of instruments in the same acoustical space. Fewer mics means fewer phase problems.

- This theory also pertains to the way speakers operate. If two speakers are in phase and they both receive the identical waveform, both speaker cones move in and out at the same time. If two speakers are out of phase, and if they both receive the identical waveform, one speaker cone moves in while the other speaker cone moves out. They don't work together. They fight each other, and the combined sound they produce is not reliable.

Harmonics, Overtones, and Partials

Harmonics are the parts of the instrument sound that add unique character. Without the harmonic content, each instrument would pretty much sound the same, like a simple sine wave. The only real difference would be in the characteristic attack, decay, sustain, and release of the individual instrument.

Because harmonics and overtones are so important to sonic character (vocal or instrumental), it's important to understand some basics about harmonics. As your experience level increases, this understanding will help you grasp many other aspects of music and recording.

When you hear middle C on a piano, you're hearing many different notes simultaneously that form together to make the sound of a piano. These different notes are called *harmonics*. Harmonics and overtones are a result of, among other considerations, vibration of the instrument; size of the instrument; acoustics; the type of material the instrument is made of; or the vibration of the string, membrane, reed, and so on. Several factors add to the harmonic content, but it's a law of physics that harmonics combine with the fundamental wave to make a unique sound that is represented by one waveform. That waveform is a result of the combination of energies included in the fundamental frequency and all of the harmonics. The fundamental is the wave that defines the pitch of the sound wave.

The frequencies of the harmonics are simple to calculate. Harmonics are whole-number multiples of the fundamental frequency. In other words, if the fundamental has a frequency of 220 Hz (A below middle C), calculate the harmonics by multiplying 220 by 1, 2, 3, 4, 5, 6, and so on.

Tone Interactions—Harmonics

This is the fundamental sine wave. Its frequency determines the note name and pitch for the waveform that's built from it.

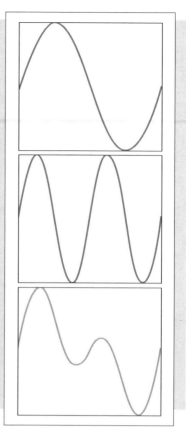

This is the second harmonic in relation to the fundamental above. Its frequency is two times the fundamental, so it completes its cycle twice in the same time period that the fundamental completes one cycle.

This is the result of combining the fundamental and the second harmonic. This new waveform has its own unique wave shape and sound. When waves combine, our ears no longer detect separate sound waves, they merely react to the one new wave that is influenced by all simultaneously occurring sounds.

- 220 x 1 = the fundamental, the frequency that gives the note its name, the first harmonic
- 220 x 2 = 440 Hz, the second harmonic
- 220 x 3 = 660 Hz, the third harmonic
- 220 x 4 = 880 Hz, the fourth harmonic
- 220 x 5 = 1,100 Hz, the fifth harmonic
- 220 x 6 = 1,320 Hz, the sixth harmonic

It's traditional to primarily consider the sonic implications of the harmonics up to about 20 kHz, because that is the typical limitation of our ears and equipment. There is a controversy regarding the importance of the upper harmonics above 20 kHz. As we understand how frequencies interact, it's not difficult to imagine that the frequencies above our audible frequency spectrum have an effect on those we can hear.

Engineers involved in archiving music and sounds for future reference carry on spirited debates about this. High-quality archival of important recordings is a big topic in the digital realm. Although digital storage seems very well suited to archiving because of its durability and long-lasting construction, the fact that CD-quality audio (at 44.1 kHz sample rate) cuts off all frequencies above 20 kHz sheds a questionable light on its long-term viability for important audio archiving. Digital sample rates of 192 kHz or higher make more sense when considering the future of audio storage.

As the harmonics combine with the fundamental, summing and canceling occurs between the fundamental and its harmonics. This summing and canceling interaction is what shapes a new and different-sounding waveform each time a new harmonic is added.

The terms *harmonic* and *overtone* are often used synonymously, but there is a difference. Whereas the harmonics are always calculated mathematically, as whole-number multiples of the fundamental, overtones are referenced to intervals and don't always precisely fit the harmonic formula. In the case of the piano, for example, the overtones are very close to the mathematical harmonics, but some are slightly off.

Some percussion sounds contain a relative of harmonics and overtones called *partials*. Like overtones, partials aren't mathematically related to the fundamental in the same simple formula as harmonics, and the effects that these sounds have can be very dramatic and interesting. Some bell-type sounds contain partials that are far removed from the true harmonics (sometimes they even sound out of tune), but the overall sound still has a defined pitch with a unique tonal character. Partials can also be lower in pitch than the *fundamental*, whereas harmonics and overtones are considered to be above the fundamental. On bells, there's generally a strong partial at about half the frequency of the fundamental, called the *hum tone*.

Harmonics, overtones, and partials extend far beyond the high-frequency limitations of our ears. For example, when we hear the lowest piano note, we're really hearing the fundamental plus several harmonics working together to complete the piano sound. If we only consider that the piano contains fundamental pitches from 27.50 to 4,186.01 Hz, it might not seem important to have a microphone that hears above 4,186.01 Hz. However, if we understand that for each fundamental there are several harmonics, overtones, or partials sounding simultaneously that go up to or above 20 kHz, we realize the importance of using equipment (mics, mixers, effects, and recorders) that accurately reproduces all of the frequencies in and/or above our hearing range. Also, if we see that the combination of these fundamentals and overtones is what shapes the individual waveform, it becomes evident that if we want to accurately record a particular waveform, we should use a microphone that hears all frequencies equally. If the mic adds to or subtracts from the frequency content of a sound, then the mic is really changing the shape of the waveform.

Shape

There are some traditional wave shapes that we refer to when describing sounds. Sine, sawtooth, square, and triangle waves each have distinct characteristic sounds. Sounding a lot like a flute, the sine wave has the simplest shape.

A sine wave is a smooth and continuous variation in energy throughout the wave cycle, steadily increasing in compression, then gently cresting over the peak to fall smoothly into rarefaction, then rising gently back to the center line. This simple waveform is also called a *pure tone*.

The fundamental is a sine wave, and each of the harmonics and overtones is also a sine wave. When the fundamental combines with its harmonics to create a new and unique waveform, they create a complex waveform.

It's important to realize that when we hear the fundamental and its harmonics, overtones, or partials, we don't hear any of the individual sine waves. Instead, we hear the result of the combination of all waves as one distinct waveform. The relative level of the harmonics determines their effect on the fundamental frequency, therefore shaping and molding the waveform and creating a sonic character. This individuality or signature character of a sound is called the *timbre*—it is also frequently called color or tone quality.

Wave Shapes

This is the third harmonic of the sine wave in the previous illustration. Notice that each harmonic is also a sine wave, but when it's combined with the fundamental and the other harmonics, an entirely new and unique waveform is created.

This is a sawtooth waveform. It's created by combining all harmonics in proper proportion. The sawtooth and triangle waveforms have a bright, edgy sound. Waveforms are given descriptive names based on the shape of their sound wave.

This is a square wave. It's created by combining the odd harmonics (1, 3, 5, 7, 9, etc.) in the proper proportion. A square wave sounds much like a clarinet.

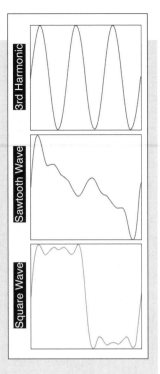

Sawtooth, square, and triangle waves get their names from the overall shapes of their unique waves. Sawtooth and triangle waves are edgy sounding and have more of a brass and bright string-type sound. A square wave sounds like a clarinet.

The complexity of the piano waveform is the result of a rich harmonic content. Piano is an instrument full of interesting harmonics. Listen carefully to a low note on the piano, and notice the complexity of the sound of a single note. If you listen closely, you can isolate and hear several pitches occurring with the fundamental. We perceive the harmonic content, along with the fundamental, as one sound. In actuality, the single piano note is constructed of many sine waves combining to give the impression of a single note with a unique timbre.

The illustration above has the fundamental wave drawn on top of the piano sound wave. This fundamental wave is very simple, yet the sound of the piano is very complex.

If you understand the theory of harmonics, you're well on your way to understanding the theory of sound. You'll also approach music and sound with a little more respect, finesse, and insight.

The Piano Waveform

This is the actual waveform of a single piano note recorded in stereo. The top waveform is the signal from the mic placed over the low strings. The wave on the bottom is the signal from the mic placed over the high strings. Notice the complexity of these waveforms compared to the sine waves in the previous illustration.

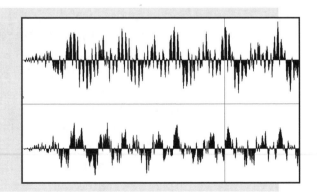

The Sine Wave Versus the Piano Waveform

Notice the purple line drawn on top of the piano waveform. This line represents the fundamental frequency of the piano note. The fundamental frequency is really nothing more than a simple sine wave.

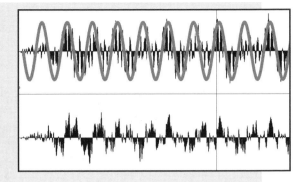

Envelope

The envelope describes the initial action, development, and diminishing of a waveform over the course of time. There are four primary phases of the envelope of any waveform: attack, decay, sustain, and release.

Attack

The way a sound is initiated is called its *attack*. The attack phase sees the amplitude rise from zero to the attack peak. Sometimes the attack rises to exactly the level of the sustain phase; other times, the attack contains a peak attack, called a *transient*, which falls slightly after the initial attack to enter the sustain phase. Most sounds with extreme attacks contain a transient, which exceeds the average level of the overall sound.

Examples of sounds with fast attack times are:
- Wood block
- Slap
- Snare drum
- Acoustic guitar played with a pick

Examples of sounds with slow attack times are:
- A violin, gently starting a long tone
- The swell of a Hammond B-3
- The sound of a crash cymbal reversed
- The approaching sound of a helicopter

Decay

When the peak attack is reached, the energy might decrease quickly following the peak, or it could diminish slowly until it reaches constant amplitude. The reaction of the amplitude after the attack is called *decay*.

Sustain

Once the sound has leveled from the attack and the decay, the period that the sound is still generating from the source is called *sustain*. Sustain is dependent on the generation of sound at the source. As long as the source continues, the waveform is sustaining. The sustain phase can remain at a constant amplitude, increase in amplitude, or decrease in amplitude.

The Envelope - ADSR

How sound develops, holds, and decays over time comprises the envelope. The envelope parameters that we use to describe a sound's amplitude characteristic over time are attack, decay, sustain, and release.

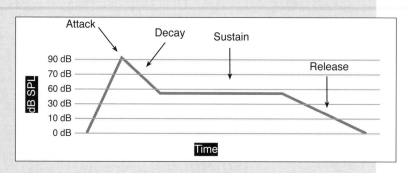

Release

Once the source stops generating the sound, the envelope enters the release phase. The easiest example to explain the release phase is reverberation (natural or simulated). When a violinist stops and removes the bow after a long, sustained note, the sound of the violin fading away in the concert hall represents the release.

INTERCONNECT BASICS

When it comes to connecting and designing audio systems, wire and connectors are very important factors. There are several aspects of wire type and cable configuration that dramatically affect sound quality. It's important that we understand the physical differences between basic cable and connector types; it's also important that we experience and realize the differences between well-designed and subpar audio cables. The cable industry is full of controversy, with an ongoing battle between manufacturers who make claims of sonic superiority based on marketing claims and those who dispute those claims. I have performed several listening tests comparing cables of various cost and design and in many of those tests, I can verify that there are sonic differences—in some, I can't. Each of us should feel compelled to perform our own tests and comparisons, especially when there is a dramatic cost difference. The home hi-fi audio industry isn't bashful about charging exorbitant prices for audio and AC cables. Are they worth it? That's a personal decision we each need to make. Keep in mind, though, that underpriced cabling gets no respect in the hi-fi industry—in many cases, the higher they price it, the better it sells.

Frequently, there are considerations other than audio quality involved in cable choice, even once the appropriate size and configuration has been determined. Some people just appreciate (and can afford) buying the best of the best—it's part of the fun they have in life. Others want to provide the best tools at each point in the signal path and they assume that several small advantages will add up to a perceptible difference in the final product. While both of these traits are understandable, it's important that we make decisions based on fact and our own perception. As functional sound operators and engineers, it's important that we practice listening analytically, assessing perceived audio quality honestly and critically. On the other hand, there are some criteria that must be met based on hard facts. These include considerations regarding resistance, inductance, and capacitance.

Resistance

In our discussion about speaker connections to amplifiers in Book 1 of the *Hal Leonard Recording Method*, we discussed ohms. An ohm (named after German physicist Georg Simon Ohm, who formulated Ohm's law) is a unit of resistance to the flow of alternating electrical current. A speaker or network of speakers provides resistance, measured in ohms, to the amplifier output; cable also exhibits resistance, also measured in ohms.

Typical speaker cable lengths add negligibly to the combined impedance load imposed by the speakers and cable. However, excessively long cable runs utilizing very thin cable can be problematic. Resistance throughout the cable length is cumulative with distance. Using identical wire, long runs exhibit greater impedance loads than short runs. Also, thicker cable provides less resistance than thinner cable.

It is generally understood that an addition of 1 ohm to the speaker/cable load should result in about a 0.1 dB change in level. By definition, 1 dB is the smallest incremental variance that should be perceptible to the human ear, although some listeners claim to perceive level changes of less than 1 dB.

From an article by Gene DellaSala (GDS), published on eCoustics.com, November 3, 2003:

> The basic purpose of a cable is to transfer the signal from point A to point B unadulterated. At audio frequencies the goal is to minimize losses by controlling the amount of resistance, inductance, and capacitance. For speaker cables, we have found the primary concerns for optimal signal transfer is to minimize resistance, followed by inductance, while also keeping capacitance in check to eliminate the possibilities of amplifier oscillation or frequency peaking. For line-level analog interconnects, it's a good idea to use cables that are low in capacitance and are well-shielded to eliminate interference and external noise sources from mitigating into the signal.

Inductance and Capacitance

Inductance, measured in henrys, describes the capability of a signal path to create magnetism. Capacitance, measured in farads, describes the capability of a circuit to store an electrical charge. In the audio frequency band, the influence of inductance and capacitance is negligible.

From an article by Dr. Eric Bogatin, published in Printed Circuit Design & Fab, July 1, 2007:

> Capacitance is a measure of the capacity of a pair of conductors to store charge at the price of the voltage between the conductors. The higher the capacity to store charge, the more charge can be stored for the same voltage between the conductors. Capacitance is a measure of the efficiency of storing charge, at the price of voltage.
>
> The efficiency of storing a charge does not depend on the absolute amount of charge that is currently on the conductors; it is just about the geometry of the conductors and the material properties. Bring two conductors closer, and the amount of charge they can store per volt between them increases.
>
> . . . Inductance is a measure of the efficiency of a conductor path to create rings of magnetic field lines, and the price you pay is the current through the conductor. You don't need current in the conductor to have an efficiency of creating rings of magnetic field lines.

Line-, Instrument-, and Mic-Level Impedance

As stated previously, impedance is the resistance to the flow of electrical current (AC). High impedance is high resistance to the flow of electrical current; low impedance is low resistance to the flow of electrical current. If you keep that simple mental picture in mind, the rest of the details should fall into place nicely.

Terminology

- Ohm (indicated by the Greek letter omega [Ω]): the unit of resistance to the flow of alternating current used to measure impedance.
- Impedance: the resistance of a circuit to the flow of alternating electrical current.
- Z: the abbreviation and symbol used in place of the word "impedance."
- Hi Z: high impedance. The exact numerical tag (in Ω) for high impedance varies, depending on whether we're dealing with input impedance or output impedance. It's generally in the range of 5,000 to 15,000 Ω for output impedance, and 50,000 Ω to 1,000,000 Ω for input impedance. It's important here to understand that hi Z is usually greater than 5,000 to 10,000 Ω.
- Lo Z: low impedance. The exact numerical tag (in Ω) varies for low impedance, as well as high impedance. It's generally in the range of 50 to 300 Ω for output impedance. It's normal for microphone output impedance to be between 50 and 150 Ω, and 500 to 3,000 Ω for input impedance. Normal input impedance for lo Z mixers is 600 Ω. Essentially, lo Z usually uses small numbers below 600 Ω.
- Output impedance: the actual impedance (resistance to the electron flow measured in Ω) at the output of a device (microphone, amplifier, guitar, keyboard). To keep it simple, realize that the output impedance is designed to work well with specific input impedance.
- Input impedance: the actual impedance (resistance to the electron flow measured in Ω) at the input of a device.

Compatibility Between Hi Z and Lo Z

To keep it simple, realize that the input impedance is designed to work well with specific output impedance. Low impedance and high impedance are substantially different in their defined ranges. They are not ideally compatible.

When connecting the output of a low-impedance mic to the input of a high-impedance amplifier, there's a problem. The lo Z microphone is designed to introduce the signal to the input of a low-impedance amplifier. Typically, when a low-impedance signal meets the high-impedance input, the signal seems so weak that, even with the input level set to maximum, an acceptable level cannot be achieved. Though the connection is made and signal enters the input, the impedances of the two devices don't match, so they can't operate to their full and intended capacity.

The other incompatible scenario involves attempting to plug a high-impedance output (microphone, guitar, keyboard, and so on) into a low-impedance input (mixer, amp, speaker, and so on). In this case, the hi Z output is expecting to meet a hi Z input; in other words, the hi Z signal is expecting to meet high resistance. With the high-impedance output signal connected to a low-impedance input, the signal meets relatively little resistance and therefore easily overdrives the input—input level controls must often be set to near minimum in order to attain a level low enough for a satisfactory level.

High-impedance outputs are supposed to meet high-impedance inputs; low-impedance outputs are supposed to meet low-impedance inputs. It's not true that the input and output impedance need to be identical. In fact, the input impedance is generally supposed to be about 10 times the output impedance, but as I mentioned earlier, we need to keep in mind that high impedance uses high ratings (above 10,000 Ω), and low impedance uses low ratings (typically below 1000 Ω).

Audio Example 5-1

High-Impedance Instrument into a Low-Impedance Input

Audio Example 5-2

Low-Impedance Mic into a High-Impedance Input

Solution: To enable high- and low-impedance devices to work together, simply use an impedance transformer—also called a *line-matching transformer* or *direct box*—to change impedance from high to low or low to high; that's the easy part. We should, however, strive to understand some of the reasons we do what we do. This simple explanation of impedance is meant to get you started toward your enlightenment. It is admittedly primary in its depth, but it functions as an excellent point of reference for further technical growth.

Speaker and Amplifier Impedance

While professional power amplifier inputs are virtually always balanced line level, connections between speakers and amplifiers are also measured in ohms. They fall into a much lower range, typically between 2 and 16 ohms.

Carefully matching the impedance of a speaker, or network of speakers, to the power amplifier output is extremely important and must match more closely than connections between mic-, instrument-, and line-level devices. This topic is covered in greater detail in chapter 10.

Balanced Versus Unbalanced

For the purposes of this course, we'll cover this topic much like we did with impedance, using simple references and, wherever possible, non-technical language. There are plenty of books about electronic circuit design and books that approach the audio world from a technical perspective; however, this book is written to help you as an operator, and with that in mind, I want to make sure you understand some of the most essential basics as they pertain to helping you be a better sound operator. Some of the differences between balanced and unbalanced wiring schemes are simple, and some are interestingly complex. As a point of reference, remember this: Almost all guitars are unbalanced and almost all microphones are balanced. Let's look at these two types of wiring.

Terminology

- **Lead** (pronounced *leed*): another term for wire.
- **Hot lead**: In a cable, the hot lead, surrounded by a layer of insulation, is the wire carrying the desired sound or signal. From a guitar, the hot lead carries the guitar signal from the magnetic pickup to the input of the amplifier.
- **Cold lead**: In a cable, the cold lead, surrounded by a layer of insulation, is the wire carrying the desired sound or signal. It is identical to the signal carried by the hot lead, although reversed in polarity. The cold lead is part of the balanced wiring scheme.
- **Braided shield**: Cables for instruments, mics, and outboard gear—pretty much anything other than speaker cable—have one or two wires carrying the desired signal. Surrounding the lead(s) are very thin strands of wire braided into a tube so that electrostatic noises and interference can be diffused, absorbed, and rejected. This braided wire tube is called the *shield*.

Unbalanced Guitar Cables

Normal guitar and keyboard cables, also called *line cables*, contain one hot lead to carry the instrument signal—a braided wire shield surrounds this hot lead. The purpose of the shield is to diffuse, absorb, and reject electrostatic noises and RF interference.

This system works pretty well within its limitations. The braided shield does a pretty good job at keeping radio signals and other interferences from reaching the hot lead—as long as the cable is shorter than about 20 feet. Once the cable is longer than 20 feet, there's so much interference bombarding the shield that the hot lead starts to carry the interference along with the audio signal. The long cable acts as a crude antenna and picks up plenty of transmissions from multiple transmitters. This fact is true even when we study balanced cables; the main difference is that the balanced wiring scheme cleverly beats the system by using the system.

Balanced Wiring

Low-impedance mics, as well as most modern outboard equipment and mixers, use balanced connections. While the length limit of unbalanced cables is about 20 feet, balanced low-impedance cables can be as long as 1,000 feet, or so, without the addition of noise or electrostatic interference and without significant degradation of the audio signal.

A cable for a balanced lo Z mic uses three conductors, unlike the unbalanced system that just uses the hot lead and the braided shield. Of these three conductors, two are used to carry the signal, and the other is connected to ground. Two-conductor shielded cables are also very common. Whether two or three leads are present, they are all twisted together throughout the length of the cable so that they are exposed to the same electrostatic noises and RF interference.

The term "balanced" is derived from the fact that two leads carry the signal and that they are perfectly balanced, sharing the exact same impedance. They receive the same noise and interferences, though one is *hot* and the other is reversed in polarity (*cold*) relative to the hot lead. The hot lead is typically connected to pin 2 on the XLR connector and the cold lead, which, again, carries the identical signal as the hot lead but reversed in polarity, connects to pin 3. As a point of interest, though not crucial to our understanding of the balanced wiring concept, some European standards and some balanced line-level connections designate pin 3 as hot and pin 2 cold.

At the source, the signal is split and sent down the hot and cold leads—the polarity of the cold lead is reversed relative to the hot lead. The hot lead is indicated by the plus sign (+) and the cold lead is indicated by the minus sign (–).

At the input stage (the end of the cable opposite the source), the hot and cold leads are recombined using a differential amplifier, which measures the difference in voltage between the leads. All noise and interference should be identical in the hot and cold leads and, since the polarity of the intended audio signal was reversed at the onset, it's the source audio that is different between the leads. The differential amplifier sees only the original audio signal as it amplifies the difference between the leads so the source audio is recombined and its amplitude is doubled. Because the electrostatic noises and RF interference have been saturating the hot and cold leads equally throughout the length of the cable, they are ignored by the differential amplifier as it recombines the hot and cold leads. Noises and interference are, therefore, eliminated.

Any three-point connector can be used on cables that connect balanced devices. As long as there's a place for the hot and cold leads and a ground to connect, the system will work. XLR connectors are the most common, but a plug such as a 1/4-inch tip-ring-sleeve configuration is also common. In large studios, a smaller version of the 1/4-inch TRS connector—the Tiny Telephone (TT) connector—is commonly used for balanced patch bay connections.

In summary, the result of balanced wiring is total cancellation of noise and interference, plus a doubling in amplitude compared to the signal in an unbalanced system.

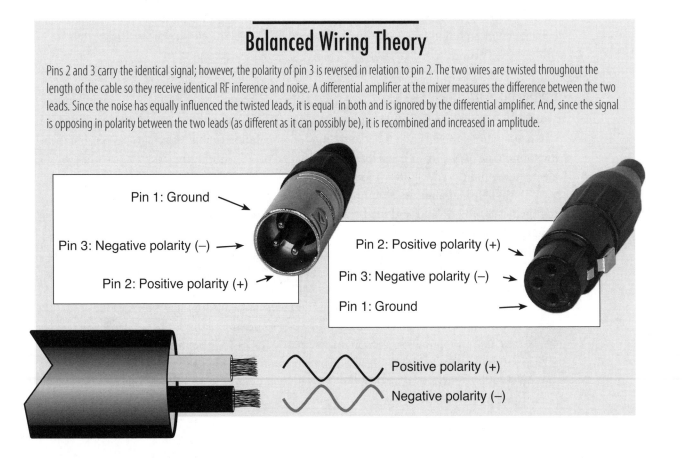

Balanced Wiring Theory

Pins 2 and 3 carry the identical signal; however, the polarity of pin 3 is reversed in relation to pin 2. The two wires are twisted throughout the length of the cable so they receive identical RF inference and noise. A differential amplifier at the mixer measures the difference between the two leads. Since the noise has equally influenced the twisted leads, it is equal in both and is ignored by the differential amplifier. And, since the signal is opposing in polarity between the two leads (as different as it can possibly be), it is recombined and increased in amplitude.

Pin 1: Ground

Pin 3: Negative polarity (–)

Pin 2: Positive polarity (+)

Pin 2: Positive polarity (+)

Pin 3: Negative polarity (–)

Pin 1: Ground

Positive polarity (+)

Negative polarity (–)

Choosing Cable

The controversy has been brewing for years in the audio world. This excerpt of "A Spat Among Audiophiles Over High-End Speaker Wire," by Roy Furchgott, was taken from the December 23, 1999 issue of the *New York Times*.

> In the last year, Lewis Lipnick has tested high-end audio cables from 28 manufacturers. As a professional musician with the National Symphony Orchestra and as an audio consultant, he counts on his exacting ear to tell him if changing cables affects the accuracy of the sound from his $25,000 Krell amplifiers.
>
> His personal choice is a pair of speaker wires that cost $13,000. "Anyone would have to have cloth ears not to tell the difference between cables," he said.
>
> "In my professional opinion that's baloney," said Alan P. Kefauver, a classically trained musician and director of the Recording Arts and Sciences program at the Peabody Institute of Johns Hopkins University. "Has the wire been cryogenically frozen? Is it flat or round? It makes no difference, unless it makes you feel better."
>
> His choice for speaker wire? Good-quality 16-gauge zip wire.

Over the years, I've found myself straddling the chasm between these two philosophies. On one hand, I swear I can hear the differences between certain cables—I've even included examples in my books to substantiate what I'm hearing. On the other hand, I appreciate the careful and scientific approach of the objective technicians who take the stand that "If I can't see it on my meter, it ain't different!"

My background is musical. I've spent the vast majority of my life performing, writing, arranging, recording, producing, and mixing music in the studio or at live shows. I care more about what it sounds like than what it looks like on a meter. However, some of the things that I hear in the tests I've done are pretty subtle, and it's not always easy to tell if it's subtly better or worse. Some of my tests have yielded substantial audible variations with the simple swapping of one cable in the setup.

The ultimate purpose of an audio cable is simple: deliver the signal to the destination in exactly the same condition that it's received at the source. Anything other than that is simply undesirable. If a cable design is overly complex, the potential for coloration increases. To complicate the process of finding trustworthy information about cable quality and sonic integrity, our industry is rampant with adamant opinion. I've heard passionate arguments by very successful audio industry professionals that no one could hear the difference between the highest-priced speaker cable and 12-2 copper electrical wire. I've heard equally impassioned professionals proclaim the benefits of rewiring their entire systems with the most upscale cable of the day.

In my mind, it always comes down to one question: Is this going to give me even a little more accuracy? If I think there's a chance of even a slight improvement—and if I can realistically afford it—I'll take the step. I'm a proponent of the theory that several minor improvements add up to a better final product. I also don't like wasting money on silly things. I've had the discussion with some of my buddies who are ardent hi-fi audio buffs, asking them to explain the validity of spending astronomical amounts of money to build systems. I like to highlight that the music they love most was very likely tracked and mixed using Yamaha NS10s as reference monitors with something like a Crown DC300 amplifier powering them—an entire system that could be purchased for much less than the price of one of their cables.

The answer to this conversation doesn't really matter—that is, if there is an answer. Music and audio stimulate our senses and emotions. Whether at a live show, in the studio, or at home, there is a huge amount of stuff going on behind the listener's or viewer's mental curtain. Music and sound are about emotion, passion, and stimulation of our senses. It's just impossible to live in each other's skin—to hear or feel the differences imposed by this or that. Our perceptions are real to us. I'm on a quest to match reality with perception. Care to join me?

Reasonable Cable Expenditure

If you consider the cost of your system, it's pretty realistic to expect that cable expenditure should fit within a reasonable percentage of the total cost. If your system were worth $2,000, it would seem silly to spend $1,000 on cables; however, it would make sense to spend between a $100 and $200. If you're using a huge live sound system with lots of monitors, power amps, and speaker cabinets, the cost of cabling is significant. If you forget to factor in the cost of cabling, you'll be surprised when it's time to connect everything together.

Even a modest live sound reinforcement system requires careful consideration when budgeting for connectors and cables. If you're spending somewhere around 10 percent of your budget on cables, you're within the bounds of acceptability—it is all very relative, though. It is a bad value to purchase really inexpensive cables because they don't withstand constant use. You might need to replace a $10 cable a few times in the same lifespan of a $25 cable, not only costing more money in the long run but also causing untold heartache and pain while sleuthing out a problem during a gig.

It's easy to see how the cost of cabling can add up quickly when you consider that 10 XLR microphone cables from Rapco or Pro Co could easily cost between $220 and $250—Monster brand cables would cost at least 50–100 percent more. Even just a 25-foot, 24-channel Pro Co snake costs a little over $300, when all is said and done. A 100-foot, 24-channel snake could easily approach or exceed $500.

The math is pretty straight ahead when you consider the ballpark cost to provide a suitable amount of mid-priced cabling for a 16-channel system:

- 100-foot, 16-channel snake: $450
- One 20-foot XLR cable per mixer channel (16): $400
- 20-foot XLR or 1/4-inch TRS cable to connect mixer to power amplifiers (2–4): $40–$80
- 50-foot speaker cable for two sides of a bi-amplified system (4): $200

So the cost of a modest cable package is quickly over $1,000, and we haven't considered insert cables, cables for the recorder and monitors, adapters, and so on. That's for a mid-priced cable package. The cost of the Cadillac cable package can go as high as you're willing to go, which probably doesn't make sense for most situations. However, if you just enjoy having the best and you're comfortable with the expense, don't feel bad when someone questions your sanity. That can be kind of fun, too.

Pre-Assembled Cables

Cables aren't inexpensive; at least cables worth owning aren't, so it is not advisable to spend a lot of time searching far and wide for the least expensive, "cheapest" cables. Typically, these low-priced "bargains" are poorly constructed and fail much sooner and more often than mid- or high-priced cables. I know with relative certainty that the cables I regularly purchase are likely to be functioning properly for years to come. I've also learned that the cables I grab out of the "super deal" bin are very likely to fail within the first several months, weeks, or days of their use. Therefore, in the long term these "super deals" are likely to be the most expensive cables.

For common cable lengths, it's usually less expensive to purchase pre-assembled cables than it is to purchase wire and assemble the cables yourself. Most of the major name-brand cable manufacturers provide a reliable product. The fact that they can purchase cable and connectors in bulk, and they have an efficient and inexpensive means of assembly, allows them to still make a profit while providing reliable cables, typically at a lower cost than the combined cost of the components.

Custom-Assembled Cables

Obviously, long cable runs and permanent installations requiring specific lengths of cable must be custom designed and meticulously assembled. For installations with custom cabling spanning long runs, Belden has been the most commonly used cable for a long time. They provide an excellent product and are very experienced in providing cabling for all types of systems.

Assembling cables requires patience, a meticulous mindset, a steady hand, the right tools, and usually some soldering skills. There are excellent solderless connectors available that are reliable and easy to assemble. Neutrik makes excellent soldered and solderless connectors.

If you want to use cables with soldered connectors, you're uncomfortable operating a soldering iron, and you rely on the dependability of your cables, you either need to develop your soldering technique or hire an experienced technician to put everything together correctly. A connector that has been skillfully soldered onto the cable will stand up to constant use and it will last a very long time before it fails. A poorly soldered connection is most assuredly going to fail at the worst possible moment.

Probable Causes for Various Cable Failures

There are a few clues as to what's going wrong when a cable begins to fail.
1. If you hear the audio signal along with a substantial amount of noise and radio interference, it typically means that, on an unbalanced cable, the hot lead is connected but the shield is disconnected.
2. If you hear no signal, it typically means the hot lead on an instrument cable is disconnected or that both leads on a balanced connection are disconnected.
3. If you're using a balanced cable and the signal is much weaker, or quieter than normal, it typically means that either the hot or cold lead is disconnected.

It's also possible, especially in a poorly constructed cable, that two or more of the leads or shield could be touching each other inside the connector, causing noise or shorting the connection. Often, cables constructed by inexperienced technicians are on the verge of these types of failures inside the connector from the moment they are put into service.

Least Common Denominator

If you decide that you want to incorporate some highly touted and more expensive cables into your setup, do so systematically. Keep in mind that any signal chain is only as good as its weakest link. Therefore, it is senseless to have an expensive cable in the same signal chain with an inexpensive cable. When you're ready to raise the bar, the best place to incorporate new and more expensive cable is between simple connection points. Connections using just a few cables make the most sense, such as:

- Between the mixer output and the recorder input.
- Between the mixer output and the powered speaker cabinet input.
- Between both the mixer output and the power amplifiers, and between the power amplifiers and the speaker cabinets.

It makes absolutely no sense to use a $100 instrument cable to connect to an inexpensive direct box that's connected to the mixer using a $10 mic cable running through an inexpensive snake to connect to the mixer, which is connected to the power amplifier with $10 XLR cables and then connected from the power amplifier to the speaker cabinets using inexpensive lamp cord. Keep all things in perspective.

Acoustical Influences on Listening Tests

Sometimes, the room you're in exposes deficiencies in cable. I've personally experienced the effect that the room has on cable assessment. After visiting Bruce Swedien (engineer for Michael Jackson's *Thriller, Bad, Off the Wall,* and so on) in the early '90s, and hearing him rave about replacing all of the wire in his studio with Monster brand cable (and then after hearing music through his studio monitors), I came away convinced that cable influences the sound we hear from our systems. His studio was amazing in every way: meticulously installed, full of the best gear, huge, and sonically out of this world. Bruce is famous for being extremely influential in the development of what came to be the standard for popular commercial audio. Through his work with Quincy Jones and Michael Jackson, he led the way to powerful and pristine audio productions that remain a sonic reference standard for many excellent engineers.

After returning home, I decide to start my own search for audio purity by testing all the cables I could find. I started with the cables between the output of my console and inputs of my powered studio monitors. It made sense that this was a place in my signal path that needed to be the best it could possibly be and that it was also the least complicated checkpoint because the test involved only one cable on each of the left and right channels. The differences I heard were stark, especially in the low band. I settled on a cable constructed from Mogami quad star cable and Neutrik connectors. Especially when compared to an inexpensive (yet popular) name brand, it sounded markedly fuller in the low band and very clear in the high band. After confirming my findings with a few of my audio geek buddies, who also enjoy listening tests and shoot-outs and other such fun activities, I was pleased to implement these new cables in my studio. However, I had just moved to a new location and, when we performed the evaluation, I hadn't finished the acoustical treatment in the studio. The differences we heard were real, audible, and pronounced, but once I had finished treating the studio with diffusers, traps, and foam, we repeated the tests. What we discovered was a much less pronounced difference in the sound between the cables. Although we could still hear a similar shift in frequency content while comparing the different cables, the tonal character of the acoustical environment had been substantially tamed—the low-mid mode was much less dominant after treatment. We concluded that the effect of the various cables was real but less of a problem in my studio once it was acoustically treated.

So what's the takeaway for each of us in this battle between high-priced cable manufacturers, audio purists, technical elite, and the ardently opinionated? We should each take the time to listen and compare. We should then make our own decisions based on the data we've compiled and the experiences we've encountered. Keep in mind that cable can make a sonic difference and the environment in which we assess the cable can influence the results.

Cable Construction Standards

Cables from a respected manufacturer can be trusted to be reliable. Plus, we can expect these cables to deliver a reasonably accurate rendition of the signal they receive. In reality, there are very few actual cable manufacturers. Many of the cables we use are imported and the name of the apparent manufacturer is printed on the cable.

In addition to making sure the actual wire is high quality, it is equally important to pay close attention to the connectors used and the quality of construction. Neutrik and Switchcraft connectors have been trustworthy for years. As I mentioned previously, if you are building your own cables, be meticulous. There are standards for cable construction which, when adhered to, produce cables that can be trusted to work for a long time.

At a local AES meeting here in Seattle, Steve Turnidge (circuit designer, mastering engineer, and author), along with Aaron Gates (system designer and installer), presented the standards to which they adhere when they build cables. Steve designs circuits for Krell, Rane, and his own company, Synthwerks. Aaron has managed and implemented many full-scale installations, including Microsoft Game Studios, MSNBC Studios, and the Experience Music Project.

Varying Levels of Soldering Technique

These two 1/4-inch connectors offer dramatically different reliability—this picture is worth 1000 words. Take great care and pride in the neatness and accuracy of your soldering tasks—spend the time it takes to get it right. You'll be much happier and you'll save a lot of heartache at the gig!

Their reference is the NASA wiring standard. It's easy to see that when these standards are met, there is an extremely high chance that there will be no cable failure in the system. Although wires break, connectors are bent, and unexplained things happen, if we build our cables according to a rigorous standard, such as that specified by NASA, we are giving our systems and our music the best chance to be heard and understood.

Visit the following websites to get a glimpse into what it takes for a cable termination to endure space travel, which is admittedly nothing compared to the same cable enduring a rock guitarist!

Speaker Cables

Use the appropriate wire to connect your speakers to your power amp. Speaker wire is not the same as a guitar cable. Use designated speaker wire. Also, choosing wire that is the wrong thickness for your situation can cause a problem with the efficiency of your amp and speakers. Speaker cable must contain two identical insulated wires and doesn't require a braided shield. The two wires in the speaker cable are typically composed of many strands of thin copper wire.

Sometimes, it is difficult to tell a speaker cable from an instrument cable because they both are commonly constructed with visually-similar wire and 1/4-inch tip-sleeve phone plugs on each end. Although speaker connections can be made with wire that looks like heavy-duty lamp wire, most professional speaker cable is round and black, like an instrument cable. These cables are not interchangeable. When a speaker cable is used as an instrument cable, the lack of a shield results in increased level of electrostatic noise and RF interference. When an instrument cable is used as a speaker cable, there is an impedance problem because the hot lead and shield are not the same impedance as the pair of identical wires required by a speaker connection. The impedance variance will cause a decrease in the efficiency of the speaker-to-amplifier connection.

Speaker Wire Connections

Always use heavy-duty wire designed specifically for use with speakers. Wire is sized according the American Wire Gauge (AWG) standard. The smaller the wire number, the thicker the wire. Thicker wire has less resistance to signal. To have minimal degradation of signal in longer runs, we use thicker wire.

Be absolutely certain that the red post on the back of the power amp is connected to the red post on the back of the speaker and that black goes to black! If these are connected backwards on one of the speakers, the speakers are said to be out of phase. When this happens, a sound wave that is sent simultaneously to both speakers moves one speaker cone out while it moves the other speaker cone in. Speakers connected out of phase work against each other instead of with each other. What you hear from them is inaccurate and unpredictable, especially in the lower frequencies.

If there is a question about the intended application for a particular cable, open up the connector and inspect the wire. If there is an insulated wire and a braided shield, it's an instrument cable. If there are two identical insulated wires, it's a speaker cable.

Wire Gauge

Speaker wire is categorized according to a few different wire gauge standards, the most common of which was established by Brown & Sharp, a division of Hexagon Meteorology, Inc. Brown & Sharp established the American Wire Gauge (AWG) specification, which divides wire into AWG gauges from 0000 (pronounced "4 aught") gauge, with a diameter of .46 inches, to 36 gauge, with a diameter of 0.0050 inches. The actual gauge spec is based on a mathematical equation and it's not uncommon to see gauge tables from 00000000 (8/0) to 50 gauge—neither extreme applies to our duties as sound operators.

Wire exhibits impedance characteristics—different sizes of wire resist the flow of electrical current by different amounts. Larger wire exhibits less resistance (lower impedance) and smaller wire exhibits greater resistance (higher impedance). It's important to note that inappropriately small wire used for excessively long runs could add substantially to the combined speaker/cable impedance load. Using the appropriate cable length and gauge should result in a small impact on the overall impedance of the connection between devices. The actual impact of the cable on a system depends a variety of factors, including the cable length and thickness, the amount of power being provided, and the total impedance of the speaker cabinet. With all of these things considered, most speaker cable runs in a live sound setting require between 16- and 10-gauge wire. Some engineers prefer smaller than typical gauge cable (larger than typical diameter wire) in an effort to minimize the impact on overall impedance. In addition, thicker cable is often associated with better transfer of low frequencies. If you're building a large system and you're not exactly sure which wire is appropriate, consult an experienced and reputable system designer who has a track record of first-class installations similar to yours.

Instrument-Level Cables

Instrument cables are almost always designed for unbalanced connections. They use one hot lead containing the instrument's output signal surrounded by a braided shield, which helps absorb and diffuse electrostatic noises and RFI (radio frequency interference). Instrument cables typically utilize 1/4-inch TS phone connectors on each end.

Some speaker cables look similar to instrument-level cables; however, they use two completely different types of wire and are not interchangeable.

Line-Level Cables

Line-level cable is designed to carry signals like those from a keyboard or guitar to a mixer or instrument amplifier. Line-level cables also connect outboard gear to the mixer or the mixer to the power amplifier or the powered monitors. Line-level cables can either be unbalanced, with a tip-sleeve connector, or balanced, using XLR or tip-ring-sleeve connectors.

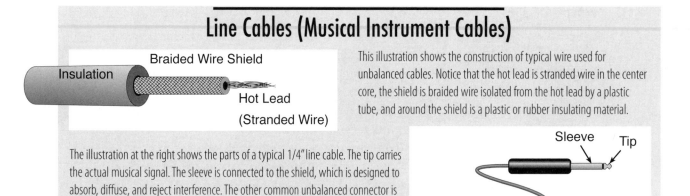

Line Cables (Musical Instrument Cables)

Insulation

Braided Wire Shield

Hot Lead

(Stranded Wire)

This illustration shows the construction of typical wire used for unbalanced cables. Notice that the hot lead is stranded wire in the center core, the shield is braided wire isolated from the hot lead by a plastic tube, and around the shield is a plastic or rubber insulating material.

Sleeve   Tip

The illustration at the right shows the parts of a typical 1/4" line cable. The tip carries the actual musical signal. The sleeve is connected to the shield, which is designed to absorb, diffuse, and reject interference. The other common unbalanced connector is the RCA phono plug.

Microphone Cables

Microphone cables typically utilize two-conductor shielded cable with XLR connectors at each end. In the professional studio, microphone cables are often used to connect line-level devices, such as mixers, effects, and power amplifiers.

Devices that connect with microphone cables at line level use a balanced wiring scheme; devices that connect with regular guitar cables use an unbalanced wiring scheme.

Do Cables Really Sound Different?

The difference between the sound of a poorly designed and a brilliantly designed cable can be noticeable. Exactly how extreme this difference is depends on the quality and type of the source material and the system through which it's being delivered. If a narrow-bandwidth signal composed of mid frequencies and few transients is compared on two vastly different cables, the audible differences might be minimal. However, when full-bandwidth audio rich in transient content, dimensionality, and depth is compared between a marginal and an excellent cable, there will typically be a noticeable difference in sound quality.

Listen for yourself. Most pro audio dealers are happy to show off their higher-priced product. When comparing equipment, listen to some fantastic recordings, with which you are very familiar, through the gear you're auditioning. It's best to use high-quality audio that receives industry praise for its excellence—after all, that's the standard you are trying to meet or beat. Right?!

Listen to Audio Example 5-3. The acoustic guitar is first miked and recorded through some common-quality cable. Then it's recorded through a microphone with some very high-quality Monster cable. Notice the difference in transient sounds, depth, and transparency.

Wire for Balanced Cables (Two-Conductor Shielded)

Most wire for balanced cables has two separate leads twisted together in the center core throughout the length of the cable. Both of these leads carry the signal and connect to pins two and three; the braided shield connects to ground.

Twisted pair carrying identical signals in opposing polarities: positive and negative.

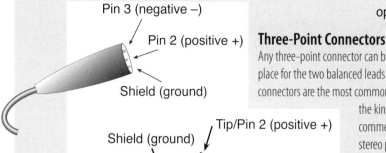

Pin 3 (negative –)
Pin 2 (positive +)
Shield (ground)

Shield (ground)
Tip/Pin 2 (positive +)
Ring/Pin 3 (negative –)

Three-Point Connectors

Any three-point connector can be used on balanced cables. As long as there's a place for the two balanced leads and a ground to connect, the system will work. XLR connectors are the most common, but the quarter-inch tip-ring-sleeve plug—like the kind on your stereo headphones—is also common. In commercial studios, a smaller version of the quarter-inch stereo plug, called the Tiny Telephone connector, is also common.

Audio Example 5-3

Mic on Acoustic Guitar Using Common Mic Cable, Then Monster Studio Pro 1000 Cable

Audio Example 5-4 demonstrates the difference in vocal sound using common mic cable first, and then a high-quality mic cable from Monster. Notice the difference in transient sounds, depth, and transparency.

Audio Example 5-4

Vocal Using Common Mic Cable, Then Monster Studio Pro 1000 Cable

Digital-Interconnect Cables

The cables you choose to make digital connections affect sound quality. You must use cable designed to carry digital data for transfers between digital devices. S/PDIF connections use RCA jacks and connectors. Even though a standard home audio RCA cable is capable of making the connection between a S/PDIF input and output, it is incapable of providing the stability and impedance required to facilitate an accurate data transfer. Similarly, an AES/EBU digital connection is made using XLR connectors. Even though a standard microphone cable is capable of completing the connection, a different type of wire is required to facilitate the most accurate data transfer.

Listen to Audio Examples 5-3 through 5-7. In each example, a different cable and format configuration is demonstrated. Listen specifically to all frequency ranges as well as transients. Also, consider the "feel" of the recording. Often, the factor that makes one setup sound better than another is difficult to explain, but it's easy to feel. The following examples are recorded using exactly the same program material and the identical transfer process.

The differences you hear on your setup depend greatly on the quality and accuracy of your monitoring system, as well as your insight and perception. Once you understand and experience subtle sonic differences, you'll realize the impact they hold for your musical expression. Constantly compare and analyze the details of your mixes. It will result in much more competitive quality. You'll realize more satisfaction, and you'll probably get more work.

I repeat: it's important to use cable designed for digital data transfer when connecting digital devices. Digital transfers require a specific and stable impedance of 75 ohms (S/PDIF) or 110 ohms (AES/EBU) to most faithfully transfer the digital data. A cable designed for analog applications transmits continuously varying voltage that mirrors the analog waveform it receives at the source. A cable designed for digital applications must efficiently and accurately transmit data, which is essentially a varying square wave indicating on or off ("1" or "0") binary bit status. These are two different tasks and they require different cables.

From the BlueJeansCable.com article "What Is Impedance, Anyway?":

> Where analog audio or video signals consist of electrical waves, which rise or fall continuously through a range, digital signals are very different—they switch rapidly between two states representing bits, 1 and 0. This switching creates something close to what we call a square wave, a waveform that, instead of being sloped like a sine wave, has sharp, sudden transitions. . . . Although a digital signal can be said to have a "frequency" at the rate at which it switches, electrically, a square wave of a given frequency is equivalent to a sine wave at that frequency accompanied by an infinite series of harmonics—that is, multiples of the frequency. If all of these harmonics aren't faithfully carried through the cable . . . then the "shoulders" of the digital square wave begin to round off. The more the wave becomes rounded, the higher the possibility of bit errors becomes. The device at the load end will, of course, reconstitute the digital information from this somewhat rounded wave, but as the rounding becomes worse and worse, eventually there comes a point where the errors are too severe to be corrected, and the signal can no longer be reconstituted. The best defense against the problem is, of course, a cable of the right impedance: for digital video or S/PDIF digital audio, this means a 75-ohm cable like Belden 1694A; for AES/EBU balanced digital audio, this means a 110-ohm cable like Belden 1800F.

Optical Digital Cables (Fiber Optics)

Fiber optic cables transmit bursts of light that represent the binary digital bits that represent audio sound waves. This type of cable is extremely well suited to the task and, conceptually, the cable should not make a difference in the data transfer. However, optical cable is fragile, doesn't reliably withstand bending, and, when subjected to harsh treatment, can't always be trusted to perform a flawless transfer. I have used a lot of optical connections in my studio, at other studios, and even in my own personal home entertainment system—I've had very few problems.

It's important to treat optical interconnect cables gently and to avoid bending them. Bending can cause cracks in the glass fiber material, which could degrade the light transfer and potentially damage, or interrupt the flow of, the data. Carefully route the cables to avoid kinks and bends; be certain there are no heavy devices setting on an optical cable. If you're careful to avoid bends and abuse to the optical cable, it should service as a very efficient and accurate means of making digital audio data transfers. In addition, optical connections are not susceptible to ground hums and RF interference.

In the audio world, there are two important optical protocols: TOSLINK and ADAT Lightpipe.

- **TOSLINK:** The term "TOSLINK" stands for Toshiba Link because Toshiba developed this two-channel stereo data protocol. TOSLINK data, carried via fiber optic cable, is essentially the same as S/PDIF data, which is carried via coaxial cable—inexpensive real-time converters are readily available. TOSLINK connections are commonly available on CD and DVD players, home entertainment systems and receivers, and most other home and professional stereo digital devices. In fact, TOSLINK connections are available on the headphone and microphone jacks on Apple's MacBook Pro. While TOSLINK can also carry multiplexed 5.1 DTS or Dolby Digital audio, it cannot carry six discrete digital audio channels.

- **ADAT Lightpipe:** Developed by Alesis for data connections between its ADAT digital recorders, ADAT Lightpipe (ADAT Optical Interface) carries up to eight tracks of 48 kHz, 24-bit audio. Alesis also developed a scheme to carry 96 kHz, 24-bit audio, but two ADAT Optical audio channels are required for each 96 kHz channel. ADAT Optical connections have been commonly adopted by the audio industry as a means to expand DAW interfaces, eight channels at a time, and on digital mixers as a means to connect to various multitrack devices.

TOSLINK and ADAT Optical formats use the same connector and cable. These devices can be easily connected together, but to no avail—their respective data formats are incompatible.

Camp Time

There is a camp in the audio world that holds steadfastly to the theory that cables don't make a difference in digital transfers—it's usually the same camp that says a clone of a digital file can't possibly deviate from the original. Both of these viewpoints can be backed up theoretically and both have been disputed artistically.

In 1993, I brought highly respected and award-winning mastering engineer Bernie Grundman to Seattle to teach a seminar on mastering. Bernie services the highest-level clients in the recording industry and has won countless awards for his mastering work. With the digital age well on its way to ramping into modern recording culture (DAT was a standard and Pro Tools was first released in 1991), the hype about digital recording centered largely on the lack of noise and the value of identical digital clones. However, Bernie, who built his career on maintaining the highest standards in everything audio, wasn't willing to simply buy into the hype—he was hearing a difference between digital generations. He brought examples of cloned files and files that had been cloned multiple times to the seminar and played them for the audience. Tannoy had provided an amazing system for the seminar and everyone in attendance could hear the image collapsing throughout the progression of digital clones. It was amazing and very controversial at the time.

My point in sharing my experience with Bernie is to inspire you to listen for yourself and make your own decisions, whether you're assessing cable, clones, software, or whatever. We function in a world where art and expression are wrapped in science and technology. Artists tend to stand on what they feel or hear. Technicians tend to stand on what they can quantify mathematically and what they can see on a meter. That can get messy.

Audio Example 5-5

AES/EBU to DAT Using Common Cable, and Then SP1000 AES Silver Digital Monster Cable

Audio Example 5-6

S/P DIF to DAT Using Common RCA Cables, and Then M1000 D Silver Digital Monster Cable

Audio Example 5-7

Analog Out to DAT Using Common XLR Cables, and Then Prolink Studio Pro 1000 XLR Monster Cables

Audio Example 5-8

ADAT Lightpipe into Digital Performer Using Common Optical Cable, Bounced to Disk

Audio Example 5-9

ADAT Lightpipe into Digital Performer Using Monster Cable's Interlink Digital Light Speed 100 Optical Cable, Bounced to Disk

Connectors

We encounter several types of connectors when hooking together audio equipment. In this section, we cover RCA connectors, 1/4-inch connectors, XLR connectors, adapters, plugging in, powering up/down, grounding, and hums.

RCA Connectors

RCA phono connectors are the type found on most home stereo equipment and are physically smaller than the plug that goes into a guitar or keyboard. RCA phono connectors are among the least expensive connectors and were very common in home-recording equipment manufactured in the mid-'80s to the mid-'90s. By today's standards, though, they are seldom used for serious audio connection. They are only appropriate for unbalanced applications and are virtually always used for high-impedance connections, such as those from consumer CD and DVD players, some computers, and other music players.

RCA Phono Connectors

For analog audio connections, RCA phono connectors are most typically used in home stereo configurations. They're small, inexpensive, and not usually used in a professional application. With one contact for the hot lead and one for the shield, this connector is used in an unbalanced application.

RCA connectors are also used for SP/DIF digital connections in home and professional applications.

1/4" Phone Plug Tip-Sleeve (Mono/Unbalanced)

The 1/4" tip-sleeve phone plug is most commonly use in an instrument cable. It carries an unbalanced, mono signal. In addition, phone plugs are often used for speaker cables, using speaker wire rather than line-level instrument cable.

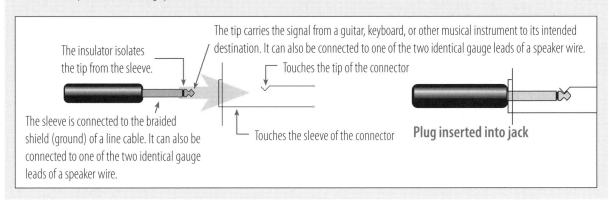

The tip carries the signal from a guitar, keyboard, or other musical instrument to its intended destination. It can also be connected to one of the two identical gauge leads of a speaker wire.

The insulator isolates the tip from the sleeve.

Touches the tip of the connector

The sleeve is connected to the braided shield (ground) of a line cable. It can also be connected to one of the two identical gauge leads of a speaker wire.

Touches the sleeve of the connector

Plug inserted into jack

1/4-Inch Phone Connectors

1/4-inch phone connectors are the type found on regular cables for guitars or keyboards. These connectors are commonly used on musical instruments and in live sound systems, at home and in professional recording studios.

Notice that a guitar cable has one tip and one sleeve on the connector. In a guitar cable, the wire connected to the tip carries the actual musical signal. The wire carrying the signal is called the *hot wire* or *hot lead*. The sleeve is connected to the braided shield that's around the hot wire. The purpose of the shield is to diffuse outside interference, such as electrostatic interference and extraneous radio signals.

1/4" Phone Plug Tip-Ring-Sleeve (Stereo/Balanced)

The 1/4" tip-ring-sleeve phone plug is most commonly seen on stereo headphones. In this application, the tip and ring connections carry the left and right channels of a stereo headphone send. This connector, like the XLR, is also commonly used to carry balanced signals, where the tip and sleeve carry the audio signal like pins 2 and 3 of the XLR connector.

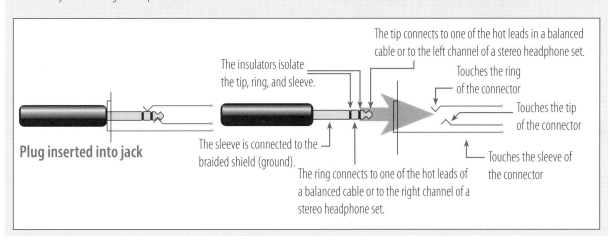

The tip connects to one of the hot leads in a balanced cable or to the left channel of a stereo headphone set.

The insulators isolate the tip, ring, and sleeve.

Touches the ring of the connector

Touches the tip of the connector

Plug inserted into jack

The sleeve is connected to the braided shield (ground).

Touches the sleeve of the connector

The ring connects to one of the hot leads of a balanced cable or to the right channel of a stereo headphone set.

The other type of 1/4-inch phone connector is the type found on stereo headphones. This plug has one tip, one small ring (next to the tip), and a sleeve. These connectors are referred to as 1/4-inch TRS (tip-ring-sleeve). In headphones, the tip and ring are for the left and right musical signals, and the sleeve is connected to the braided shield that surrounds the two hot wires. The 1/4-inch TRS connector is also commonly used for balanced line-level connections. This connector can be used for other devices that require a three-point connection.

XLR Connectors

XLR connectors are the type found on most microphones and the mic inputs of most mixers. Two of the three pins on this connector carry the signal, and the third is connected to the shield. A cable that uses XLR connectors typically carries a balanced signal, utilizing two hot leads and a shield.

It's not uncommon to find cables with an XLR on one end and a 1/4-inch phone plug on the other, or cables that have been intentionally wired in a nonstandard way. These are usually for specific applications and can be useful in certain situations. Check wiring details in your equipment manuals to see whether these will work for you.

There are other types of connectors, but RCA phono, 1/4-inch phone, and XLR are the most common. It's okay to use adapters to go from one type of connector to another, but always be sure to use connectors and adapters with the same number of points. For example, if a plug has a tip, ring, and sleeve, it must be plugged into a jack that accepts all three points in order to maintain a consistent function. In some cases, a balanced source (XLR or TRS) delivers an unbalanced signal to an input. This is accomplished by simply connecting one hot lead at the input.

Female Connector   **Male Connector**

XLR Connectors

XLR connectors are most commonly seen on standard microphone cables, professional outboard gear, and power amplifiers. The connectors offer a secure way to run balanced microphone and line-level signals.

The Dual Banana Connector

Dual banana connectors are very common speaker wire connectors, either at the amplifier output or the speaker box input. They are quickly connected and simple to use. One side has a tab to help keep track of consistent phasing between speaker boxes. When phase reversal is necessary, they can be flipped upside down. They're also available in various colors, so color

coordination of frequency splits or amplifier runs is easy. This connector is not used for line-level connections.

Speakon Connectors

Speakon connectors are used for speaker wire termination. Prior to common adoption of the Speakon connector, the dual banana connector was most common. European regulatory requirements outlawed the use of the dual banana connector, forcing the user to terminate with spade lugs or bare wire ends. This prompted the use of the Speakon connection, which has become a very common speaker box termination. This connector typically utilizes a four-point connection, letting the user simply utilize two of the points for a standard speaker connection or all four points to pass both high and low frequencies of a biamplification scheme down the same speaker cable.

These connectors are also available in an eight-point version, providing numerous options for multiple-amplification sends. Speakon connectors offer a secure, twist-lock connection; they're fast, easy, and cost-effective. They are not used for line-level sends.

Plugging In

The output of your mixer might have multiple outputs for connection to different amplifier inputs. The Main output is the correct output from the mixer to be plugged into the power amplifier. This might also be labeled Mains, Mix Out, Out to Amp, or Stereo Out. If your mixer has XLR outputs available, and if your power amp has XLR inputs, patch these points together as your first choice. This output typically provides the cleanest and most noise-free signal. Many mixers and amplifiers use 1/4-inch connectors that are capable of producing or receiving either unbalanced or balanced signals. Use balanced connections wherever possible. If you need to use unbalanced connections, remember to use the shortest cables possible to avoid radio interference and extraneous noises.

Electrical Power

Spikes and surges are fluctuations in your electrical current that rise well above the 120-volt current that runs most of your equipment. Surges generally last longer than spikes, but both usually occur so quickly that you don't even notice them. Because power surges and spikes can seriously damage delicate electronic circuits, protection is necessary for any microprocessor-controlled equipment (computers, synthesizers, mixers, processors, sequencers, printers, and so on). Oh yeah—I guess that's virtually everything we use these days!

Inexpensive power strip surge suppressors are not the ideal solution to protection against spikes and surges, although for a small system they are better than no protection at all. For reliable electronic system protection, dedicated power conditioners guard against spikes, and some conditioners even maintain a constant voltage level. Furman offers several basic affordable options that are functional and efficient. They also offer more expensive systems that provide RMS voltage regulation, surge protection, noise filtration, voltage protection, voltmeter/ammeters, isolated outlet banks, USB chargers, and diagnostic lights. Especially when using a large complex system, it is worthwhile to invest in high-quality power supply and distribution systems.

Powering Up

- Turn on the mixer and outboard gear (such as delays, reverberation devices, and compressors) before the power amps.
- Always turn power amps on last to protect speakers from pops and blasts as the rest of the electronic gear comes on.

Powering Down

- Turn power amps off first to protect speakers, and then turn the mixer and outboard gear off.

Ground Hum

Aside from causing physical pain, grounding problems can induce an irritating hum into your audio signal. If you have ever had this kind of noise show up mysteriously and at the worst times in your sound system, you know what true frustration is.

Audio Example 5-10

60-Cycle Hum

Sixty-cycle hum is the result of a grounding problem in which the 60-cycle electrical current from the wall outlet is inducing a 60-cycle-per-second tone into your musical signal.

To make matters worse, this 60-cycle tone isn't just a pure and simple 60-Hz sine wave. A sine wave is the simplest waveform and, in fact, is the only waveform that has a completely smooth crest and trough as it completes its cycle. We could easily eliminate a 60-cycle sine wave with a filter. Sixty-cycle hum has a distinct and distracting waveform, which also includes various harmonics that extend into the upper frequencies.

It's very important to have your setup properly grounded in order to eliminate 60-cycle hum and for your own physical safety while operating your system.

Grounding

Grounding is a very important consideration in any setup! The third pin, called the *ground pin*, on an AC power cable exists for your safety. It is a key component in a scheme designed to keep you from being electrocuted. If there's an electrical short or a problem in a circuit, the electricity may search out a path other than the one intended. Electricity is always attracted to something connected to the ground we walk on (the earth). The ground pin gives an electrical problem such as this somewhere to go. It provides a low-resistance path for the dissipation of fault current.

If a piece of equipment has a ground connection that is intermittent or disconnected and if you happen to touch that equipment, you might become just the path to the ground that the electricity is looking for. This could be painful, at the very least, or even fatal. Properly grounding a piece of equipment gives potentially damaging electrical problems a path to ground other than you.

Ground Loops

Grounding, while providing essential safety, offers potential signal noise through the formation of ground loops. Any time we connect multiple electrical components together (outboard equipment, mixers, amplifiers, and so on), we run the risk of creating a ground loop. This is actually a loop between multiple ground circuits, which essentially acts as an antenna, receiving and absorbing interference and noise that becomes infused in our intended audio signal.

If a system is designed properly, there shouldn't be a grounding problem and, in fact, the grounding scheme will help eliminate noise from entering the system; however, a sound operator on the road doesn't always have the luxury of integrating with a properly designed audio or AC system.

In the audio world, ground loops cause hums and buzzes. In the video world, they cause interference bars in the picture. Functionally, ground loops often cause erratic operation or even damage to audio and video equipment.

Solutions to Grounding Problems

Let's look at some practical solutions to the persistent hums and buzzes that vex so many setups. Once your sound system is on the right path electrically, your frustration level should drop significantly.

Seasoned designer/installers typically have the luxury of designing all aspects of a system and implementing a meticulously wired and impeccably integrated electrical, audio, and media system. In these installations, there is time to sleuth out buzzes and hums, and then there is time to implement a plan to eliminate them. That's not quite the same thing as showing up an hour before a gig and finding yourself faced with an unbelievably loud hum in the entire system.

As a live sound operator, count your blessings if you're in a fixed location as a house engineer. There still might be occasional issues, but they should be infrequent. If you do come up against a problem, consult with a respected and experienced system designer. He or she has likely already faced a similar problem and come up with a workable solution.

If you are a sound operator on the road, traveling from venue to venue, you'll encounter a vast range of systems, from extremely well-designed megasystems to extremely questionable and makeshift house systems. If you're using a house system, you'll still need to integrate your equipment with theirs. If you're carrying your own system, you'll need to integrate with the house AC power. All of these scenarios offer opportunity for problems.

Noise Doesn't Necessarily Indicate a Grounding Problem

A noise in the system might be coming from a bad cable. One of the possible scenarios for cable failure is that the shield becomes disconnected, which results in a noisy hum along with the intended signal. If you hear a noise in the system, first look at the channel, group, and main meters. Frequently, the offending noise will show itself clearly on one of the meters. If you see a channel meter that looks like it is receiving a constant input level, turn that channel off to see if the problem disappears. If the noise goes away as the channel is muted, replace the cable between the mixer or snake channel and the offending device. Or, if the noisy channel is connected to a direct box, push the ground button on the DI. If the noise disappears as a result of either of these procedures, celebrate and enjoy the gig!

Connect All Equipment to the Same Outlet

This is a simple and safe solution to this common problem but it is only viable with small or mid-sized system that doesn't require multiple 15- or 20-amp AC circuits. Because the ground loop is caused by the differences in electrical potential (accumulated voltage) between outlets in your building, plugging into one outlet with all of your equipment drastically reduces the likelihood of grounding problems.

Reminder: Always use high-quality AC distribution. Plastic electrical power strips are known to present a fire threat, so use metal power boxes capable of handling your equipment. It's a good idea to hire an electrician or a system designer/installer to design and build an appropriate electrical distribution system.

Hum Eliminators

Ebtech makes both AC- and line-level hum eliminators. These devices isolate the ground without disconnecting it. In many cases, they will rid the system of an annoying hum and the show can go on. Some designers are not fans of these devices and, for a permanent installation, hum eliminators are not the best solution. The best answer is to find the source of the noise and get rid of it.

Lighting Dimmers

Often, a defective or just inherently noisy lighting dimmer causes the noise. If a dimmer is causing a noise, it is usually while the lights are being dimmed. If you spot an inexpensive wall dimmer and you're struggling to find a noise, try turning the dimmer up all the way or turning the light off.

Hire a Pro

The best solution to a persistent grounding problem is to hire a qualified and experienced system designer to rewire or to supervise the rewiring of your venue. He or she can verify that all of the available electrical outlets are properly grounded and that the electrical supply systems are suitable for powering audio and media systems.

For large audio installations, it's ideal to have the power company provide a completely separate electrical feed for all audio connections. Your qualified designer can instruct an electrician in the best scheme for providing clean AC power to all outlets designated for audio equipment power needs. Ideally, these circuits should be filtered, regulated, and relayed. When designed properly, if there is a loss of power, circuits will come back on in an order determined by a relay network.

Auxiliary Power Supply

It's also a great idea to have any computer-based gear on a power backup system. These backup systems have battery power that will continue the flow of current to your equipment if there's a momentary power loss or failure. You only need to be saved once by one of these systems to be a firm believer in their use. If, during a show, the power flips off for just a second, all of your analog equipment will typically pop right back on; however, your microprocessor-controlled devices will need to reboot and might be down for a few minutes. That's a long time during a show!

Circuit Tester

Most electrical stores should stock a circuit tester, designed to verify that the AC circuit from the wall has been wired correctly. If you see any deviation from standard wiring when the tester is plugged into the wall outlet, don't use the circuit.

Danger, Danger

Remember, the human body can conduct the flow of 20 to 30 amps of 110-volt alternating current (AC). Because this can be very painful or even lethal, great care needs to be taken that all safety systems are functional and in place. Be extremely careful in the operation of any electrical equipment.

To get rid of ground hums, many sound operators have gotten into the habit of lifting the AC ground, using a 99-cent adapter between the AC outlet and the power cable. This is an extremely dangerous practice and should not be done. It can result in damage to your equipment and it can be deadly. Don't do it!

Proper grounding can be the single most important factor in keeping your system safe, quiet, and buzz-free. A poorly designed system can have many hums, noises, and other unwanted sounds and it can be dangerous.

THE FRONT-OF-HOUSE MIXER

You must understand everything about your mixer! Most modern mixers offer ample headroom and abundant features. The better your understanding of the primary mixer functions, the better your chances of experiencing creative freedom. The mixer is also called a *console, desk, board, front-of-house mixer,* or simply *FOH* in certain writings.

Some manufacturers are offering digital mixers designed specifically for live use. These mixers provide previously unheard of amounts of processing and ultimate control along with the capacity to recall mix snapshots and to integrate seamlessly with multimedia presentations and other new technology—it's hard to imagine that all mixers won't be digital someday.

Whether you're using a digital or analog mixer—of the simplest or most complex variety—the concepts and principles of sound reinforcement and sonic shaping stand true. If you develop a solid foundation of knowledge about how these devices work, your question won't be, "What do I do to this sound?" Instead it will be, "Where is the controller that lets me do what I know I want to do?"

Signal Path

Signal path is simply the route that a signal takes from point A to point B. For speed and efficiency in any situation, it's essential that you're completely familiar with the signal paths involved in your setup. Any good maintenance engineer knows that the only surefire way to find a problem in a system is to follow the signal path deliberately from its point of origin (point A, for example, the microphone) to its destination (point B, the speakers).

There are several possible problem spots between point A and point B. A thorough knowledge and understanding of your signal path lets you deal with any of these problems as quickly as possible.

Many owner's manuals give a schematic diagram of exactly what the signal path is in a mixer. You may not be totally into reading diagrams, but there's a lot to be learned by simply following the arrows and words. The basis of electronics is logic. Most complex electronic tasks can be broken down into small and simple tasks. That's exactly how live sound is. What seems like an impossible task at first isn't so bad when you realize it consists of several simple tasks performed in the right order.

Signal Path

Many owner's manuals give a schematic diagram of the mixer's signal path. You may not be totally into reading diagrams, but there's a lot to be learned by simply following the arrows and words.

A thorough understanding of your signal path will help you troubleshoot most audio problems. Review your mixer's block diagram—follow the path from the mic to the main output. These drawings are easy to follow and, along the way, you'll see exactly how you mixer routes signals.

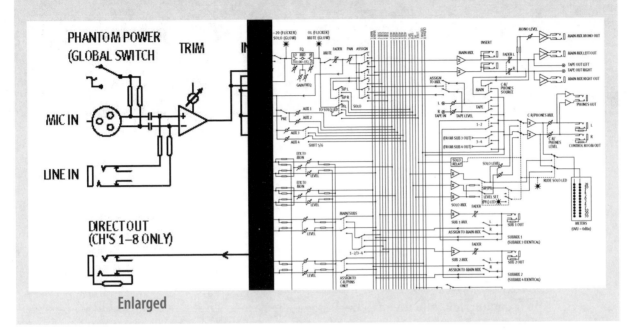

Enlarged

An example of a typical signal path is as follows: The microphone goes into the microphone input, which goes to the attenuator, which goes into the preamp, which goes into the equalizer, which goes to the group assignment, which goes to the processors or gets routed to the main mix, which often goes through a processor, which goes to the master volume fader, which goes to the main stereo output of the mixer, which goes to the power amp in, which goes to the speakers, which go to your ears, which go to your brain, which makes you laugh or cry. A thorough understanding of your signal path is the answer to most trying circumstances you'll come across.

Input Levels

Audio Example 10-1 is mixed three different ways—same music and board but different mixes. Notice the dramatic differences in the effect and feeling of these mixes. Even though they all contain the same instrumentation and orchestration, the mixer combined the available textures differently in each example.

Audio Example 10-1

Mix Comparison

Comparison of Three Different Versions of the Same Mix

The mixer is where your songs are molded and shaped into commercially and artistically palatable commodities. If this is all news to you, there's a long and winding road ahead. We'll take things a step at a time, but for now you need to know what the controls on the mixer do. Unless a manufacturer has blatantly copied a successful mixer design, no two mixers are set up in exactly the same way; however, the concepts involved with most mixers are essentially the same.

In this section, we'll cover those concepts and terms that relate to the signal going to and coming out of the mixing board. These concepts include:

- High and low impedance
- Direct boxes and why they're needed
- Phantom power
- Line levels

A mixer is used to combine, or mix, different sound sources. These sound sources might be:

- On their way to the front-of-house mix
- On their way to effects from instruments or microphones
- On their way to the stereo or multitrack service recording device
- On their way to the monitor system

We can control a number of variables at a number of points in the pathway from the sound source to the house mix and monitors. This pathway is called the *signal path*. Each point holds its own possibility for degrading or enhancing the audio integrity of your music.

Input Stage

Let's begin at the input stage, where the mics, instruments, playback, and presentation devices plug into the mixer. Mic inputs come in two types: high impedance and low impedance. There's no real difference in sound quality between these two as long as each is used within its limitations.

In practical application, microphone connections are almost always low impedance, whereas instrument outputs are typically high impedance. In order to plug a guitar or keyboard into a microphone input, you must incorporate a line matching transformer.

The main concern when considering impedance is that high-impedance outputs go into high-impedance inputs and low-impedance outputs go into low-impedance inputs.

Output Impedance

There is a difference between input and output impedance. In days when everything was centered on the vacuum tube (an inherently high-impedance device), it was most efficient and financially feasible to match input and output impedances. In the early 1900s, Bell Laboratories found that to achieve maximum power transfer in long-distance telephone

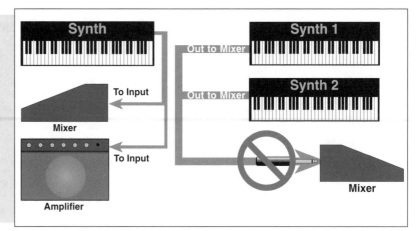

circuits, the impedances of interconnected devices should be matched. Impedance matching reduced the number of vacuum tube amplifiers needed. Because these amplifying circuits produced a lot of heat and were expensive and bulky, there was sufficient motivation to do whatever it took to match impedances.

Bell Laboratories invented a small cheap amplifier, called the *transistor*, in 1948. With the advent of the transistor (an inherently low-impedance device), everything changed. The transistor utilizes maximum voltage transfer where the destination device (called the load) should have an impedance at least 10 times that of the sending device (the source). This concept, known as bridging, is the most common circuit configuration used to connect audio devices.

Because of the load-source relationship, it is possible to simply split one output several times for connection to multiple inputs. Conversely, summing multiple sending signals (sources) to one destination device (load) is not recommended. It's necessary to utilize a summing circuit to combine multiple sources to a load.

The Preamp

One of the first things your signal from the mic sees as it enters the mixer is the mic preamp (sometimes called the *input preamp* or simply the *preamp*). The preamp is actually a small amplifier circuit, and its controls are generally at the top of each channel. The preamp level controls how much a source is amplified and is sometimes labeled as the *Mic Gain Trim*, *Mic Preamp*, *Input Preamp*, *Trim*, *Preamp*, or *Gain*.

A signal that's been patched into a microphone input has entered the mixer before the preamp. The preamp needs to receive a signal that is at mic level. *Mic level* (typically 30–60 dB below line level) is what we call the strength of the signal that comes out of the mic as it hears your music. A mic level signal must be amplified to a signal strength that the mixer wants. Mixers work at line level, so a mic level signal needs to be amplified by the preamp to line level before it gets to the rest of the signal path.

Best results are usually achieved when the preamp doesn't need to be turned up all the way. A preamp circuit usually recirculates the signal back through itself to amplify. This process can add noise, then amplify that noise, then amplify that noise, and so on. So, use as little preamplification as possible to achieve sufficient line level.

Some mixers have an LED (light-emitting diode, or red light) next to the preamp control. This is a peak level indicator and is used to indicate peak signal strength that either is (or is getting close to) overdriving the input. The proper way to adjust the preamp control is to turn it up until the peak LED is blinking occasionally, then decrease the preamp level slightly. It's usually okay if the peak LED blinks a few times.

Many modern mixers have very clean and transparent input preamps with ample headroom. Headroom, in any circuit, is the operational range between normal and maximum signal level. These mixers rarely utilize a peak LED at the input stage to help set the preamp level. In this case, the channel fader and master stereo are optimized at unity gain. (Unity gain is the state where a circuit outputs the same level it receives at its input.) When the channel fader is set at unity (U), the preamp level is adjusted to produce the proper level for the mix.

Attenuator

It's a fact that sometimes the signal that comes from a microphone or instrument into the board is too strong for the preamp stage of your mixer. This can happen when miking a very loud instrument, such as a drum or electric guitar amp, or when accepting the DI of a guitar, sound module, or bass with particularly strong output levels. Some microphones produce a stronger signal than others. This is a particular problem when miking drums or loud guitar and bass amplification systems. If the signal is too strong going into the preamp, then there will be unacceptable distortion. When this happens at the input, there's no fixing it later.

This situation requires the use of an attenuator, also called a *pad*. This is almost always found at the top of each channel by the preamp level control. Sometimes, especially on condenser microphones, there is a pad between the mic capsule and the microphone circuitry. Try this attenuator before using the mixer attenuator.

An attenuator restricts the flow of signal into the preamp by a measured amount or, in some cases, by a variable amount. Most attenuators include 10-, 20-, or 30-dB pads, which are labeled –10 dB, –20 dB, or –30 dB. Listen to Audio Example 10-1 to hear the sound of

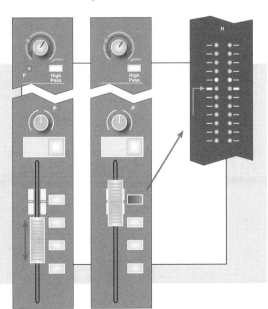

Adjusting Channel Gain

There are two basic methods to adjust initial gain settings.

1. The channel on the left demonstrates the use of channel peak LED to adjust the input level. With this method, while the source is active, turn the trim up until the peak LED blinks, then back the trim off slightly.
2. The channel on the right demonstrates a unity setting on the channel fader. Set the fader to unity and then adjust the trim for the proper mix level. Most boards that use this system offer a means to meter the input signal. Many mixers display the channel level on the main left-right meter when the channel solo button is selected.

an overdriven input. This example would sound clean and clear if only the attenuator switch were set correctly!

Video Example 10-1

Demonstration of Trim Adjustment

Audio Example 10-2

The Overdriven Input

If there's noticeable distortion from a sound source, even if the preamp is turned down, use the pad. Start with the least amount of pad available first. If distortion disappears, all is well. If there's still distortion, try more attenuation.

Once the distortion is gone, use the preamp level control to attain sufficient input level. Listen to Audio Example 10-3 to hear the dramatic difference this adjustment can make in the clarity of an audio signal.

Audio Example 10-3

Attenuator Adjustment

Again, if the input stage of your mixer has a red peak LED by the input level control, it's desirable to turn the input up until the peak LED blinks occasionally, then back the level off slightly. This way we know we have the signal coming into the mixer as hot as possible without distortion. This is good.

Ideally, we'll always run electronic instruments through the system with their outputs set at maximum. This procedure results in the best possible signal-to-noise (S/N) ratio and provides a more surefire way to get the instrument back to its original level for a future use.

If you don't have an attenuator and if you're reinforcing an instrument such as a bass, keyboard, or guitar through a direct box, you can turn the output of the instrument down slightly to keep from overdriving the input preamp. Be sure to mark or note the position of the instrument's controls (especially volume) so you can duplicate levels in the future.

Direct Box

It's possible, acceptable, and standard procedure to use a direct box to match a high-impedance output to a low-impedance input or vice versa. A direct box is also called a *line-matching transformer*, *impedance-matching transformer*, *impedance transformer*, or *DI* (*direct injection*). Its sole purpose is to change the impedance of the instrument or device plugged into its input.

Impedance transformers work equally well in both directions—low to high or high to low. Using the same transformer, you can plug a high-impedance instrument into the high-impedance input and then patch the low-impedance output into a low-impedance input, or, if necessary, you can plug low-impedance into the low-impedance end and come out of the transformer high-impedance.

Direct Box

The hi Z input is a Y. One side of the Y sends the signal to the transformer; the other side of the Y sends the signal to the Out to Amp jack. This makes it possible for instruments to plug into the direct box, then into the amplifier from the Out to Amp jack.

If you hear a loud ground hum after plugging into the direct box, it will usually go away if you press the Ground switch.

Most DIs have a pad to help keep strong signals from overdriving the console inputs.

Passive Versus Active DIs

There are two main types of direct boxes: passive and active. Passive direct boxes are the least expensive and generally do a fine job of matching one impedance in to another impedance out. Active direct boxes are usually more expensive and contain amplifying circuitry that requires power from a battery or other external power supply. These amplifying circuits are used to enhance bass and treble. An active direct box typically gives your signal more punch and clarity in the high frequencies and low frequencies.

Audio Example 10-4

Bass Guitar through a Passive Direct Box

Audio Example 10-5

Bass Guitar through an Active Direct Box

Audio Example 10-6

Direct Box Comparisons

The difference between these two examples can be subtle, but it's often the nuances that make the difference between okay and brilliant! A 10-percent improvement of each channel really impacts the final product. Optimize every part of the mix. It makes a noticeable difference.

Direct boxes typically have a ground lift switch. Try flipping this switch if you can hear a noticeable 60-cycle ground hum along with the instrument sound. There is usually one position that eliminates hum.

Meters

We must use meters to tell how much signal is getting to the console, recorders, outboard gear, and so on. The meters commonly used today are: the volume unit (VU) meter, the full scale (FS) meter, and the peak program meter (PPM).

Volume Unit (VU)

In the modern sound reinforcement world, VU meters are the least common type of meter; however, they're commonly used on analog tape recorders.

VU stands for *volume unit*. This meter, with an average rise and fall time of about 0.3 seconds, reads average signal levels, not peak levels or fast attacks. A VU meter has a needle that moves across a scale from about –20 VU up to about +5 VU. The act of moving the needle's physical mass across the VU scale limits the potential metering speed.

The VU meter enjoys continued popularity in the broadcast world because its readings correspond closely to the perceived loudness of speech signals.

VU meters are commonly used in outboard equipment to monitor input and output strength, as well as parameters such as gain reduction and other dynamic controls. In these applications they are easy to read and provide an excellent visual representation of average signal levels.

Full Scale (FS) and Peak Program Meters (PPM)

Full scale and peak program meters accurately display transient attacks. Transients are nearly instantaneous, sharp attacks, such as those from any metal or hard wood instrument that is struck by a hard stick or mallets. Peak program meters—available on analog consoles and recorders since well before the digital revolution—help engineers recognize, meter, and monitor this transient component that typical VU meters can't.

Peak program meters contain a series of small lights that turn on immediately in response to a predetermined voltage. Because there's no movement of a physical mass (such as the VU meter's needle), peak meters are ideal for accurately indicating transients. Because

Peak and VU Meters

Zero on a peak meter registers maximum level (full scale); zero on a VU measures optimum average level, with several dB headroom before maximum level is attained. There is an art to using a VU meter, where your knowledge and insight into the operational characteristic influence level assessment. Peak meters are typically used for digital metering; they provide black and white assessment of level—you've reached overload or you haven't. Many engineers prefer to meter levels, especially mixdown levels, on both peak and VU meters. VU meters equate more closely to actual loudness. If you control peak levels and maintain strong VU levels throughout, you can be assured that your mix is strong and clean.

Typically, digital peak meters and VU meters are compared using a 1-kHz tone. Originally, 0 VU when fed by the identical 1-kHz tone was equated to –18 on the digital peak meter. This provided plenty of headroom for transients and overloads—it also produced very conservative recording levels.

0 VU is typically equated to a peak level between –15 and –18 dBFS.

A level of 0 dBFS should be approached at some point in the recording, but it's most important to avoid exceeding full scale.

the reaction of the meter to an incoming signal is electronic, rather than physical, the speed of the meter is adjustable by simply changing the speed of the electronic rise and release. Typically, a PPM has a potential rise time of about 10 milliseconds and a fall time potential of about four seconds. On many peak program meters, you can select between peak and average readings. Average signal level readings are still valuable, especially considering that average signal strength offers a close correlation to the loudness of speech signals.

Digital mixers and recorders reference a full scale meter (FS), which is also a peak meter. However, in setting digital levels, the avoidance of peak levels above full scale, called overs, is fundamentally important. It's important when setting levels on a digital device (whether a mixer, recorder, plug-in, or outboard gear) to set levels high enough to realize the full amplitude resolution—to use as many bits as possible to define the amplitude status at each sample—while still leaving ample headroom to avoid any overs. It's important to realize that excessively low-level digital levels sound grainy and harsh, while keeping in mind that digital overs can annoy any listener in a live show and/or destroy a portion of an otherwise excellent recording of a brilliant performance.

Video Example 6-2

VU and Peak Meter from the Same Sound Source

Adjusting VU Levels for Transients

When monitoring a signal with an analog VU meter, there are unique considerations that require an understanding of the meter and the way it differs from a peak or full scale meter. As previously mentioned, a transient attack is the percussive attack present in percussion instruments. Virtually any time one hard surface is struck with another hard surface, such as a hard stick, mallet, or beater, the attack component of the waveform contains a transient. The following instruments contain transient attacks and their input and/or record levels should be adjusted accordingly:

- Cymbals
- Cowbell
- Tambourine
- Tom, Kick
- Shakers
- Maracas
- Acoustic Piano

- Hi-Hat
- Snare Drum
- Triangle
- Claves
- Conga, Bongo
- Guiro
- Acoustic Guitar

When metering instruments that contain transients with a standard VU meter, adjust levels so that the loudest sounds register between −9 VU and −7 VU. This approach results in much more accurate and clean percussive sound. The transient is usually at least 9 VU hotter than the average level, so when the standard VU meter reads −9 VU, the input is probably seeing 0 VU. If you meter 0 VU on a transient, the input might see +9 VU!

A peak LED is normally just one red light that comes on when the signal is about to oversaturate tape, overdrive a circuit, or exceed maximum digital level. It's usually okay for the peak LED that lives in one corner of a VU meter to blink occasionally, but if it's on continuously, back off the input level until it blinks less frequently. When a peak LED comes on, it means that even though the VU is registering well within acceptable limits, the actual level that's reaching the signal path is getting pretty hot.

In a mix, if the average or VU level is conservative but the peak LEDs are always on, there's probably a percussion instrument in your mix that's too hot, and even if it doesn't sound too loud on your system, it'll probably sound too loud on other systems. When reinforcing most instruments, adjust the level so the VU meters read between 0 and +2 VU. For percussive instruments that have transient peaks, the VU meter should read around –9 VU to –7 VU at the peaks.

Phase

If two signals are electronically out of phase, their waveforms are mirror images of each other. The crest of one wave happens simultaneously with the identical trough of the other wave. When this happens, there is phase cancellation—in other words, no signal. If two waveforms are in phase, they crest and trough together. This results in a doubling of the amount of energy from that waveform, or twice the amount of air being moved.

When phase problems occur, they're typically a result of two basic scenarios:

- An incorrectly wired microphone or balanced line cable
- Multiple microphones in the same acoustic space, spaced so that undesirable phase interactions occur as the same sound arrives at the different mics in destructive phase relations

This problem doesn't show up as much in a stereo mix, but any time your mix is played in mono or any time you are combining multiple microphones to the mix center, this can be the worst problem of all. To hear the effect of combining a sound with itself in and out of phase, listen to the guitar in Audio Example 10-7.

At the beginning of Audio Example 10-7 the original sound is playing into one channel of the mixer. Next, the signal is split and run into another channel of the mixer. Notice the volume increase as the two channels are combined. Finally, the Audio Example shows the sound difference as the phase is reversed on the second channel. The channels combined are obviously thin and reduced in level.

Audio Examples 10-7

Phase Reversal

The nature of combining sounds dictates that there is always phase interaction. We wouldn't want to hinder that because good phase interaction gives our music depth and richness. However, we do want to be particularly aware of phase interactions that can have an adverse effect on the quality of our music.

If your mixer has a phase switch on each channel, it's probably at the top of the channel near the preamp and attenuator controls. Its purpose is to help compensate for phase interaction problems. These problems are most dramatic in a mono mix or when channels are panned to the same position.

Phase Relationships When Using Multiple Mics

Whenever two or more microphones pick up the same waveform, there is a likelihood that the mics will hear the waveform at different points in the energy cycle. Notice that the two waveforms compared at the bottom of this illustration represent the portions of the waveform that were simultaneously heard at mics A and B. Even though our example uses a simple sine wave source, it's easy to see that the combination of the two waveforms will not result in the smooth and even sine wave. Considering a more complex waveform, some frequencies will combine in a destructive manner when picked up by multiple mics, and other frequencies won't.

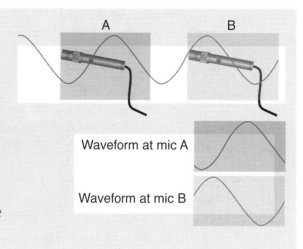

Waveform at mic A

Waveform at mic B

Short delay times, chorus, and phasing effects can also cause these kinds of problems in mono, so you might also need to change some delay times to help even things out. There will be more about this when we cover building a mix.

It's a good idea to check for phase problems when setting up your mix. In any situation in which multiple mics are used in close proximity, be on alert for a hollow or thin sound. This is different from a sound that simply has a reduced frequency content in the low frequencies—it literally has a hollow and thin sound. When you recognize this sound, look at the microphones and how they are set up and aimed. Any mics that are pointed at each other, on- or off-axis, could be causing the problem. One at a time, invert the phase on those mics. There's a good chance that you'll hear the sound solidify and fill out.

Input Level Comparison

These initial variables (preamp, attenuator, meters, and phase) are very important points for us to deal with. Any good sound operator has a solid grasp of these crucial parts of the signal path. These are the basics; you'll continually return to them for clean, quality, professional audio.

Listen to Audio Examples 10-8 through 10-10. If your signal isn't clean and accurate at the input stage, it won't be clean and accurate anywhere. Improper settings at the input stage could easily result in a mix that is distorted and sonically irritating at any volume.

Audio Example 10-8
Proper Input Levels

Audio Example 10-9
Low Input Levels Resulting in a Noisy Mix

Audio Example 10-10
High Input Levels Causing Distortion

We must have proper levels coming into the mixer before we can even begin to set levels to tape. Any distortion here is magnified at each point. Any noise that exists here is magnified at each point. Listen to the effects of improper level adjustment at the input. Audio Examples 10-8, 10-9, and 10-10 use the same song, the same mixer, and the same tracks with different input levels. All three of these Audio Examples are recorded at the same mix level; the only variable is the input trim level.

Notice in Audio Example 10-8, the sound is clean and strong. This was recorded with the input levels set properly. In Audio Example 10-9, the input level is set very low. When this happens the levels at the end of the signal path must be elevated unnaturally in order to achieve proper mix levels. Along with raising levels at the final stages of the signal path comes the noise that inherently resides in the mixer circuitry.

Audio Example 10-10 is simply too hot coming into the signal path. The distortion here happens immediately. There's no way to fix the problem when the audio is distorted at the input stage.

Phantom Power

Condenser microphones and active direct boxes need power to operate. If they don't receive it, they won't work. This power can come from a battery in the unit or from the phantom power supply located within the mixer.

Phantom power (a very low amperage 48-volt DC current) is available at any mic input that has a phantom power switch. Because amperage is the actual punch behind the voltage and because phantom power has a very low amperage, there's little danger that this power will cause you any physical harm, even though the power travels to the mic or direct box through the same mic cable that the musical signal travels to the mixer.

Phantom power requirements can vary from mic to mic, so check your mic specifications to ensure that the mic is getting the power it needs. Voltage requirements are typically between 12 and 52 volts. Most mics that require low voltages have a regulatory circuit to reduce higher voltages so that normal 48-volt phantom power can be used without damaging the mic. Microphones that require higher voltages won't usually sound all that great until they get the power they require. These mics often come with their own power supply.

Your mixer might not have phantom power built in. Most microphone manufacturers offer external phantom power supplies for one or more mics. Simply plug the phantom power supply into an AC outlet, and then plug the cable from the mic or direct box into the phantom power supply. Finally, patch from the XLR output of the phantom power supply into the mixer mic input.

Phantom power is preferred over battery power because it is constant and reliable, whereas batteries can wear down, lose power, and cause the mic or direct box to operate below its optimum specifications (even though it might still be working).

If the mic or direct box doesn't need phantom power, it's good practice to turn the power off on those channels, though it isn't absolutely essential. Many consoles have phantom power on/off switches. Some mixers have phantom power that stays on all of the time. This is okay, but if there's an on/off switch, turn it on when you need it and off when you don't.

Mic Level

The effective output of each microphone is typically quantified in relation to line level. Most microphones output a signal that is between 30 and 60 dB below line level. This means that the signal from the microphone must be boosted between 30 and 60 dB before the signal strength is at line level and ready to move through the mixer.

Line Level

Line in and *line out* are common terms typically associated with mixer, outboard equipment, and tape recorder inputs and outputs. The signal that comes from a microphone has a strength that's called *mic level*, and a m°ixer needs to have that signal amplified to what is called *line level*. The amplifier that brings the mic level up to line level is called the *mic preamp*.

Instrument inputs on mixers are line level. An input that is line level enters the board after the microphone preamp and is, therefore, not affected by its adjustment.

Most mixers have attenuators on the line, and mic inputs compensate for the difference in equipment output levels. As we optimize each instrument or voice signal, we must optimize the gain structure at each point of the signal path. When all the levels are correct for each mic preamp, line attenuator, fader, EQ, bus fader, and so on, we can reproduce the cleanest, most accurate signal. When one link of this chain is weak, the overall sonic integrity crashes and burns.

Mixers that have only one 1/4-inch phone input on each channel typically have a Mic/Line switch. Select the appropriate position for your situation.

+4 dBm Versus –10 dBV

You might have heard the terms plus 4 or minus 10 (+4 or –10) used when referring to a mixer, signal processor, computer, or recording device. This is another consideration for compatibility between pieces of equipment, aside from the low-impedance/high-impedance dilemma. Different equipment can have different relative line level strength. This level comparison, tagged in dB, is specified as either +4 dBm or –10 dBV.

When we use the term dB, it's useful to keep in mind that it is a term that expresses a ratio between two powers and can be tagged to many different types of power that we encounter.

In our option of +4 dBm, dB is tagged to milliwatts; and –10 dBV is tagged to volts. Without going into the math of it all, let's simply remember that +4 equipment only works well with other +4 equipment, and –10 equipment only works well with other –10 equipment.

Some units provide a switch between +4 and –10, so all you do is select the level that matches your system. There are also boxes made that let you go in at one level and out at the other. In many situations, a tool like this is a necessary solution.

Most +4 gear is balanced low-impedance. This is the type that's used in professional systems and uses either XLR connectors or some other type of three-pin connector, such as a quarter-inch tip-ring-sleeve connector. Gear can also be of the unbalanced variety and still operate at +4.

We most often think of −10 dBV gear as unbalanced, although this is not always the case. This type of gear is considered semi-professional. Most semi-professional gear equipment operates at −10 dBV. Gear that uses RCA phono connectors or regular mono guitar plugs operates at −10 dBV.

When used properly and with shorter cable runs, there should not be a noticeable difference in sound quality from a unit operating at −10 as opposed to +4, even though +4 is the professional standard.

+4 dBm balanced equipment works especially well when longer cable runs are necessary, such as in a large concert hall, an arena, or a large recording studio—especially when radio interference and electrostatic noises are a particular problem.

If you are considering signal processors (reverbs, compressors, gates, and so on), mixers, or recording devices, you must always maintain compatibility between +4 and −10 equipment.

Units are available that allow +4 dBm and −10 dBV gear to work perfectly together. Plug in one end at +4, and the signal comes out the other end at −10, or vice versa.

Channel Insert

Most modern mixers have what is called a *channel insert*. This is the point where a piece of outboard signal processing can be plugged into the signal path on each individual channel. If your mixer has inserts, they're probably directly above or below the microphone inputs.

A channel insert lets you access only one channel at a time and is used to include a signal processor in the signal path of that specific channel. The processor you insert becomes a permanent part of the signal path from that point on; therefore, its level affects the remainder of the signal path. An insert is especially useful when you are using a compressor, gate, or other dynamic processor.

A channel insert utilizes a send to send the signal (usually as it comes out of the preamp) to the signal processor. The signal processor output is then patched into the return of the channel insert. This completes the signal path, and the signal typically continues on its way through the EQ circuit and on through the rest of its path.

Channel Insert

Many mixers provide channel inserts—the point where an outboard signal processor can be plugged into the signal path. If your mixer has inputs, they're probably directly above or below the mic inputs.

A channel insert will have a send that sends the signal, usually as it comes out of the preamp, to the processor. The output of the signal processor is then patched into the return of the channel insert. This completes the signal path. The signal then continues on its way through the EQ circuit and on through the rest of its path.

A channel insert and an effects bus are similar in that they can both deal with signal processing. An insert affects one channel only. Inserts are ideal for patching dynamics processors, such as compressors, limiters, and gates, into a signal path.

Effects Bus

An effects bus (such as aux 1 or aux 2) lets you send a mix from the bus to an effect (typically outside the mixer, but often simply routed internally to an effect, headphones, and so on), leaving the master mix on the input faders without effects. The output of the effect is then plugged into the effects returns or open channels on the mixer. This is good for reverbs and multi-effects processors.

When we discuss the input faders as a group, we're talking about a bus. The term *bus* is confusing to many, but the basic concept of a bus is simple—and very important to understand. A bus usually refers to a row of faders or knobs.

If you think about a city bus, you know that it has a point of origin (one bus depot) and a destination (another depot), and you know that it picks up passengers and delivers them to their destination. That's exactly what a bus on a mixer does. For example, in concert the fader bus has a point of origin (the instruments and voices on stage) and a destination (the amplification system). Its passengers are the different ingredients of the mix. Not all channels (passengers) necessarily get on the bus, but whoever rides goes to the same destination.

Mixers also have auxiliary buses or effects buses. Aux buses (also called *cue sends*, *effects sends*, or *monitor sends*) operate in the same way as the fader bus. An aux bus (another complete set of knobs or faders) receives its signal from the channel inputs. There is typically a switch that lets the sound operator choose whether the signal originates toward the beginning of the signal path (Pre Fader Listen) or near the end of the signal path (After Fader Listen). It picks up its own set of the available passengers (channels) and takes them to their own destination (usually an effects unit or the monitor system).

When a bus is used with an effect, such as a reverb, delay, or multi-effects processor, the individual controls on the bus are called *effects sends* because they're sending different channels to the effects unit on this bus. The entire bus is also called a *send*.

Return is a term that goes with send. The send sends the musical ingredients to the reverb or effect. The return accepts the output of the reverb or effect as it returns to the mix.

The Difference Between a Live Mixer and a Recording Mixer

Most of the controls on a live mixer are identical in concept to the controls on a recording mixer. Inputs, trims, preamps, inserts, equalization, sub-group assignments, and so on do the same things on both types of mixers.

With the growth of multichannel audio, many live boards offer several configurations to supply mono, stereo, left-center-right, and surround outputs, a feature previously reserved for recording consoles.

Mutes

- A typical recording console implements mutes differently than a typical live console. When in a live sound reinforcement application, it is crucial that channel mutes affect the monitor sends. In other words, whether the monitors are receiving a pre- or post-fader send, the Mute button should interrupt the send into the monitor bus.

- This type of configuration is much more effective anytime feedback is an issue—when the sound operator hears feedback and mutes channels, the feedback will stop more quickly if the mutes interrupt the send to the house and the monitors.

- It also helps to clean up the monitors if the operator is able to easily mute unnecessary microphones. If a mic is not being used, the entire mix will sound better if it is muted in the house and the monitors.

Recording

- A recording console does not interrupt the auxiliary (monitor) send when the channel is set to pre-fader. In a recording session, pre-fader auxes often feed the headphones, and the recording engineer needs to be able to set up a separate pre-fader mix for the musicians that is completely independent of the control room mix. With this configuration, the engineer is free to move the faders and mute channels without affecting the send to the phones.

- Like the live mixer, the recording console interrupts the post-fader send when the Mute button is pushed. During recording, post-fader auxes are typically used for effects sends. In this application, the effect send must mute at the same time as the channel so that any effects returns will naturally stop or fade away in response to the muted sends. If the post-fader effects sends don't mute along with the channel, the mix engineer could mute the dry channel but still hear the return from effects, such as reverberation and delays.

Outputs

Live

- Most live consoles allow the operator to send each channel to center, left, right, or a combination of four or eight buses.

- The bus assignments on each channel are primarily used to feed sub-groups that, in turn, feed the main mix.

- Many live mixers provide direct outputs for each channel that conveniently feed headphone monitor systems or multitrack recorders.

Recording

- Recording consoles typically provide 24 or more options for routing from each channel. These outputs are typically connected to the line inputs of a multitrack recorder. This feature is unnecessary in most live settings.

- Recording consoles often have a dedicated set of outputs to the mixdown recorder. These outputs are constant and are unaffected by most other monitoring functions, such as PFL, AFL, and the selection of various monitor buses, such as mix, recorder output, or aux solos.

Inputs

Live

- Live console inputs are typically more simple and straightforward than recording mixers. If you use a combination of 24 instrument inputs and 8 vocal inputs, each input is connected to the channel mic or line in. That's pretty much it with the exception of some monitor, effects, and mains FOH system connections.
- There is typically only one set of mic and line inputs per channel on a live mixer and the sound operator has access to either input, but not both.

Recording

- Like the live mixer, recording consoles facilitate mic and line inputs for each channel; however, recording mixers also provide another set of up to 48 or more inputs for returns from the multitrack recorder.
- During mixdown, the recording mixer is often configured to receive all returns from the multitrack while also accepting returns into the main input faders from extra effects, MIDI instruments, acoustic ambiences, and so on to blend with the multitrack outputs for creation of the final mix.

Solo Functions

Live

- Most live mixers provide AFL (*after fader listen*) and PFL (*pre fader listen*). These solos let the sound operator listen to any channels through the headphones to verify signal integrity or simply to check whether the channel controls what he or she thinks it controls.
- AFL and PFL on the live mixer don't affect the house mix.

Input Faders

The input faders are the primary mix level tools. In a live sound reinforcement application, the input faders are often set to the unity gain position, noted by the "U" on the fader hash marks, and then the gain trims are adjusted for approximate volume settings.

Some recording mixers offer level control via a rotary knob, some via a fader. No matter which configuration, the primary difference between a live sound console and a recording console is this concept, in which the input channel adjusts the recording level, yet the multitrack might or might not return to the same or a different channel. You must constantly keep the recorder track in mind for monitoring purposes.

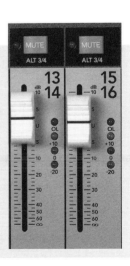

Recording

- Like the live mixer, most recording consoles provide AFL and PFL In addition, the recording mixer provides SIP (*solo in place*), which is also called *mix solo*. SIP is active in the main mix and is often used to immediately break the mix down to minimal instrumentation, such as drums and bass alone, vocals only, and so on.

- AFL and PFL affect what is heard from the control room monitor bus, but not what is being sent to the mix recorder. SIP also affects the main mix outputs to the mix recorder.

Portability

Live

- Live mixers are designed to be transported, set up, and put away on a daily basis.

- Their frames are more rigid, to handle transporting, and they are much lighter.

- It is easy for a couple people to put most live mixers gently into the road case.

Recording

- Recording mixers aren't designed to be transported more than a few times during their years of service.

- Their frames are typically much more flexible, resulting in increased wear and tear, and risk of failure during transport.

Managing the Signal Path

This section covers input faders, gain structure, buses, group assignments, pan, solo, and mute.

Input Faders

When everything is set properly at the preamp, use the input faders to set the channel mix levels. Mixing live sound is different than recording. In a live mix, the main channel faders are used to set the mix volumes for each channel, whereas in a recording session, these faders might be used during tracking to set the recording level, with another knob or fader controlling the actual mix. In the recording world, the input faders are used to control mix levels during mixdown in the same way they're used in live sound reinforcement to build the mix. Many consoles provide the flexibility to configure the layout according to your preference.

Pre and Post

Aux buses often include a switch that chooses whether each individual point in the bus hears the signal before it gets to the EQ and fader (indicated by the word *Pre*) or after the EQ and fader (indicated by the word *Post*).

Selecting Pre lets you set up a mix that's totally separate from the input faders and EQ. This is good for headphone sends. Once the headphone mix is good for the musicians, it's

Aux Buses with Pre and Post

On Aux 1, the Pre-Post switch is set to Pre. This lets the Aux 1 bus hear the signal before it gets to the EQ and fader.

On Aux 2, the Pre-Post switch is set to Post. This lets the Aux 2 bus hear the signal after it has gone through the EQ and fader circuitry.

best to leave it set. You don't want changes you make for your listening purposes to change the musicians' mix in the phones.

Selecting Post is good for effects sends. A bus used for reverb sends works best when the send to the reverb decreases as the instrument or voice fader is turned down. Post sends are perfect for this application because the send is after the fader. As the fader is decreased, so is the send to the reverb, maintaining a constant balance between the dry and affected sounds. If a Pre send is used for reverb, the channel fader can be off, but the send to the reverb is still on. When your channel fader is down, the reverb return can still be heard loud and clear.

Using the Aux Bus

Imagine there's a guitar on channel 4, and it's turned up in the mix. We hear the guitar clean and dry. Dry means the sound is heard without effect. If the output of aux bus 1 is patched into an effects processor input, and the aux 1 send is turned up at channel 4, we should see a reading at the input meter of the effect when there is signal present. This indicates that we have a successful send to the processor.

The effects processor won't be heard until we patch its output back into the mix—either into an available, unused channel or into a dedicated effects return. If your mixer has specific effects returns, it's often helpful to think of these returns as simply extra channels on your mixer.

Once the effects outputs are patched into the returns, raise the return levels on the mixer to hear them in the mix. The signal coming from a reverberation setting should be 100-percent wet—that means it's providing only reverberated sound and none of the dry sound. Maintain separate control of the dry channel. Get the reverberated sound only from the completely wet returns. With separate wet and dry control, you can blend the sounds in the mix to produce just the right sonic blend. Listen to Audio Examples 10-11, 10-12, and 10-13 to hear the dry and wet sounds being blended in the mix.

Audio Example 10-11

Dry Guitar

Audio Example 10-12

Reverb Only

Group Assignment

The group assignment, also called the *bus assignment*, *track assign*, or *switching matrix*, is used to send whatever is received at the input of the mixer (mic, instrument, or tape) to any one or a combination of output buses. In a live sound application, these bus outputs are normally routed to the main mix output, whereas in a recording application, they're also typically connected to the inputs of the multitrack.

If your mixer has eight sub-groups, the channel faders will typically have a bus assignment nearby that lets you assign each separate channel to one or more groups. These group assignments are very powerful and convenient. In most cases they're routed back into the main mix and used to control the levels of groups of similar instruments, such as drums, percussion, backing vocals, guitars, keys, and so on.

Always remember to remove the assignments of any sub-grouped channels into the main mix. A channel should not be assigned to the main mix via both the individual channel and the group master.

Sometimes the group masters are used to supply mixes to recording devices, or any specific location that requires an output from a specific group of channels. A typical mixer designed for live sound application uses sub-groups to help the operator organize the mix into intelligently devised groups of like instruments or voices.

Summing

Sub-group assignments and, in the recording world, track assignments are used to combine two or more mix ingredients to one output. These summing circuits are able to combine multiple signals and send them to a single input circuit without overdriving or distorting.

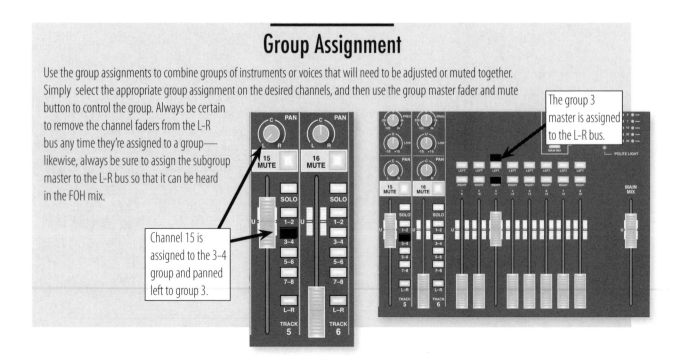

Group Assignment

Use the group assignments to combine groups of instruments or voices that will need to be adjusted or muted together. Simply select the appropriate group assignment on the desired channels, and then use the group master fader and mute button to control the group. Always be certain to remove the channel faders from the L-R bus any time they're assigned to a group—likewise, always be sure to assign the subgroup master to the L-R bus so that it can be heard in the FOH mix.

The group 3 master is assigned to the L-R bus.

Channel 15 is assigned to the 3-4 group and panned left to group 3.

They act like a big Y cable with summing circuitry included to make everything as clean as possible.

Avoid patching the outputs of two or more instruments (or other devices) together through a Y cable into one input. This typically overdrives and distorts the input. Anytime you sum (combine) multiple outputs to one input, use a circuit such as the group assignment circuit on your mixer. This is designed specifically to maintain proper impedance and signal strength for its destination input. This type of circuit is also called a *combining bus*, *combining matrix*, *summing bus*, *summing matrix*, *switching matrix*, *track assignment bus*, or *track assignment matrix*.

Three Practical Applications for Splitting a Signal Using a Y

1. When sending a guitar to the direct input of a mixer and simultaneously to an amplifier: You can use a Y cable out of the guitar or the guitar effects setup

2. When sending a microphone signal to a live system and simultaneously to a recording system: A Y cable is also called a *splitter*. A splitter box usually has a snake that plugs into the recording board; it also has outputs (sometimes in the form of a snake) that connect to the live mixing board.

3. When plugging the final output of a mixer into two or more devices, such as power amp inputs, recording devices, or additional room feeds (overflow, nursing mothers, and so on).

Audio Examples 10-14, 10-15, and 10-16 demonstrate the sound of splitting the guitar signal with a Y cord straight out of the guitarist's effects. One side of the Y goes directly to the mixer through a direct box. The other side is sent to an amplifier. The amp is miked, and the microphone is plugged into the mixer. This setup works well with a guitar, synth, drum machine, or any other electronically generated sound source.

Audio Example 10-14

Direct Guitar

Audio Example 10-15

The Miked Amp

We can combine the direct and miked guitar to one group with the group assignment bus. Listen to Audio Example 10-16 as the guitar sounds combine in different levels to create a new and interesting texture.

Audio Example 10-16

Combining Direct and Miked Signals

Pan

The pan control, sometimes called the *pan pot* (for *panoramic potentiometer*), is used to move a channel in the stereo panorama. Sounds are positioned at any point in the left to right spectrum (between the left and right speakers). Some pan controls are either all the way left or all the way right with no position in between, but panning is usually infinitely sweepable from full left to full right or anywhere in between. Often the pan control is used for selecting odd or even bus assignments on the multitrack bus assignments. Odd is left and even is right.

You can use the pan control along with the group assignment bus to combine multiple instruments, such as several keyboards, to a stereo pair of faders (the sub-group masters). This can give you a very big sound while letting you focus your mixing attention during the program.

Gain Structure

It's necessary to consider gain structure as we control different levels at different points in the signal path. *Gain structure* refers to the relative levels of the signal as it moves from the source to the destination. At each point where level changes are possible, you must monitor the signal strength and, if possible, solo the signal.

We've already discussed the proper method for adjusting the input preamp level, and we've heard some examples of music with the input stage too cold and too hot. These examples give an obvious demonstration of the importance of proper level adjustment at this primary stage. Each stage with user-controlled levels carries its own importance to the integrity of your signal. Ideally, you'll be able to adjust each stage to be as hot as possible, with minimal distortion.

Unity Gain

Some mixers have a suggested unity setting for input faders and group master faders. They are usually indicated by a grayed area near the top of the fader's throw or numerically by a zero indication. Try placing the input and group assignment bus faders to their ideal settings. Then adjust the input preamp for a proper level reading on the channel fader. This is a safe approach and works well much of the time.

Experiment with different approaches to find what works best with your setup. No approach works every time, so remember to trust your ears. If your sound is clean and punchy but the settings don't seem to be by the book, you're better off than if you have textbook settings on your mixer with substandard sound.

Confidence in your control of the gain structure can take time and experience, so start practicing. See what happens when you try a new approach.

If you adjust the input level properly, if the input fader is somewhere close to the ideal setting, and if the group fader is also close to the ideal, all should be well. If one or more of these settings is abnormally high or low, you might have a problem with your gain structure.

Potential Problems

If the input level is abnormally high (even if the signal isn't noticeably distorted), the channel fader might be abnormally low. Faders work more smoothly and are easier to control in the upper part of the fader throw. When the fader is abnormally low, it's much more difficult to fade an instrument down or to fine-tune the mix levels.

If the preamp level is destructively high, the signal will overdrive the mixer, and the integrity of the entire signal path will be jeopardized. The preamp adjustment is very important. An improper setting here will result in surefire failure.

If the preamp level is abnormally low, the input fader and/or the channel, group, and main faders might be abnormally high. When this happens, the inherent noise that resides in your mixer is turned up further than it needs to be. Therefore, we end up with more noise in relation to signal (an undesirable signal-to-noise ratio). This is bad.

Solo

A Solo button turns everything off except the soloed channel. A live mixer routes the soloed channel(s) to the headphone output. A recording mixer routes the soled cahnnel(s) to the control room monitors. This lets you hear one channel or instrument by itself, as if it were a solo. This feature is very useful in evaluating a channel for signal integrity and quality of sound. It's often impossible to tell what's really going on with a channel when listening to it in the context of the rest of the arrangement. Headphones should only be used to assess soloed channels, not as a reference to evaluate the sound of the mix in the room, hall, or arena.

There are three main types of soloing: PFL, AFL, and Mixdown. Each manufacturer chooses exactly where the signal is intercepted for each solo function on their consoles. They might or might not conform to the norm, so it's up to you to read your mixer manual to find out exactly how the solo functions work.

PFL (Pre Fader Listen)

PFL stands for *Pre Fader Listen*. The PFL button solos a channel immediately before the fader. This provides an accurate picture of how a particular channel is sounding just as it's going into your mix or just before it gets to the multitrack. The position of the channel fader has no effect on the PFL solo because the signal is tapped prior to its arrival at the fader. Often, the PFL solo isn't affected by EQ or pan settings either.

PFL is usually the best way to verify signal integrity because it is closest to the source. If the signal is clean at the PFL position and unsatisfactory at the channel fader, a patching or console problem is likely.

The PFL solo button doesn't affect the main outputs, aux sends, or mix sends. Because it is non-destructive to the mix output, the PFL solo provides an excellent way to quickly verify the signal on a specific channel. A mixer designed for live sound application routes both the PFL and AFL signals to its headphone output. A recording mixer routes PFL and AFL to the control room monitors, through the monitor section, but does not interrupt the main mix connection to the mixdown recording device.

AFL (After Fader Listen)

The AFL (*After Fader Listen*) solo is, as its name indicates, affected by the fader position. The signal is soloed immediately after it leaves the fader. In addition, AFL is typically affected by the EQ and mute settings, plus processing that has been inserted in the signal path. It provides a convenient way to monitor a channel or channels just before they enter the main mix.

Solo in Place/Mixdown Solo

Solo in Place, also called *Mixdown Solo*, is similar to AFL. It solos the selected channel or channels as they are in the mix and, in fact, typically affects the main mix output—all panning, EQ, inserts, mutes are in effect. Because this feature typically interrupts the house mix, it is not as common as AFL and PFL—it can be a potential disaster in the hands of a rookie sound operator.

Mutes

A Mute button is an off button—a channel mute turns the channel off. Use the mutes instead of the faders to turn a channel off, especially when setting up a mix. Beginning sound operators often pull the faders down instead of using the mutes. Once the channel level is set in relation to the other channels, you'll save time by simply pressing and releasing the Mute buttons. The levels will remain the same, and you'll avoid continual rebalancing.

The Equalizer

The equalizer or EQ section is usually located at about the center of each channel and is definitely one of the most important sections of the mixer. EQ is also called *tone control*; *highs and lows*; or *highs, mids, and lows*. Onboard EQ typically has an in/out or bypass button. With the button set to in, your signal goes through the EQ. With the button set to out or bypass, the EQ circuitry is not in the signal path. If you're not using EQ, it is best to bypass the circuit rather than just set all of the controls to flat (no boost and no cut). Any time you bypass a circuit, you eliminate one more possibility for coloration or distortion.

From a purist's standpoint, EQ is to be used sparingly, if at all. Before you use EQ, use the best mic choice and technique. Be sure the instrument you're miking sounds its best. Trying to mike a poorly tuned drum can be a nightmare. It's a fact that you can get wonderful sounds with just the right mic in just the right place on just the right instrument. That's the ideal.

From a practical standpoint, there are many situations in which using EQ is the only way to get a great sound on time and on budget. This is especially true if you don't own a wide array of mics. Proper use of EQ is fundamental to an outstanding mix.

Proper control of each instrument's unique tone (also called its *timbre*) is one of the most musical uses of the mixer, so let's look more closely at equalization. There are several different types of EQ on the hundreds of different mixers available. What we want to look at are some basic principles that are common to all kinds of mixers, as well as outboard equalizers.

We use EQ for two different purposes: to get rid of (cut) part of the tone that we don't want, and to enhance (boost) some part of the tone that we do want. Boosting and cutting frequency ranges are both very important. A young sound operator typically reaches for equalization to add highs or lows, but rarely listens to a sound to critically locate a frequency range to cut.

Frequency Selection (Hertz)

Boosting and cutting at a specified frequency number on any equalizer (for example, 100 Hz) alters more than just one frequency. It alters a frequency band that is sometimes adjustable in width. So when we say, "Boost the bass guitar at 100 Hz," we're really indicating a frequency range with its center at 100 Hz.

The ability to hear the effect of isolating these frequency bands provides a point of reference from which to work. Try to learn the sound of each frequency band and the number of Hertz that goes with that sound.

To understand boosting or cutting a frequency, picture a curve with its center point at that frequency.

Listen to the effect that cutting and boosting certain frequencies has on Audio Examples 10-17 to 10-25.

Audio Example 10-17
Boost, Then a Cut at 60 Hz

Audio Example 10-18
Boost, Then a Cut at 120 Hz

Audio Example 10-19
Boost, Then a Cut at 240 Hz

Audio Example 10-20
Boost, Then a Cut at 500 Hz

Audio Example 10-21
Boost, Then a Cut at 1 kHz

Audio Example 10-22
Boost, Then a Cut at 2 kHz

Audio Example 10-23
Boost, Then a Cut at 4 kHz

Audio Example 10-24
Boost, Then a Cut at 8 kHz

Audio Example 10-25
Boost, Then a Cut at 16 kHz

Our goal in understanding and recognizing these frequencies is to be able to create sound pieces that fit together. The frequencies in Audio Examples 10-17 to 10-25 represent most of the center points for the sliders on a 10-band graphic EQ.

If the guitar channel has a balance of the entire frequency range, it might sound great all by itself. If the bass channel has a very broad-range sound with lots of highs and lows, it might sound great all by itself. If the keyboard channel has a huge, broadband sound, it might sound great all by itself. However, when you put these instruments together in a song, they'll probably get in each other's way and cause problems for the overall mix.

Ideally, find the frequencies that are unnecessary on each channel, cut those, then enhance or boost the frequency ranges you like. Keep the big picture in mind while selecting frequencies to cut or boost. Boost and cut different frequency ranges on the different instruments and fit the pieces together like a puzzle.

For instance, if the bass sounds muddy and needs to be cleaned up by cutting at about 250 Hz, and if the high end of the bass could use a little attack at about 2500 Hz, that's great. When we EQ the electric guitar channel, it's very possible that we could end up boosting the 250-Hz range to add punch. That works great because we've just filled the hole that we created in the bass EQ. Audio Example 10-26 demonstrates a bass recorded without EQ (flat).

Audio Example 10-26

Bass (Flat)

Listen to Audio Example 10-27 as I turn down a frequency with its center point at 250 Hz. It sounds much better because I've turned down the frequency range that typically clouds the sound.

Audio Example 10-27

Bass (Cut 250 Hz)

Audio Example 10-28 demonstrates a guitar recorded flat.

Audio Example 10-28

Guitar (Flat)

Audio Example 10-29 demonstrates the guitar with a boost at 250 Hz. This frequency is typical for adding punch to the guitar sound.

Audio Example 10-29

Guitar (Boost 250 Hz)

Audio Example 10-30 demonstrates the guitar and bass blending together. Notice how each part becomes more understandable as the EQ is inserted.

Audio Example 10-30

Guitar and Bass Together

In a mix, the lead or rhythm guitar doesn't generally need the lower frequencies below about 80 Hz. You can cut those frequencies substantially (if not completely), minimizing interference of the guitar's low end with the bass guitar.

If the guitar needs a little grind (edge, presence, and so on) in the high end, select from the 2- to 4-kHz range. Because you have already boosted 2.5 kHz on the bass guitar, the best choice is to boost 3.5 to 4 kHz on guitar. If these frequencies don't work well on the guitar, try shifting the bass high-end EQ slightly. Find different frequencies to boost on each instrument—frequencies that work well together and still sound good on the individual channels. If you avoid equalizing each instrument at the same frequency, your song will sound smoother, and it'll be easier to listen.

Frequency Ranges

The range of frequencies that the human ear can hear is roughly from 20 Hz to 20 kHz. This broad frequency range is broken down into specific groups. It's necessary to know and recognize these ranges.

- Highs – above 3.5 kHz
- Mids – between 250 Hz and 3.5 kHz
- Lows – below 250 Hz

These are often broken into more specific categories:

- Brilliance – above 6 kHz
- Presence – 3.5–6 kHz
- Upper midrange – 1.5–3.5 kHz
- Lower midrange – 250 Hz–1.5 kHz
- Bass – 60–250 Hz
- Sub-bass – below 60 Hz

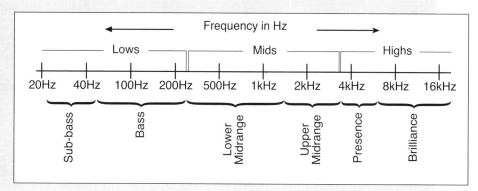

Definition of Frequency Ranges

As I stated before, the range of frequencies that the human ear can hear is roughly from 20 Hz to 20 kHz. Individual response may vary, depending on age, climate, and how many rock bands the ears' owner might have heard or played in. This broad frequency range is broken down into specific groups. It's necessary for us to know and recognize these ranges.

Listen to Audio Examples 10-31 to 10-34. I'll isolate these specific ranges.

Audio Example 10-31

Flat

Audio Example 10-32

Highs (Above 3.5 kHz)

Equalization Curve - Bandwidth

Boosting or cutting a particular frequency also boosts or cuts the frequencies nearby. If you boost 500 Hz on an equalizer, 500 Hz is the center point of a curve being boosted. Keep in mind that a substantial range of frequencies might be boosted along with the center point of the curve. The exact range of frequencies boosted is dependent upon the shape of the curve.

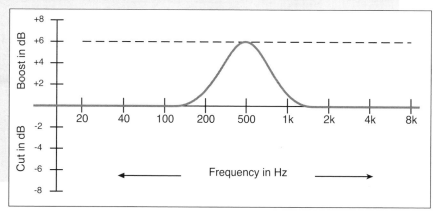

Audio Example 10-33

Mids (250 Hz to 3.5 kHz)

Audio Example 10-34

Lows (Below 250 Hz)

These are often broken down into more specific categories. Listen to each of these more specific ranges.

Audio Example 10-35

Flat (Reference)

Audio Example 10-36

Brilliance

Audio Example 10-37

Presence

Audio Example 10-38

Upper Midrange

Audio Example 10-39

Lower Midrange

Audio Example 10-40

Bass

Audio Example 10-41

Sub-Bass

Some of these ranges may be more or less audible on your system, though they're recorded at the same level. Even on the best system, these won't sound equally loud because of the uneven frequency response of the human ear throughout the volume spectrum.

Bandwidth

Bandwidth is simply the width, quantified in octaves, of a frequency spectrum. A human being hears a bandwidth of about 10 octaves—from 20 Hz to 20 kHz.

Most equalizers contain controls for at least three bands, with each band about one octave wide. This means that the boost or cut is centered on the defining frequency but contains frequencies that extend 1/2 octave below the center point and 1/2 octave above the center point. A one-octave bandwidth is specific enough to enable us to get the job done, but not so specific that we might create more problems than we eliminate. Bandwidth is sometimes referred to as the *Q*.

Bandwidth - The Q

Many equalizers let you control the width of the curve being manipulated. Notice that the differing bandwidths in this illustration refer to bandwidth in octaves or fractions of an octave. Parametric equalizers provide variable bandwidth control.

- Band #1 is about one octave wide.
- Band #2 is about two octaves wide.
- Band #3 is about half an octave wide.

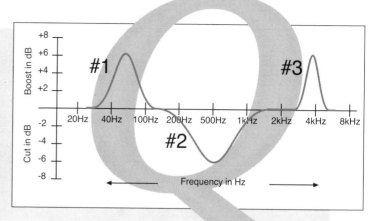

As the frequency band is raised or lowered, the frequencies on either side follow along in the overall shape of a bell curve centered on the given frequency. Bandwidth has to do with pinpointing how much of the frequency spectrum is being adjusted. A parametric equalizer is unique in that it has a bandwidth control.

A wide bandwidth (two or more octaves) is good for overall tone coloring. A narrow bandwidth (less than half an octave) is good for finding and fixing a problem.

Sweepable EQ

A lot of mixers have sweepable EQ (also called *semi-parametric* EQ). Sweepable EQ dramatically increases the flexibility of sound shaping. There are two controls per sweepable band:

1. A cut/boost control to turn the selected frequency up or down
2. A frequency selector that lets you sweep a certain range of frequencies

This is a very convenient and flexible EQ. With the frequency selector, you can zero in on the exact frequency you need to cut or boost. Often, the kick drum has one sweet spot where the lows are warm and rich or the attack on the guitar is at a very specific frequency. With sweepable EQ, you can set up a boost or cut, then dial in the frequency that breathes life into your music.

Mixers that have sweepable EQ typically have three separate bands on each channel: one for highs, one for mids, and one for lows. Sometimes the highs and lows are fixed-frequency equalizers, but the mids are sweepable.

Sweepable EQ

A sweepable equalizer has a Boost/Cut control to determine the severity of EQ. The frequency selector lets you slide the band throughout a specific range.

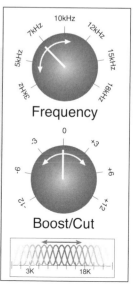

Frequency

Boost/Cut

Bandwidth (Q)

Narrow   Wide

Parametric EQ

The width of the selected frequency band is controlled by the Q adjustment (also called bandwidth). Curve A (below) is a very broad tone control. Curve B is a very specific pinpoint boost. The Q varies infinitely from its widest bandwidth to its narrowest. Frequency and Boost/Cut operate like the sweepable EQ.

Parametric equalization is the most flexible and powerful tone control.

A ← → B

1 kHz

Parametric EQ

This is the most flexible type of EQ. It operates just like a sweepable EQ, but gives you one other control: the bandwidth, or Q.

With the bandwidth control, you choose whether you're cutting or boosting a large range of frequencies or a very specific range of frequencies. For example, you might boost a four-octave band centered at 1,000 Hz, or you might cut a very narrow band of frequencies, a quarter of an octave wide, centered at 1,000 Hz.

With a tool like this, you can create sonic pieces that fit together like a glove. Parametric equalizers are a great addition to any mixer. They are readily available in outboard configurations, and some of the more expensive consoles provide built-in parametric equalization.

Video Example 10-3

Demonstration of Parametric Q and Sweep

Graphic EQ

This is called a graphic equalizer because it's the most visually graphic of all EQs. It's obvious, at a glance, which frequencies you've boosted or cut.

A graphic equalizer isn't appropriate to include in the channels of a mixer, simply because of the space required to contain 10 or more sliders, but it is a standard type

Graphic Equalizers

The 10-band graphic EQ provides good general sonic shaping. Each slider controls a one-octave bandwidth.

A 31-band graphic equalizer provides control over a third of an octave with each slider.

+12

30Hz 60Hz 120Hz 250Hz 500Hz 1kHz 2kHz 4kHz 8kHz 16kHz

-12   Frequency in Hz

10-Band Graphic Equalizer
One Octave Per Slider

+12 dB

-12 dB

20 25 31 40 50 60 80 100 125 160 200 250 315 400 500 630 800   1 1.25 1.6 2 2.5 3.15 4   5 6.3 8 10 12.5 16 20

Hz   kHz

31-Band Graphic Equalizer
1/3 Octave Per Fader.

of outboard EQ. The graphic EQs that we use in sound reinforcement have 10, 31, or sometimes 15 individual sliders that each cut or boost a set frequency with a set bandwidth. The bandwidth on a 10-band graphic is one octave. The bandwidth on a 31-band graphic is one third of an octave.

Graphic equalizers were very popular in the studio a number of years ago. Today, graphic equalizers are used mostly in live sound reinforcement applications because they are convenient and very visual, and they work well in conjunction with acoustical measurement devices. It is typical for a traditional sound system design to include a 31-band equalizer between the main output and the house amplifier inputs.

Notch Filter

A notch filter is used to seek and destroy problem frequencies, such as a high-end squeal, a ground hum, or possibly a noise from a heater, fan, or camera.

Notch filters have a very narrow bandwidth and are often sweepable. These filters generally cut only.

Peaking Filters

All the equalizers we've covered so far are peaking filters because they cut or boost a band in a bell-curve shape to a peak that is centered on the defining frequency. These are by far the most common types of equalizers.

High-Pass Filter

A high-pass filter lets the high frequencies pass through unaffected but cuts the low frequencies. In previous study of equalizers, we used an image of a bell curve with a center point at the selected frequency moving up or down as the frequency range was boosted or cut. Band-pass, high-pass, and low-pass filters don't fit that picture. With these filters, we specify a frequency at which the cut begins. The selected frequency is called the *cutoff frequency*, or sometimes the *knee*. Once the filter point is defined, the severity of the cut (the steepness of the filter) is calibrated in dB per octave. This rate of the cut is called the *slope*. In their normal use, these filters cut at a rate between 6 and 12 dB per octave.

A high-pass filter can help minimize 60-cycle hum on a particular channel by filtering, or turning down, the fundamental frequency of the hum. High-

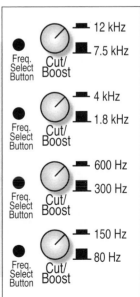

Selectable Frequencies

Two bands of EQ are available on each knob, enabling access to eight frequency bands. Pressing the Frequency Select button determines which frequency is boosted or cut. Each knob adjusts one frequency or the other, not both at the same time.

This type of equalizer is very functional, but the sound operator must be careful not to stack too many boosts or cuts at a specific selectable frequency. For example, increasing the highs by boosting several channels at 4 kHz can cause a mix that is unnaturally edgy-sounding.

Types of Filters

Equalization comes in many forms. The first types of equalizers we covered were peaking filters, where a range of frequencies are cut or boosted in the form of a bell curve. In addition there are several forms of band-adjusting filters, where an entire range of frequencies is adjusted uniformly. The icons typically used to indicate these filters accurately depict their functions.

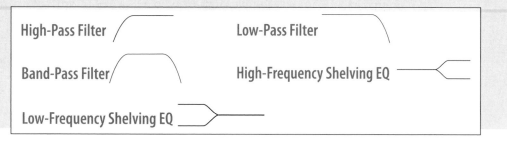

pass filters function very well when you need to eliminate an ambient rumble, such as a furnace in the background or street noise that leaks into a vocal mic.

Most modern high-pass filters provide a sweepable frequency selector. In the context of creating a mix, a high-pass filter is traditionally used to trim away unused or unnecessary low-frequency information. Simply listen to the channel and sweep the high-pass filter upward from the low-frequency range until you hear the low end thin out a little. Using filters in this manner is a good way to clean out unnecessary sonic ingredients. However, always be careful to critically assess the sonic impact on the overall mix. This is music, after all, and sometimes ingredients we don't consciously hear are indeed affecting the emotional or physical impact of the overall sound.

The majority of mixers found in churches around the country offer fixed frequency high-pass filters that eliminate frequencies below about 100 Hz; some include a fixed frequency low-pass filter. Many equalizer sections provide a sweepable high-pass filter on the low band, and some include a sweepable low-pass filter on the high band.

Low-Pass Filter

A low-pass filter lets the low frequencies pass through unaffected and cuts the highs, usually above about 8 to 10 kHz. Low-pass filters have many uses. For instance, they can help minimize cymbal leakage onto the tom channels, filter out a high buzz in a guitar amp, or filter out string noise on a bass guitar channel.

Sweeping a filter lets the sound operator fine-tune the channel so that only the desired frequencies pass through. Listen to a channel with the low-pass filter set with cutoff frequency as high as possible, then lower the cutoff frequency until you can hear the highs diminish. Then, raise the cutoff frequency slightly for a natural sound, while filtering out extraneous high frequencies. As with the high-pass filter, always be careful to critically assess the sonic impact on the overall mix. You might be filtering frequencies that are combining in the mix to create a sound or a feeling, even though you don't think you can hear them.

These filters, whether high-, low-, or band-pass, should be used to filter specific unwanted mix ingredients; they shouldn't be used to trim away at every channel on the mix, just in case there's a problem in a frequency range.

Band-Pass Filter

A band-pass filter lets us select a frequency range (a band) and lets it pass through unaffected. In other words, all frequencies above and below a specified frequency range are filtered out. The band-pass filter is just like the marriage of a high-pass and a low-pass filter. With a band-pass filter, it's easy to create a lo-fi sound, such as that projected from a small transistor radio, or to zero in on any specific frequency range for a special effect.

Shelving EQ

A shelving EQ leaves all frequencies flat to a certain point, then turns all frequencies above or below that point down or up at a rate specified in dB per octave. As with high- and low-pass filters, shelving equalizers roll off the highs or lows, at a slope between 6 and 12 dB per octave; however, past the slope, all frequencies remain boosted or cut to the end of the frequency spectrum.

Shelving equalizers are a convenient way to add *air* (the high frequencies we can't necessarily hear as much as feel) to a mix. Simply sweep the cutoff frequency into the highs, above 12 kHz or so, and raise the shelf slightly. This is a common technique, especially with the advent of the ultra-quiet gear available today.

Combined Equalizers

Software-based equalizers emulate each type of hardware EQ. Many contain identical controls and even emulate the look and feel of highly respected classic equipment.

Adjusting Gain Structure

Modern mixers provide plenty of headroom and they are virtually noise-free. It's important that the sound operator always control the gain structure. There shouldn't be any point in the signal path where levels are unnaturally hot or cold. The following gain setting routine is very effective—it puts each gain stage in an acceptable range and provides an excellent staring point for most mixes. Any inserts into the signal path must be verified as to proper input level and nearly identical overall output level.

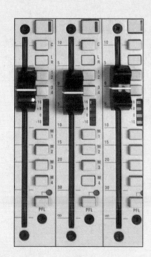

Set faders to unity

Adjust gain trims for appropriate level

Adjust master output control for desired volume in the house

Additionally, several software-based equalizers offer all the features we've discussed, all in one equalizer. In fact, most offer multiple options at each band.

The Equalizer's Sound

Equalizing circuitry does affect the sound of the source. In fact, there are equalizers that sound good and equalizers that don't sound so good. The quality, manufacturer, and design of any audio tool matters. Always listen to your music with and without the equalizer; be very discerning. It's better to avoid EQ rather than to use an EQ with lots of flexibility and a crummy sound.

Well-respected outboard gear gets that way because it works well and sounds good. Value reputation, yet always assess for yourself. Very inexpensive equalizers usually sound bad and are noisy—but not always. Very expensive equalizers usually sound great and are very clean and noise-free—but not always. It's up to you to listen and select the equipment that works for you.

Remember, even with the multitude of available equalizers, don't use EQ first to shape your sounds. First, get as close to the sound you want using mic choice and mic technique, then use EQ if it's necessary.

Stereo Master

The stereo master control is the final level adjustment out of the mixer. Whereas the level adjustment to the mixdown recording device is very important, the stereo master in a live sound application serves as a master volume control. If the system is set up properly, the output level within the acceptable volume range will be somewhat normal—typically peaking at 0 VU with plenty of headroom remaining.

The live sound operator should set the channel faders on, around, or near unity gain; adjust the gain trims for acceptable readings on the channel faders; and then use the stereo master fader for overall volume control.

Talk Back to the Band

Most mixers provide a talkback system designed to facilitate efficient communications from the sound operator to the musicians. These systems typically let the operator choose which musicians to talk to by allowing him or her to select an aux bus, subgroup, or a combination of auxes and groups to activate. A microphone is usually connected to the talkback jack in the rear of the mixer for this purpose.

Using the talkback system is much easier on the sound operator than yelling back and forth from the FOH mixer to the stage. This system is a must when the musicians are using in-ear monitors.

Talkback/Communications

The talkback button lets the sound operator talk into multiple buses—typically, into each aux or group of auxes, or into the main output. Talkback is the sound operator's link to the musicians.

This is especially convenient when the band and singers are using in-ear monitor systems. If the drummer needs to play louder or softer, the operator must simply choose a bus that feeds the in-ear system in the talkback section of the console and speak while pressing the talkback button.

The sound operator frequently serves as the announcer for certain types of presentations, so the ability to plug a microphone into the talkback bus and turn the mic on by simply pushing the Mains button in the talkback section is very convenient.

Test Tones

Your mixer might have a section marked *tones*, *test tones*, *osc*, or *oscillator*, especially if you own a large-format or vintage recording console. This section contains a frequency generator that produces different specific frequencies in their purest form—sine waves. These frequencies are used to adjust input and output levels of your mixer, recorders, and outboard equipment. Tones are used for electronic calibration and level setting, whereas pink and white noise are used for acoustical adjustments.

Consider the stereo master output from your mixer to the mixdown recorder. Raise the level of the reference tone (between 500 and 1,000 Hz) until the VU meter reads 0 VU on the stereo output of the mixer. Do this with the stereo master output faders set at the point where your mix level is correct. Adjust the tone's level to the meters with the tone's output control.

Fine-tune the left/right output balance. If one side reads slightly higher than the other from the same 1-kHz tone, balance the two sides. Many mixers have separate level controls for left and right stereo outs. Proper adjustment of the left/right balance ensures the best accuracy in panning and stereo imaging.

Console Layout

The modern digital mixer, in theory, could be constructed with one fader and a knob. The knob would select the target parameter, and the fader would vary the setting. If you want to decrease the lows on channel 61, just turn the knob until you get to CH61-EQ-LOW, then use the fader to turn the lows down. This would be very nice and compact, but it wouldn't be very convenient in the heat of battle. During a performance, the sound operator must be able to move quickly through mix changes and, even more importantly, he or she must be able to react with lightning speed to catastrophic events, such as a dead mic, feedback, or an FOH system malfunction.

Mixers—both digital and analog—are laid out specifically for the operator's convenience. Recording mixers are laid out in a couple different formats that serve the needs of the recording engineer: split and inline. However, almost all live mixers are set up with the channel faders to the left, the sub-group controls just to the right, and the main mix output level faders furthest to the right. Mixers with 24 or fewer channels typically keep all of the

Comparison of Console Layouts

The mixers below illustrate three different mixer layouts.

- Mixer A is the Yamaha M7CI-48 digital mixer. Being a fully digital console, it provides instant access to a lot of faders at once; however, to access EQ, dynamics, effects, and so on, a channel must be selected—adjustments are made using the on-screen interface and the controllers located below it.
- Mixer B is a Mackie ONYX 1620. It is a small format 2-bus mixer with 16 channels and a simple stereo output.
- Mixer C is a Soundcraft GB4, 32-channel, 4-bus mixer. It has 32 channels with the control center, subgroups, and master faders between the first 24 and the last 8 channels.

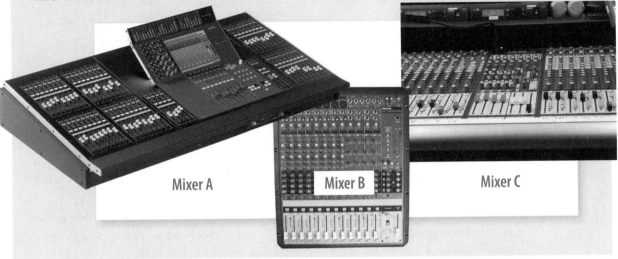

Mixer A   Mixer B   Mixer C

channel faders to the left of the sub-groups and masters. Mixers with more than 24 channels usually position channels 25 and above to the right of the sub-groups and masters.

Each mixer is configured according to the needs of the sound operator; however, they're not all the same. Even though mixer layouts are similar, certain manufacturers devise layout schemes that are very comfortable for some and very uncomfortable for others. Those who are used to working in recording might favor a particular mixer style, whereas someone who has worked primarily in live sound might prefer another style.

Matrix

Many modern front-of-house mixers include a matrix section. The matrix is a combining bus that lets the sound operator set up a unique balance of the group and main outputs. This is a very convenient feature that is often used to feed alternate zones, such as an overflow room, or to set up a separate mix for the service-recording device.

The mix that is set up to reinforce instruments and amplifiers in the sanctuary is dependent on the acoustic sounds. Any time you need to send the FOH mix to a recording device or to a separate room, bear in mind that they won't have the advantage of the acoustic room sounds to fill in the mix. The mix that such devices or rooms receive will require a different mix—a mix that includes elevated levels from the instruments that provide substantial acoustic volume. Typically, the instruments with the loudest acoustic

The Matrix

The matrix is a summing bus that lets the sound operator combine the subgroups and main outputs into a different blend. It typically provides separate level controls for all groups and masters—the new mix is routed to the matrix output.

The matrix lets the sound operator set up an optional relative mix for recording, broadcast feeds, or any other destination. This is very useful when feeding a device or location that is separated from the acoustic stage volume and acoustic ambience.

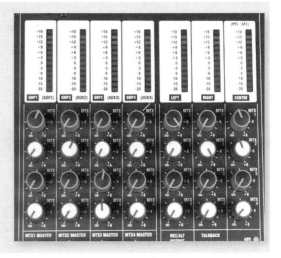

volume need to be boosted most in the recording or alternate location—if not, they will be weak in the mix.

Direct Outs

Many mixers provide direct outs on each channel. These are separate from inserts, group outputs, and other rear panel outputs. These direct outputs access the signal without interrupting any other routing. With the increased popularity of in-ear monitor systems and affordable multitrack recorders, these outputs have become a necessary feature.

- Feed the inputs of the in-ear monitor core device with the direct outputs to give mix control to each musician without depending on the sound operator. This makes the sound operator's job a little easier, freeing him or her up to focus on other details.

- Connect the direct outs to the line inputs of a multitrack recorder. This type of setup is ideal for recording. The mixer trims should be set for optimal levels in the mix, which is ideal because the trims affect the direct output levels.

- If the direct outs are connected to an in-ear monitor system and you want to record, simply split the direct outs and send the same signal to a multitrack.

Analog Versus Digital Mixers

If you possess a thorough understanding of the mixer functions we've just studied, you'll be able to operate efficiently on an analog or digital mixer. With any new mixer, the real test is knowing where all the features are located—you should already have a good idea of the features you can expect to be available.

The logistical advantage of an analog console is its simplicity. Each channel is the same—the EQ knob and every other knob is at the same spot on the console each time. Digital mixers are so flexible that it's not always easy to guess where the controls are; they might be hidden somewhere in a tangled web of menus.

The really nice feature of nearly any digital mixer is its ability to store snapshots of the entire mixer layout for retrieval later. It's a simple matter to set up a tracking session

Combining Digital Mixers

New technology enables digital consoles to link together as one. These two PreSonus StudioLive 24-channel mixers connect via FireWire and function as if they were a single 48-channel mixer. They are extremely affordable and they sound great, not to mention that they provide full dynamics and effects control on each channel.

Digital mixers incorporate traditional design and layout concepts for quick and easy use. They also provide layers of control to access multiple parameters at different key locations.

Designed for either studio or live applications, these mixers can also be controlled remotely from an iPad—the interface is incredibly flexible. When held in portrait position, each channel can be accessed. Controls are readily available for compression, gating, EQ, 10 aux sends, two FX levels, and the channel mix level. Simply turn the iPad to landscape position and the screen automatically changes to one of three basic views: overview for the entire mix, aux levels for all channels and subs, or four stereo pairs of graphic equalizers that can be assigned to the mains, subs, and auxes. The Masters section is also available in this view for level and dynamics control of the mains and subs.

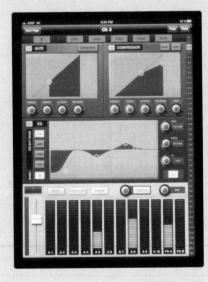

iPad landscape and portrait views

and then just save the snapshot of that session. The snapshot contains all EQ, level setting, routing, effects, etc., so resetting for tracking on another day is a simple matter of a button push to recall your previously saved settings.

In addition, digital consoles provide automated control of virtually every knob, button, and parameter. When it comes to mixdown, the modern engineer can finely craft each small segment of a production a little at a time, writing the data into the automation system. Once the mix is completed and it's time to render it to the mixdown format of choice, the mix engineer simply starts playback and sits back and enjoys his or her creation.

Control Surfaces

A control surface, like the Avid VENUE, feels and acts just like a digital console. There are, however, a couple of primary differences between the two:

- The digital console has audio inputs and outputs as well as A/D and D/A converters.
- The control surface does not pass audio signals and does not have A/D and D/A converters. The control surface has all of the faders, knobs, and switches that control all the EQ, dynamics, effects, level, pan, surround, busing, routing, inserting, sending, and so on, but must connect to the audio interface to have anything to control.

It's hard to beat the flexibility and control provided by modern control surfaces. From the small and powerful Euphonix Artist Control and Mix Control devices to the massively flexible Avid D-Show series, these devices let the engineer control all parameters of the mixer and plug-ins in many different ways.

The Virtual Mixer

A virtual mixer is the mixer built into your digital audio recording software. Again, like the digital mixer, there aren't really any new functions on these mixers, just onscreen representations of the same functions. Sometimes the virtual mixer is more cumbersome to operate, especially if you're not using a very large monitor. However, although there aren't necessarily new functions for most basic processing tasks, the virtual mixer and plug-ins make possible many combinations and variations of parameters that are either very difficult or impossible to get from a standard hardware device.

The software mixer built into the modern DAW is capable of performing essentially all mixing tasks. The primary limitation is the mouse. As digital workstations have grown to dominate the home and professional studio worlds, the trend has been to provide an effective work environment, normally control-surface-based. While control surfaces are able to simulate the look and feel of a classic recording console, a patient and diligent recordist can accomplish amazing mixing feats with a mouse, some software, and a computer.

FOH Mixer Position

The ideal position for the front-of-house mixer depends on whether you are connected in stereo or mono. Whereas many sound operators prefer a mono mix, there is something very nice about the stereo effects that come from keyboards and instrument effects processors, so consider connecting in stereo.

Mono Mix

If the mix is mono, it is typically best for the sound operator to sit directly in front of one side and between two-thirds and three-quarters of the way toward the back wall. Avoid placing the mixer against (or nearly against) the back wall—the combination of the reflections off the back wall and the direct sound can add up to a very unreliable representation of the overall FOH mix.

It is best if the mixer is out in the open so there are minimal reflections off close walls and other physical structures around it.

Stereo Mix

If the sound system is running in stereo, the sound operator should be in the center of the room, equidistant from the left and right speakers.

Again, it is best if the mixer is out in the open so there are minimal reflections off close walls and other physical structures around it.

Even though the system is running in stereo, avoid panning instruments hard left and right—the congregation needs to hear the full mix no matter where they are in the sanctuary.

FOH Mixer Position

Position the mixer in line with one of the speaker stacks or arrays for a mono mix. For a stereo mix, position the FOH mixer equidistant from the two speaker stacks or arrays. Also, be sure the mixer isn't too close to the back wall. Depending on the size of the venue, the mixer should usually be placed between halfway and three quarters of the way toward the back of the room. Positioning the mixer close to the back wall provides an unreliable perspective for the sound operator because of the strong and immediate reflections coming from behind.

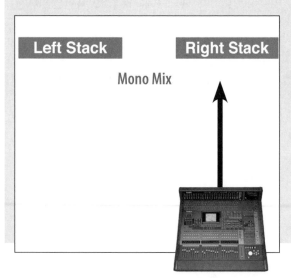

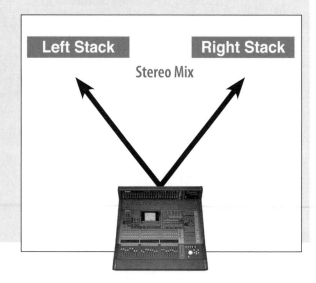

chapter 11

SIGNAL PROCESSORS

Signal processors play an important role in shaping the live sound into a polished, professional sound. When they're over-used they can easily destroy the power and intelligibility of an otherwise good mix. When they're used tastefully and musically, they can add a powerful dimension, supporting the communication of heart and soul that the musicians are trying so hard to portray.

There are two overriding categories of signal processors:

- Dynamics processors, which manipulate signal strength
- Effects processors, which use delays to build natural- and unnatural-sounding effects

Dynamics Processors

Dynamics processors in the effects rack function in the same way as the dynamics processors we studied previously in relation to system dynamics control. Whereas the more general limiting and compression on the system functions to control the mixed audio and to protect the speakers and the audience against peaks and audio surges, individual dynamics processors provide a different function. They function to decrease the dynamic range of the audio source.

Dynamic range is simply the distance in dB from the loudest to the softest portions of the track. The true effect of the dynamic processor, when inserted into the channel signal path, is to reveal portions of the audio source at the lowest levels. The fact that the level-control circuit turns down the loudest segments is somewhat incidental in itself—the value of the process is seen when the signal is turned back up to regain the lost level. Once the signal is turned up, the nuance in audio source is revealed by the amount of the made-up gain. Words and notes that would have been buried in the mix without the compressor are heard clearly as a result of the compression procedure.

Compressor

The compressor is an automatic volume control that turns loud parts of the musical signal down when they exceed a certain level (voltage). The circuit that actually controls the level is the VCA (*voltage controlled amplifier*), the DCA (*digitally controlled amplifier*), or another externally controlled amplifier circuit. Because an audio waveform is alternating current with changing amplitude over time, it is essentially voltage. The amplitude can be used to trigger the amplifying circuit to adjust in direct correlation to the energy level.

Imagine yourself listening to the mix, and every time the vocal track starts to get too loud and read too hot on the meter, you turn the fader down and then back up again for the rest of the track. That is exactly how a compressor works.

When used correctly, compression doesn't detract from the life of the original sound—in fact, it can be the one tool that helps life and depth to be heard and understood in a mix. Imagine a vocal track. Singers perform many nuances and licks that define their individual style. Within the same second, they may jump from a subtle, emotional phrase to a screaming-loud, needle-pegging, engineer-torturing high note. Even the best of us aren't fast enough to catch all of these changes by simply riding the input fader. In this situation, a compressor is needed to protect against excessive levels.

As the loudest parts of the source are turned down, we're able to bring the overall level of the channel up. In effect, this brings the softer sounds up in relation to the louder sounds. The subtle nuance becomes more noticeable so the individuality and style of the artist is more easily recognized, plus the understandability and audibility of the lyrics are greatly increased.

Video Example 11-1

Demonstration Compression and Makeup Gain

Patching the Dynamic Processor

Dynamic processors are typically inserted into a channel. The output of the processor, when patched into the channel return, supplies the dynamically altered signal back into the channel signal path—it's a permanent part of the sound from that point.

It's also common to patch a source directly into a dynamics processor before it reaches the mixer.

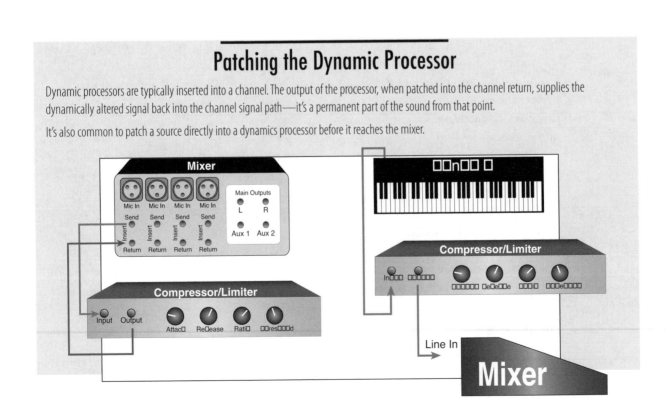

To Compress or Not to Compress—That Is the Question

Some engineers use a lot of compression and limiting; other engineers don't use any. In commercial popular music, the use (and overuse) of dynamic processors is common, especially in the recording world. Because commercial recordings use compressors on many of the mix ingredients and because the mastering process often overuses peak limiters, some use of compressors and limiters is almost necessary in order to achieve a polished sound.

The dramatic difference between a recording mix and a live mix is feedback. If the live sound operator used the same amount of compression as the recording engineer, feedback would be extreme even in the best and most finely tuned sound system.

Compression Parameters

There are five controls common to most compressors: threshold, attack time, release time, ratio, and output level. Once you see how these work, you can operate any compressor, anywhere, anytime. To make it even better, these controls are easy to understand, and they do just what they say they do.

The End Result of Compression

Ideally, the end result of compression is that the loudest portion of the signal sounds about the same as normal, but the softest portions seem louder.

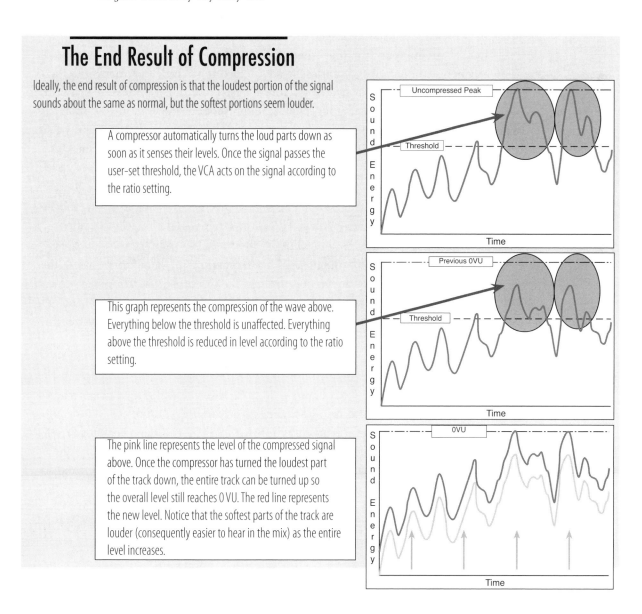

A compressor automatically turns the loud parts down as soon as it senses their levels. Once the signal passes the user-set threshold, the VCA acts on the signal according to the ratio setting.

This graph represents the compression of the wave above. Everything below the threshold is unaffected. Everything above the threshold is reduced in level according to the ratio setting.

The pink line represents the level of the compressed signal above. Once the compressor has turned the loudest part of the track down, the entire track can be turned up so the overall level still reaches 0 VU. The red line represents the new level. Notice that the softest parts of the track are louder (consequently easier to hear in the mix) as the entire level increases.

Controls on the Compressor/Limiter

Almost all compressor/limiters contain the same control options, whether hardware or software. Once you understand the functions on one compressor/limiter, you'll find seamless transition to another. The unit pictured here contains the basic controls: attack time, release time, threshold, ratio, output level, peak/RMS, knee, and meter function.

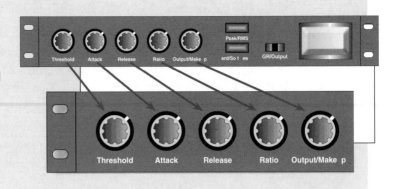

Threshold

As amplitude increases, voltage increases. The threshold is the point where the compressor begins to recognize the signal amplitude. Once the compressor recognizes the signal—when the amplitude rises above a certain voltage—it begins to act in a way that is determined by the attack time, release time, and ratio controls.

There are two different ways that compressors deal with the threshold:

- One way boosts the signal up into the threshold. Picture yourself in a room with an opening in the ceiling directly overhead. You represent the signal, with your head being the loudest sounds. The opening represents the threshold of the compressor. Imagine that the floor moves up, and you begin to go through the opening. That's the way that some compressors move the signal into the threshold—they turn it up until it goes through the threshold.

- The other way compressors deal with the threshold is by moving it down into the signal. Picture yourself in a room with an opening directly overhead. Now the ceiling moves down until you're through the opening. This is the other way the threshold control works—the signal level stays the same, but the threshold moves down into the peaks.

No matter which way the threshold works, it's the part of the signal that exceeds the threshold that's processed. Once the signal is through the threshold, the VCA turns down just the part of the signal that's gone through, leaving the rest of the signal unaffected. The portion that's above the threshold will be turned down according to how you have set the remaining controls (attack time, release time, and ratio).

Attack Time

The attack time controls the amount of time it takes the compressor to turn the signal down, once it has passed the threshold. If the attack time is too fast, the compressor will turn down the transients. This can cause an instrument to lose life and clarity. On a vocal, for instance, if the attack time is too fast, all of the "t" and "s" sounds will start to disappear. On the other hand, if the attack time is too slow and the vocal is very compressed, the T's and S's will fly through uncompressed and sound exaggerated.

Audio Example 11-1

S's and T's

Variations in the attack time setting help diminish or accentuate the relative attack of instruments such as guitar, bass, piano, or drums. Long attack times adjust average levels; short attack times adjust peak levels.

Specific attack time limitations vary between processors, though they typically range from 0.1 ms to 200 ms. One characteristic of an expensive compressor is fast attack time capability. Also, some compressors have the attack time fixed for a specific purpose, such as vocals.

Release Time

Release time is the time that it takes for the compressor to let go, or turn the signal back up, once it's below the threshold. The release time might be as fast as 50 ms or as slow five seconds.

Fast release times work well with fast attack times to control peak levels. Slow release times work well with slow attack times to control average levels. There is no practical value to adjusting the attack time so it's slower than the release time.

Long release times with severe compression can result in increased sustain. With the proper setting of the threshold, release, and attack time, a guitar, for example, can benefit by increased sustain. Over time, as the VCA turns the signal back up to its original level, an otherwise quickly decaying signal maintains its sustain longer.

Attack Time Settings Control Understandability and Punch

The attack setting provides a means to adjust the relative level of the initial portion of an audio source. In a vocal passage, the initial transient sounds—especially the sounds "s," "t," and "k"—offer two possible complications for recording:

1. If the vocalist has a natural abundance of sibilance, the vocal sound might take on a harsh character. These transient sounds, called sibilance, can cause irritating effects when reverberated, and they can even overdrive electronic circuitry. In this case, a fast attack time during compression helps smooth out the sound—the track will settle into the mix better.

2. If the instrumental bed is very percussive, and if the vocal sound contains understated sibilance, the lyrics might be lost in the mix because they're not understandable. In this case, try compressing the vocal channel, using a slower attack time; the compressor will let the sibilance pass through unaltered, yet the rest of the word will be compressed according to the control settings. The length of attack time varies with each vocal sound and application, but settings between 5 and 50 ms typically work well. Listen to the sound as you make the attack time adjustment; once you find the right setting, the vocal will seem more alive and understandable.

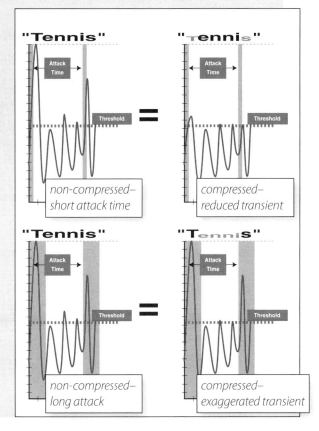

For a natural and unobtrusive sound, set the attack time relatively fast and the release time relatively slow. Each instrument or voice is different, so there's still importance placed on listening while you adjust these controls.

Ratio

Once the compressor starts acting on the signal, the *ratio* control determines how extreme the VCA action will be. The ratio is simply a comparison between the level that goes through the threshold and the output of the VCA; it's expressed as a mathematical ratio (10:1, 3:1, and so on). The first number in the ratio indicates how many dB of input increase will result in 1 dB of output increase. The higher the ratio, the greater the compression.

If the threshold is adjusted so that the loudest note of the song exceeds the threshold by 3 dB, and the ratio is 3:1, the 3-dB peak is reduced to a 1-dB peak—the gain is reduced by 2 dB. Using that same 3:1 ratio, if you input a 12-dB peak, the unit would output a 4-dB peak—still a ratio of 3:1, and gain is reduced by 8 dB.

Output Level

The output level control makes up for reduction in gain caused by the VCA. If the gain has been reduced by 6 dB, for example, the output level control is used to boost the signal back up to its original level.

The Difference Between a Compressor and a Limiter

The ratio setting determines the difference between a compressor and a limiter. Ratio settings below 10:1 result in compression. Ratio settings above 10:1 result in limiting. That explains why most manufacturers offer combined compressor/limiters. Extreme compression becomes limiting.

Hard Knee Versus Soft Knee Compression/Limiting

Hard knee/soft knee selection determines how the compressor reacts to the signal once it passes this threshold and the amplifier circuitry engages. Whereas the ratio control determines the severity of compression, the knee determines how severely and immediately the compressor acts on that signal.

When the compressor is set on soft knee and the signal exceeds the threshold, the amplitude is gradually reduced throughout the first 5 dB or so of gain reduction. When the compressor is set on hard knee and the signal exceeds the threshold, it is rapidly and severely reduced in amplitude. The hard knee/soft knee settings are still dependent on the ratio, attack, release, and threshold settings. The knee setting specifically relates to how the amplifier circuitry reacts at the onset of compression or limiting.

The difference between hard knee and soft knee compression is more apparent at extreme compression ratios and gain reduction. Soft knee compression is most useful during high-ratio compression or limiting. The gentle approach of the soft knee setting is least obvious as the compressor begins gain reduction. Hard knee settings are very efficient when extreme and immediate limiting is called for, especially when used on audio containing an abundance of transient peaks.

Hard Knee Versus Soft Knee Compression/Limiting

Once the compressor senses signal above the threshold and the attack time has passed, the level-control circuitry begins to respond. The dynamic action matches the word picture—soft knee creates a gently rounded level adjustment; hard knee creates a sharp angle. A hard knee setting activates the dynamic process immediately; a soft knee setting gradually engages dynamic control during the first 5 dB or so. Typically, soft knee compression is more gentle and less audible than hard knee —try this setting on a lead vocal or lyrical instrument for inconspicuous level control.

Hard knee dynamic control is more extreme and much less sonically forgiving. Try hard knee limiting when absolute level control is necessary.

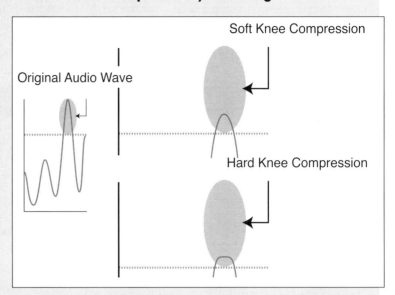

Typically, soft knee compression is gentler and less audible than hard knee compression. Try this setting on a lead vocal or lyrical instrument for inconspicuous level control.

Hard knee dynamic control is more extreme and much less sonically forgiving. Try hard knee limiting when absolute level control is necessary.

Peak/RMS Detection

RMS refers to average signal amplitude, based on the mathematical function of the root mean square. Peak refers to immediate and transient amplitude levels, which occur frequently throughout most audio recordings. The Peak/RMS setting determines whether the compressor/limiter responds to average amplitude changes or peak amplitude changes. RMS compression is more gentle and less obtrusive than peak compression. Peak compression is well suited to limiting applications. It responds quickly and efficiently to incoming amplitude changes containing transient information.

Side Chain

The side chain provides an avenue for activating the level-control circuitry from a source other than the audio signal connected to the device's main input.

Any audio source can be patched into the side chain input for creative applications. However, it's common to run a split from the audio signal through an equalizer then back into the side chain. In this way, the equalizer can be boosted at a specific frequency and cut at others, allowing the user to select a problem frequency to trigger gain reduction. This technique works very well when low-frequency pops or thumps must be compressed while the rest of the audio signal is left unaffected, or when certain high-frequency transients must be controlled.

Listen to the following Audio Examples highlighting the sonic impact of different compressor/limiter settings. The acoustic guitar is often compressed, and in these examples it provides an excellent comparison. With its clean, clear sound and transient attack, the parameter adjustments are very apparent.

Audio Example 11-3

Acoustic Guitar—No Compression

This acoustic guitar, without compression, has clean sound; however, it has a wide dynamic range. Notice the difference between the level of the loudest sound and the softest sound.

Audio Example 11-4

Acoustic Guitar—Variations in Attack Time Settings

Notice the change in the attack of each note. By increasing and decreasing the attack time, intimacy and sonic impact change dramatically.

Audio Example 11-5

Acoustic Guitar—Long and Short Release Times

With the release time set too short, the processor is continually active, risking sonic degradation. With the release time set too short, the level changes become very noticeable.

Audio Example 11-6

Acoustic Guitar—Ratios: From Compression to Limiting

Each ratio setting provides a different result, from gentle gain control to the brick wall.

VU Meter Versus LEDs Versus Onscreen

Each dynamic processor provides a method to measure the amount of gain reduction occurring at any given time. Whereas a typical meter reads from left to right to indicate the amount of signal present, a compressor/limiter meter typically moves from right to left to indicate the amount of signal decrease (in dB).

When a traditional VU meter indicates gain reduction, there is no level change as long as the meter is resting at the far right side. As the level is decreased, the meter moves to the left—the numbers on the meter represent decibels of gain reduction.

When a series of LEDs are used to indicate gain reduction, each LED that illuminates indicates more gain reduction. The numbers under each LED show the amount of gain reduction.

Computer-based compressor/limiters use an onscreen version of either of these metering systems.

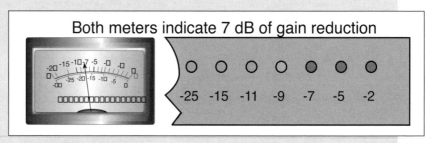

Both meters indicate 7 dB of gain reduction

Acoustic Guitar—Adjusting the Threshold for Optimum Sonics

Listen to the changing sound as the threshold moves down into the signal amplitude. If the threshold is too high, there is no dynamic compression. If the threshold includes too much of the amplitude, the level-changing circuitry (VCA, DCA, Optical Amp, and so on) is always working; this typically produces a thin, weak, or strained sound.

Acoustic Guitar—Adjusting the Knee and Peak/RMS Settings

With the ratio set at 7:1, attack time at 10 ms, release time at .5 seconds, and the threshold set for 6 dB of gain reduction, notice the sonic difference as I switch from soft to hard knee and from peak to RMS detection.

Meters on the Compressor/Limiter

Compressor/limiters utilize various systems for metering gain reduction, input levels, and output levels. Some devices offer separate meters for each function, whereas several units utilize a multipurpose meter that switches between functions. Either system is functionally simple, and it's important to use these meters to help ensure optimal use of the device.

Input Level Meter

Ideally, the input level meter verifies the proper signal strength as it enters the device; however, many compressor/limiters don't have one. Because compressor/limiters are usually patched inline directly or through an insert, you can take advantage of the meters on your mixer.

When the compressor limiter is in bypass mode (most units have a bypass switch), the input is probably at unity gain with the output, especially when the output level is set to "U" or zero (no boost or no cut). Therefore, as you increase gain reduction, you should be able to simply boost the output level to maintain the original level or, if your device doesn't have an output level control, you can usually make up the gain by increasing the channel gain trim.

Output Level Meter

The output level meter is simply fed by the output level control. Use it to verify that the signal level is correct at the output of the device. It's common to set the output level so that it matches the input level—both at unity gain.

Gain Reduction Meter

Gain reduction refers to the amount that the VCA has turned the signal down once it crosses the threshold.

To meter gain reduction, some compressor/limiters use a series of LEDs, and others use a VU meter. Typically, LEDs light up from right to left, indicating how far the unit has turned the signal down. Each LED represents two or more dB of gain reduction.

If your compressor has a VU meter, 0 VU is the normal (rest) position on a meter used to indicate no gain reduction. As the compressor turns down, the needle moves backward from 0 to indicate the amount of gain reduction. A –5 reading on the VU indicates 5 dB of gain reduction.

The Limiter

A limiter and compressor perform the same basic task, although a compressor controls level and amplitude in a soft and gentle manner, whereas a limiter controls level and amplitude in an extreme way.

Limiters are often used to control the level of the entire mix. An excellent mix typically contains several transient peaks (levels that exceed the average level of the entire mix). Although the limiter ignores the majority of the program material (audio that doesn't exceed the threshold), a peak that exceeds the threshold will be turned down quickly. Through the use of limiters, the mix maintains a constant and aggressive level and amplitude. A master mix might peak at 0 VU; while the limited mix also peaks at 0 VU, the difference is that the limited mix sounds louder. A good limiter operates in a way that is imperceptible to most listeners. It reacts quickly to transient peaks and maintains a full, impressive, aggressive sound throughout the limiting process.

Setup Suggestions for the Compressor/Limiter

Use the following procedure to set virtually any compressor/limiter. Even though it seems like there are countless dynamics processors available, they all work essentially the same. Whether digital or analog, vintage or new technology, this procedure will work.

- Adjust Ratio to determine function. Settings below 10:1 produce compression; settings 10:1 to infinity:1 produce limiting.
- Set the attack time to fast or slow, depending on the audio source and desired effect.
- Set the release time to about .5 seconds for general use.

Limiting

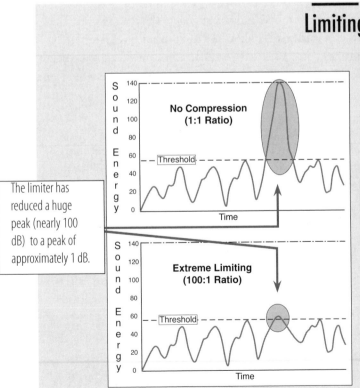

The limiter has reduced a huge peak (nearly 100 dB) to a peak of approximately 1 dB.

This graph represents a signal with a huge peak energy. Use a limiter on this kind of signal so the majority of the sound is unaffected (by the limiter's VCA) but the trouble spot is nearly eliminated.

The limiter can keep nearly any peak from overdriving the tape or from blasting through the mix. The results of effective limiting are often dramatic. If you start with a mix that has been level-impaired by a few quick blasts of energy, then you essentially remove those blasts, the entire mix level can be increased substantially, resulting in a much more powerful sound.

- Select soft knee for gentle compression or hard knee for limiting applications.
- Select RMS for most compression applications or peak for most limiting applications.
- Adjust the threshold for the desired amount of gain reduction. You should typically have 3 to 6 dB of reduction at the strongest part of the track, and there should be times when there is no gain reduction.
- Consider all rules carefully, then break them at will, and intentionally, any time the music demands.

Gate/Expanders

Gates and expanders are functionally opposite in relation to compressors and limiters. Whereas the compressor/limiter decreases the signal level when it exceeds the user-set threshold, the expander/gate decreases the signal level when it is below the user-set threshold. For example, most guitarists include their own complement of processor, which they use to build the perfect guitar sound for each song. The problem is that guitar sounds often consist of extreme compression use along with one or two types of overdrive—another type of compression. This typically results in a signal containing substantial noise, which is then routed through other potentially noisy processors (chorus, delay, EQ, and so on).

All this results in a great sound, as long as the guitarist is playing; however, whenever the guitarist is idle, there is a very audible hissing, grinding noise coming from his or her equipment. In this scenario, it's the expander/gate that comes to the rescue. The sound operator must simply adjust the threshold so that it is just above noise level—when the guitarist plays, the gate opens up, allowing the full sound to be heard, but when the guitarist

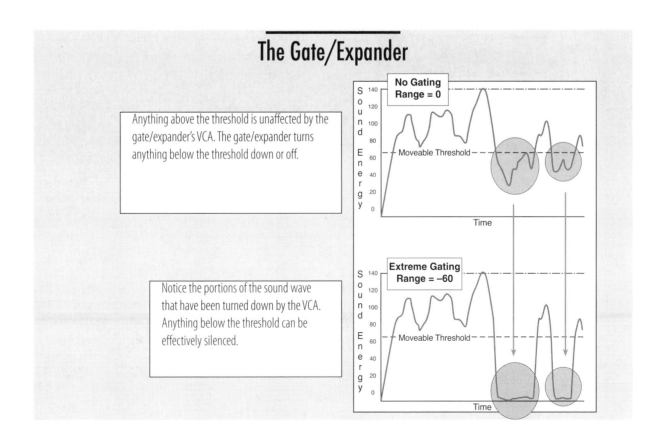

The Gate/Expander

Anything above the threshold is unaffected by the gate/expander's VCA. The gate/expander turns anything below the threshold down or off.

No Gating Range = 0

Sound Energy — Moveable Threshold — Time

Notice the portions of the sound wave that have been turned down by the VCA. Anything below the threshold can be effectively silenced.

Extreme Gating Range = –60

Sound Energy — Moveable Threshold — Time

Patching Effects Processors

It's best to connect the output of your mixer's aux bus or effects send bus to the input of the effects unit. Next, connect the output of the effect to the mixer's effects return or into an available mixer channel.

When using effects, keep the original track dry, blending the 100% wet return with it for the best musical impact. Wet/dry adjustments are made from within the effects device.

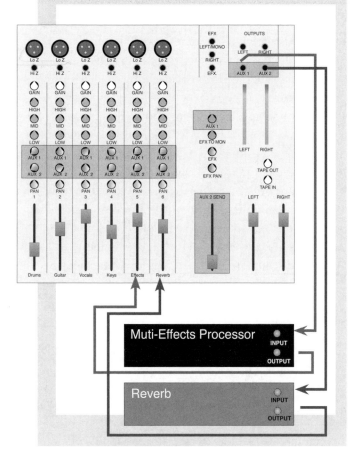

isn't playing, the gate turns the channel off, eliminating the noise.

WARNING: If the threshold is not set correctly, portions of the musical ingredient could be completely lost. The threshold must be set so that it opens anytime the musician plays or sings; however, if the threshold is set too high, some of the more quiet and subtle passages might not be strong enough to open the gate.

Effects Processors

Effects processors add the third dimension to a mix. Room size and complexity is indicated by the way sound reacts in an acoustical space. The echoes and delays that happen after the original sound emanates from the source tell the brain what the surrounding environment is like. All of the effects processors (echoes, reverberation, and chorus effects) revolve around one thing: the delay. In their simplest form, all of these effects are merely combinations of single or multiple delays combined together in proper proportion with the original signal.

Wet Versus Dry

Wet and *dry* are two terms that refer to the amount of effected signal that is blended with the original dry signal. The relationship between wet and dry is quantified in a percentage; 100% wet refers to a signal that contains none of the original (dry) signal. A sound that is completely dry has none of the effect return combined with it (0% wet). An equal combination of the wet and dry signals is referred to as 50% wet.

Patching Effects Devices

It's best to connect the output of your mixer's aux bus or effects send bus to the input of the effects unit. Next, connect the output of the effect to the mixer's effects return or into an available mixer channel.

Most effects processors have a meter on the input for proper level adjustment, and many effects processors have a final output level adjustment.

When using effects it's always desirable to keep the channel track dry and blend the 100% wet return with it for the best musical impact and control. In a small setup you might have to run the effects inline, doing all of the blending from dry to wet within the effects unit. This can work well, but it's best to keep the dry and wet controls separate.

Simple Delay Effects

A delay does just what its name says: It hears a sound and then waits for a while before it reproduces it. Current delays are simply digital recorders that digitally record the incoming signal, and then play it back with a time delay selected by the user. Delay parameters vary from unit to unit, but most delays have a range of delay length from a portion of a millisecond up to one or more seconds. This is called the *delay time* or *delay length* and is typically variable in increments of a millisecond.

Almost all digital delays are much more than simple echo units. Within the delay are all of the controls you need to produce slapback, repeating echo, doubling, chorusing, flanging, phase shifting, some primitive reverb sounds, and any hybrid variation you can dream up.

Slapback Delay

The simplest form of delay is called a *slapback*. The slapback delay is a single repeat of the signal. Its delay time is anything above about 35 ms. Any single repeat with a delay time of less than 35 ms is called a *double*.

To achieve a slapback from a delay, simply adjust the delay time and turn the delayed signal up, either on the return channel or on the mix control within the delay.

Calculating Delay Times

Delays are an important part of creating a professional-sounding mix. It's usually best if the delays are in time with the music—it helps reinforce the groove. Most modern delays let the sound operator tap a button in time with the music to set the delay speed.

Calculating the delay time per beat is simple, especially if you're recording to a Digital Audio Workstation, using the built-in sequencer as a click. Use the formula 60,000 ÷ bpm to find the length of one beat. Here's the logic. If you keep it tucked away in your memory banks, you'll never need to look at a sheet of numbers in a grid again.

- There are 1000 ms/second.
- There are 60 seconds/minute.
- Therefore, there are 60,000 ms/minute.
- Tempos are stated in beats per minute (bpm).
- Therefore, the total number of ms/minute (60,000) divided by the number of beats in a minute derives the number of ms per beat.

There are typically four beats per minute. Delays that work well, in support of the musical groove, are in time with the quarter note, eighth note, sixteenth note, or eighth- and sixteenth-note triplets.

To calculate these subdivisions of the beat, divide the ms/beat:

- By 1.5 to calculate the quarter note triplet value.
- By 2 to calculate the eighth note value.
- By 3 to calculate the eighth note triplet value.
- By 4 to calculate the sixteenth note value.
- By 6 to calculate the sixteenth note triplet value.

For a single slapback delay, feedback and modulation are set to their off positions. Slapback delays of between 150 ms and about 300 ms are very effective and common for creating a big vocal or guitar sound.

Audio Example 11-9 demonstrates a track with a 250-ms slapback delay.

Slapback delays between 35 and 75 ms are very effective for thickening a vocal or instrumental sound.

Audio Example 11-10 demonstrates a track with a 50-ms delay.

Slapback delay can be turned into a repeating delay. This smoothes out the sound of a track even more and is accomplished through the use of the regeneration control, which is also called *feedback* or *repeat*.

Regeneration feeds the delay back into the input of the delay unit, so we hear the original, the delay, and then a delay of the delayed signal. The higher you turn up the feedback, the more times the delay is repeated. Practically speaking, anything past about three repeats gets too muddy and does more musical harm than good.

The vocal track in Audio Example 11-11 starts with a simple single slapback, then the feedback increases until we hear three or four repeats.

Why does a simple delay make the sound so much bigger and better? Delay gives the brain the perception of listening in a larger, more interesting environment. As the delays combine with the original sound, the harmonics of each part combine in interesting ways. Any pitch discrepancies are averaged out as the delay combines with the original signal. If a note was sharp or flat, it's hidden when heard along with the delay of a previous note that was in tune. This helps most vocal sounds tremendously and adds to the richness and fullness of the mix.

The human brain gets its cue for room size from the initial reflections, or repeats, that it hears off surrounding surfaces. Longer delay times indicate to the brain that the room is larger. The slapback is really perceived as the reflection off the back wall of the room or auditorium as the sound bounces back (slaps back) to the performer. Many great lead vocal tracks have used a simple slapback delay as the primary or only effect. Frequently, this delay sounds cleaner than reverb and has less of a tendency to intrusively accumulate.

Slapback delay, often called *echo*, is typically related in some way to the beat and tempo of the song. The delay is often in time with the eighth note or sixteenth note, but it's also common to hear a slapback in time with the quarter note or some triplet subdivision. The delay time affects the rhythmic feel of the song. A delay that's in time with the eighth note can really smooth out the groove of the song, or if the delay time is shortened or lengthened just slightly, the groove may feel more aggressive or relaxed. Experiment with slight changes in delay time.

It's easy to find the delay in milliseconds for the quarter note in your song, especially when you're working from a sequence and the tempo is already available on screen. Simply divide 60,000 by the tempo of your song (in beats per minute). 60,000 ÷ bpm = delay time per quarter note in milliseconds (in common time).

Most modern delay effects devices provide a tap tempo button so the sound engineer merely taps along with the beat of the song to set the delay time. More complex devices actually let the user tap in several notes in a unique rhythm, which the delay then mimics on all incoming audio signals.

Regenerating the delay often smooths out the slapback effect, essentially creating multiple echoes or repeats. It's common to use a delay with two to five (or more) delays. This has a blending effect on most mixes.

Doubling/Tripling

Combining a single delay of less than 35 ms with the original track is called *doubling*. Combining two separate delays of less than 35 ms with the original track is called *tripling*. The short delay(s) can combine with the original track to sound like two people (or instruments) on the same part.

When doubling and tripling, use prime numbers for delay times. A prime number can only be divided by one and itself (for example, 1, 3, 5, 7, 11, 13, 17, 19, 23, 29, and so on). Using prime numbers minimizes the potential hollowness caused by phase problems between the original and delayed signals, typically providing a warmer sound that sounds better in stereo and mono.

Slapback Delay and Reflections

Sound travels at the rate of about 1120 ft./sec. To calculate the amount of time (in seconds) it takes for sound to travel a specific distance, divide the distance (in feet) by 1120 (ft./sec.):

time = distance (ft.) ÷ speed (1120 ft./sec.).

In a 100′ long room, sound takes about 89 ms to get from one end to the other (100÷1120). A microphone at one end of this room wouldn't pick up the slapback until it completed a round trip (about 178 ms after the original sound).

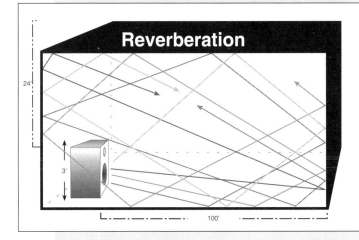

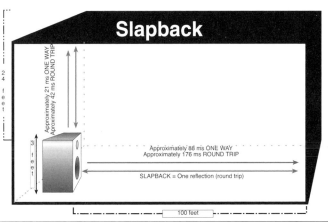

Modulation

The modulation control on a delay is for creating chorusing, flanging, and phase shifting effects. The key factor here is the LFO (*low-frequency oscillator*); its function is to continually vary the delay time. The LFO is usually capable of varying the delay from the setting indicated by the delay time to half of that value and back. Sometimes the LFO control is labeled *modulation*.

As the LFO is slowing down and speeding up the delay, it's speeding up and slowing down the playback of the delayed signal. In other words, modulation actually lowers and raises the pitch in exactly the same way that a tape recorder does if the speed is lowered and raised. Audio Example 11-12 demonstrates the sound of the LFO varying the delay time. This example starts subtly, with the variation from the original going down slightly, then back up. Finally, the LFO varies dramatically downward, then back up again.

Audio Example 11-12

The LFO

On most usable effects, these changes in pitch are slight and still within the boundaries of acceptable intonation, so they aren't making the instrument sound out of tune. In fact, the slight pitch change can have the effect of smoothing out any pitch problems on a track.

As the pitch is raised and lowered, the sound waves are shortened and lengthened. We know that when two waveforms follow the same path, they sum together. The result is twice the amount of energy. We also know that when two waveforms are out of phase, they work against and cancel each other either totally or partially.

When the modulation is lengthening and shortening the waveform and the resulting sound is combined with the original signal, the two waveforms continually react together in a changing phase relationship. They sum and cancel at varying frequencies. The interaction between the original sound and the modulated delay can simulate the sound we hear when several different instrumentalists or vocalists perform together. Even though each member of a choir tries his or her hardest to stay in tune and together rhythmically, choir members are continually varying pitch and timing. These variations are like the interaction of the modulated delay with the original track. The chorus setting on an effects processor is simulating the sound of a real choir by combining the original signal with the modulated signal.

Delay Settings for Various Delay Effects

Effect	Delay A	Delay B (Stereo)	LFO	Speed	Regeneration	Phase
Slapback	35–350ms		No	No	No	No
Echo (Repeats)	35–350ms		No	No	2–10	No
Reverb	15-35ms	15-35ms	No	No	Several	No
Doubling	1-35ms		No	No	No	No
Tripling	1-35ms	1-35 ms	No	No	No	No
Phase Shifter	0.5-2ms	0.5-2 ms	Yes	Low	Medium	Yes/No
Flanger	10-20ms	10-20ms	Yes	Low	Medium	Yes/No

The speed control adjusts how fast the pitch raises and lowers. These changes might happen very slowly, taking a few seconds to complete one cycle of raising and lowering the pitch, or they might happen quickly, raising and lowering the pitch several times per second.

Audio Example 11-13 demonstrates the extreme settings of speed and depth. It's obvious when the speed and depth controls are changed here. Sounds like these aren't normally used, but when we're using a chorus, flanger, or phase shifter, this is exactly what is happening, in moderation.

Audio Example 11-13
Extreme Speed and Depth

Phase Shifter

Now that we're seeing what all these controls do, it's time to use them all together. Obviously, the delay time is the key player in determining the way that the depth and speed react. If the delay time is very, very short, in the neighborhood of 1 ms or so, the depth control will produce no pitch change. When the original and affected sounds are combined, we hear a distinct sweep that sounds more like an EQ frequency sweeping the mids and highs. With these short delay times, we're really simulating waveforms, moving in and out of phase, unlike the larger changes of singers varying in pitch and timing. The phase shifter is the most subtle, sweeping effect, and it often produces a swooshing sound.

Audio Example 11-14 demonstrates the sound of a phase shifter.

Audio Example 11-14
Phase Shifter

Flanger

A flanger has a sound similar to the phase shifter, except it has more variation and color. The primary delay setting on a flanger is typically about 20 ms. The LFO varies the delay from near 0 ms to 20 ms and back continually. Adjust the speed to your own taste.

Flangers and phase shifters work very well on guitars and Rhodes-type keyboard sounds. They provide a rich blend and interesting harmonic motion.

Audio Example 11-15 demonstrates the sound of a flanger.

Audio Example 11-15
Flanger

Chorus

The factor that differentiates a chorus from the other delay effects is, again, the delay time. The typical delay time for a chorus is about 15 to 35 ms, with the LFO and speed set for the richest effect for the particular instrument voice or song. With these longer delay times, as the LFO varies, we actually hear a slight pitch change. The longer delays also create more of a difference in attack time. This also enhances the chorus effect. Since the chorus gets its name from the fact that it's simulating the pitch and time variation that exists within a choir, it might seem obvious that a chorus works great on background vocals. It does. Chorus is also an excellent effect for guitar and keyboard sounds.

Audio Example 11-16 demonstrates the sound of a chorus.

Audio Example 11-16

Chorus

Phase Reversal and Regeneration

The regeneration control can give us multiple repeats by feeding the delay back into the input so that it can be delayed again. This control can also be used on the phase shifter, chorus, and flanger. Regeneration, also called *feedback*, can make the effect more extreme or give the music a sci-fi feel. As you practice creating these effects with your equipment, experiment with feedback to find your own sounds.

Most units have a phase reversal switch that inverts the phase of the affected signal. Inverting the phase of the delay can cause very extreme effects when combined with the original signal (especially on phase shifter and flanger effects). This can make your music sound like it's turning inside out.

Audio Example 11-17 starts with the flanger in phase. Notice what happens to the sound as the phase of the effect is inverted.

Audio Example 11-17

Inverting Phase

Stereo Effects

The majority of effects processors are stereo, and with a stereo unit, different delay times can be assigned to the left and the right sides. If you are creating a stereo chorus, simply set one side to a delay time between 15 and 35 ms, then set the other side to a different delay time, between 15 and 35 ms. All of the rest of the controls are adjusted in the same way as a mono chorus. The returns from the processor can then be panned apart in the mix for a very wide and extreme effect. Listen as the chorus in Audio Example 11-18 pans from mono to stereo.

Audio Example 11-18

Stereo Chorus

For a stereo phase shifter and flanger, use the same procedure. Simply select different delay times for the left and right sides.

Understanding what is happening within a delay is important when you're trying to shape sounds for your mix.

Sometimes it's easiest to bake a cake by simply pressing the Bake Me a Cake button, but if you are really trying to create a meal that flows together perfectly, you might need to adjust the recipe for the cake. That's what we need to do when building a mix; we must be able to custom fit the ingredients.

Reverberation Effects

As we move from the delay effects into the reverb effects, we must first realize that reverb is just a series of delays. In fact, modern reverberation devices are capable of all delay effects. However, some devices are limited to producing either delay or reverberation effects. Also, many software plug-ins specifically focus on a single function, often in emulation of classic hardware.

Reverberation is simulation of sound in an acoustical environment, such as a concert hall, gymnasium, or bedroom. No two rooms sound exactly alike. Sound bounces back from all the surfaces in a room to the listener or the microphone. These bounces are called *reflections*. The combination of the direct and reflected sound in a room creates a distinct tonal character for each acoustical environment. Each one of the reflections in a room is like a single delay from a digital delay. When it bounces around the room, we get the effect of regeneration. When we take a single short delay and regenerate it many times, we're creating the basics of reverberation.

Reverb must have many delays and regenerations working together in the proper balance, combining to create a smooth and appealing room sound.

Envision thousands of delays bouncing (reflecting) off thousands of surfaces in a room and then back to you, the listener. That's what's happening in the reverberation of a concert hall or any acoustical environment. There are so many reflections happening in such a complex order that we can no longer distinguish individual echoes.

Accurate and believable digital simulation is accomplished by producing enough delays and echoes to imitate the smooth sound of natural reverb in a room. The reason different reverb settings sound unique is because of the different combinations of delays and regenerations. The mathematic calculation and relations of the delays involved in a reverberation sound is called an *algorithm*.

A digital reverb is capable of imitating a lot of different acoustical environments and can do so with amazing clarity and accuracy. The many different echoes and repeats produce a rich and full sound. Digital reverbs can also shape many special effects that would never occur acoustically. In fact, these sounds can be so much fun to listen to that it's hard not to overuse reverb.

Keep in mind that sound perception is not just two-dimensional, left and right. Sound perception is at least three-dimensional, with the third dimension being depth (distance). Depth is created by the use of delays and reverb. If a sound (or a mix) has too much reverb, it loses the feeling of closeness or intimacy and sounds like it's at the far end of a gymnasium. Use enough effect to achieve the desired results, but don't overuse effects.

Most digital reverberation devices offer several different sounds. These are usually labeled with descriptive names such as halls, plates, chambers, rooms, and so on.

Hall Reverb

Hall indicates a concert hall sound. These are the smoothest and richest of the reverb settings, with complex, long delay times that blend together to form a smooth decay over time. Typical hall algorithms have a decay time longer than two seconds, although user-adjustable controls allow for unnatural settings on hall sounds or any of the basic sounds.

Chamber Reverb

Chambers imitate the sound of an acoustical reverberation chamber, sometimes called an *echo chamber*. Acoustical chambers are fairly large rooms with hard surfaces. Music is played into the room through high-quality, large speakers, and then a microphone in the chamber is patched into a channel of the mixer as an effects return. Chambers aren't very common now that technology is giving us great sounds without taking up so much real estate. The sound of a chamber is smooth, like the hall's, but it has a few more mids and highs.

Plate Reverb

Plates are the brightest-sounding of the reverbs. These sounds imitate a physical plate reverb. A true plate is a large sheet of metal (about 4' by 8') suspended in a box and allowed to vibrate freely. A speaker attached to the plate itself induces sound onto the plate. Two contact microphones are typically mounted on the plate at different locations to provide a stereo return. The sound of a true plate reverb has lots of highs and is very clean and transparent.

Room Reverb

A room setting imitates many different types of rooms that are typically smaller than hall or chamber sounds. These can range from a bedroom to a large conference room or a small bathroom with lots of towels to a large bathroom with lots of tile.

Rooms with lots of soft surfaces have little high-frequency content in their reverberation. Rooms with lots of hard surfaces have lots of high-frequency content in their reverberation.

Reverse Reverb

Most modern reverbs include reverse or inverse reverb. These are simply backward reverb. After the original sound is heard, the reverb swells and stops. It is turned around. These can actually be fairly effective if used in the appropriate context.

Gated Reverb

Gated reverbs have a sound that is very intense for a period of time, and then closes off quickly. They offer a very big sound without overwhelming the mix.

Though at one time this was a trendy, popular sound, the technique has been around for a long time. The original gated reverb sound actually used a room mic, distant from the source, in a large room patched through a gate. The trigger for the gate to open was set to the side chain, where a mic close to the source was patched. This was common on snare drums at one time. There was a close mic on the snare patched to the mixer and also patched into the trigger input of the gate, so when the snare was hit, the gate opened and you could hear the large sound of the room microphone(s). When the snare wasn't being hit, the room mics were off.

Other Variations of Reverberation

There are many permutations of the reverberation sounds. You might see bright halls, rich plates, dark plates, large rooms, small rooms, or bright phone booths, but they can all be traced back to the basic sounds of halls, chambers, plates, and rooms.

These sounds often have adjustable parameters. They let us shape the sounds to our music so that we can use the technology as completely as possible to enhance the artistic vision. We need to consider these variables so that we can customize and shape the effects.

Audio Example 11-19

Reverberation Variations

Video Example 11-2

Various Acoustic Space Recordings

Reverberation Effects Parameters

The sound operator must be familiar enough with the available parameters in any effects device to be able to customize the effects during sound check or even during the performance.

Predelay

Predelay is a time delay that happens before the reverb is heard. This can be a substantial time delay (up to a second or two) or just a few milliseconds. The track is heard clean (dry) first, so the listener gets more of an up-front and close feel, then the reverb comes along shortly thereafter to fill in the holes and add richness.

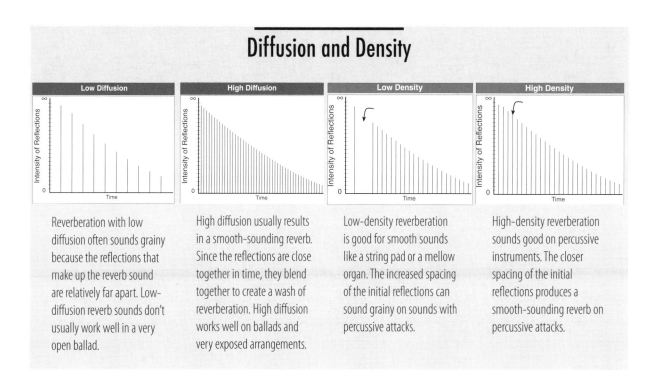

Diffusion and Density

Low Diffusion

Reverberation with low diffusion often sounds grainy because the reflections that make up the reverb sound are relatively far apart. Low-diffusion reverb sounds don't usually work well in a very open ballad.

High Diffusion

High diffusion usually results in a smooth-sounding reverb. Since the reflections are close together in time, they blend together to create a wash of reverberation. High diffusion works well on ballads and very exposed arrangements.

Low Density

Low-density reverberation is good for smooth sounds like a string pad or a mellow organ. The increased spacing of the initial reflections can sound grainy on sounds with percussive attacks.

High Density

High-density reverberation sounds good on percussive instruments. The closer spacing of the initial reflections produces a smooth-sounding reverb on percussive attacks.

Diffusion

Diffusion controls the space between the reflections. A low diffusion is equated with a very grainy photograph. We might even hear individual repeats in the reverb. A high diffusion is equated with a very fine-grain photograph, and the sound provides a very smooth wash of reverb.

Decay Time

Reverberation time, reverb time, and decay time all refer to the same thing. Traditionally, reverberation time is defined as the time it takes for the sound to decrease to one-millionth of its original sound pressure level. In other words, it's the time it takes for the reverb to go away.

Decay time can typically be adjusted from about 1/10 of a second up to about 99 seconds. We have ample control over the reverberation time.

Density

The density control adjusts the initial short delay times. Low density is good for smooth sounds, such as strings or organs. High density works best on percussive sounds.

Audio Example 11-20

Reverberation Parameters

MICROPHONE PRINCIPLES AND DESIGN

There's much more to mic choice than finding a trusted manufacturer that you can stick with. There's much more to mic placement than simply setting the mic close to the sound source. The difference between mediocre audio quality and exemplary audio quality is quite often defined by the choice and placement of microphones.

The microphone is our most fundamental tool. You can have $100,000 worth of esoteric, vintage, high-tech gear in your signal path, but if the microphone doesn't capture the desired sonic essence, it's all a waste. Each microphone offers a sonic personality and offers the potential to be much more than just an archival tool. For instance, if you and a buddy test 10 mics on Joe to see how they'll work for his new song, nine of the microphones might evoke agreement that, yeah, that sounds like Joe. However, chances are that one of the 10 mics might get the response, "Wow! Joe sounds great on this mic!"

Once you find the microphone that sounds great for whatever the sound source, it's time to compare other options in the signal path, such as preamplifiers, compressors, or equalizers—the fewer additions to the signal path between the mic and the speakers, the better. Musically, you need to do what you need to do. As long as it feeds the passion and emotion of the music, it's all right to include a hundred processors in the signal path; however, if you want to capture the true essence of the original sound at the source, find the perfect mic and the perfect preamp, and keep it simple.

If you need to use compression, make it a conscious choice, and be sure the compressor is enhancing the musical impact.

Audio Example 12-1

Six Different Types of Microphones on the Same Source

Video Example 12-1

Five Different Microphones on the Same Source

Using a mic to capture sound is not as simple as just selecting the best mic. There are two other critically important factors involved in capturing sound:

- Where we place the mic in relation to the sound source
- The acoustical environment

As you'll see in the Audio Examples in this book, the sound of the acoustical environment plays a very important role in the overall sound quality.

Selecting a microphone involves more than doing a simple random search for "the sound." Our understanding of mics and how they work, along with the ability to read and understand their specifications, allows us to make educated predictions regarding which mics to consider. You should be able to listen to a source, and then make an intelligent decision about which mics to audition.

Technically, we must consider a set of factors when choosing a microphone: directional characteristic, operating principle, response characteristic, and output characteristic. In addition, the real-world considerations are always cost, durability, and appearance. If you intend to make great music and are serious about your craft, always be in search of great-sounding microphones.

Directional Characteristic

Anytime you mic a source, you must be aware of which way to aim the mic, and whether that particular mic is sensitive only to the intended source. Understanding the microphone *polar response pattern*, also called the *pickup pattern*, is fundamental to capturing the essence of the sound you want. There are three basic polar patterns we consider when comparing and choosing microphones: omnidirectional, bidirectional, and unidirectional.

We typically refer to the overall pickup in our discussions; however, each polar response will vary dramatically when frequency ranges are compared. In addition, there are variations of the directional characteristics. Because of the way most directional characteristics are created, high frequencies are the most directional, and low frequencies are least directional.

The polar response pattern is sometimes a result of the microphone capsule's inherent operational principle—more often, it is the result of the physical housing.

Polar Response Graph

The polar response graph provides a visual image of the microphone's sensitivity to sound coming from different directions. This circular graph indicates sensitivity in a 360-degree circular scope and is interpreted as a three-dimensional image, even though it's drawn two-dimensional for the sake of simplicity.

On-Axis

The position at zero degrees on the graph represents the front of the microphone (the portion designed to pick up the sound). This position is referred to as *on-axis*.

Off-Axis

Any position on the polar response graph that isn't on-axis is called *off-axis* and is quantified in degrees. 180 degrees on the graph represents the back of the mic (the part directly in back of the portion designed to pick up the sound); it's referred to as *180 degrees off-axis*.

A microphone that demonstrates a decrease in sensitivity at a certain point on the polar graph is said to discriminate from sound at that point. We indicate that decrease in sensitivity

Polar Patterns

Polar graphs are seen in two dimensions, but they imply three dimensions. All patterns should be visualized in a spherical three-dimensional plane.

Omnidirectional

Cardioid

by denoting its degree marker. A microphone with reduced sensitivity at 180 degrees on the graph is said to exhibit 180-degree *off-axis discrimination*.

Additionally, for the sake of indicating discrimination, the graph is considered symmetrical. Indicating that a microphone exhibits 150 degree off-axis discrimination indicates that the point of least sensitivity is at 150 degrees. Because the graph is symmetrical, there would also be a point of least sensitivity at 210 degrees. This also indicates that there is sensitivity, to some degree, in the region centered on 180 degrees off-axis.

Each microphone uses a design that exhibits a unique polar response throughout the frequency spectrum.

Because of the symmetrical aspect of the polar graph, sometimes polar patterns are indicated as 0–180 degrees left and right instead of 360 degrees around. This is actually a simpler system; because off-axis discrimination is always referenced as 180 degrees or less, symmetry across the 0-degree axis is implied.

Polar Response Graph

The polar response graph plots the spatial sensitivity as it relates to the position of the sound source in relation to the microphone capsule. These graphs are considered symmetrical in relation to the plotted sensitivity and, in addition, should be considered three-dimensionally spherical.

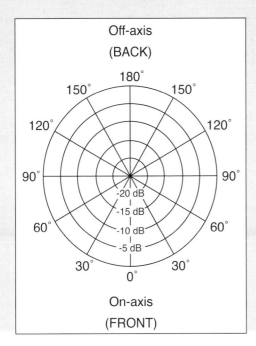

Sensitivity Scale

The polar response graph calibrates spherical sensitivity through a series of concentric circles. Each concentric circle is consecutively smaller. The outer circle indicates full sensitivity (0 dB decrease in sensitivity), and there are typically four or five consecutively smaller circles between

Polar Shapes

The two most basic polar response patterns are omnidirectional (doesn't discriminate against sounds from any direction) and cardioid (discriminates against sounds that are 180 degrees off-axis). The other two polar shapes in this illustration are bidirectional (an omnidirectional pattern on each side of the mic) and hypercardioid (a bidirectional pattern with a large half and a small half).

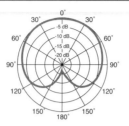

Sometimes the on-axis position is noted at the top of the graph; other times it's noted at the bottom. In either case, the 0° position is always the front of the mic.

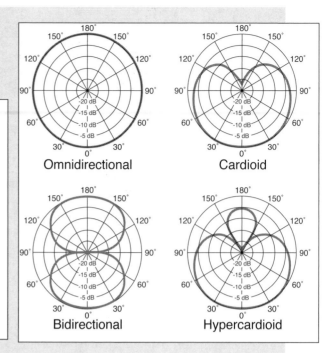

Omnidirectional

Cardioid

Bidirectional

Hypercardioid

the outer circle and the center point, each typically indicating a decrease in sensitivity of 5 dB. Plotting the decreases of sensitivity around the polar graph is what creates the polar pattern.

Some polar graphs include an additional outer circle, at +5 dB, for the rare instance that certain frequencies sum, creating a hypersensitivity that exceeds the normal full sensitivity.

Normally, the pattern on the graph, which is visually dominant and uses a solid, bold line, is the average overall pattern and is typically measured from a 1-kHz sine wave. Many electronic measurements consider a 1-kHz sine wave as the reference, or average, signal across the audible spectrum.

Omnidirectional

An omnidirectional mic, sometimes referred to simply as *omni*, hears equally from all directions. It doesn't reject sound from anywhere. An omnidirectional pickup pattern provides the fullest sound from a distance. Omni microphones are very good at capturing room ambience, recording groups of instruments that you can gather around one mic, and capturing a vocal performance while still letting the acoustics of the room interact with the sound of the voice.

Omnidirectional microphones are inappropriate in a live setting because they produce feedback more quickly than any other pickup pattern.

Omnidirectional Pickup Pattern

Mics with an omnidirectional pickup pattern pick up sound equally from all directions and don't reject sound from any direction.

Multiple Frequencies and Symmetry on the Polar Graph

The polar graph often displays the directional characteristic for multiple frequencies. To accomplish this in the least cluttered manner, all patterns are assumed to be symmetrical across the Y axis. In addition, to help clarify the results, various line styles are incorporated on each frequency. Sometimes the polar graph is split, like the graph on the left, to highlight the variations in frequency response; other times the graph is whole, like the graph on the right, with the pattern variations simply changing between left and right.

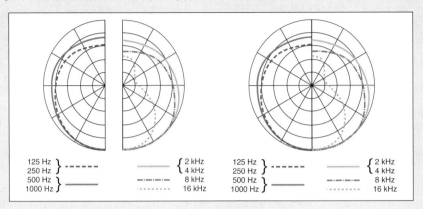

| 125 Hz | 250 Hz | 500 Hz | 1000 Hz | | 2 kHz | 4 kHz | 8 kHz | 16 kHz |

Bidirectional Pickup Pattern

Bidirectional microphones hear equally well from both sides, but they don't pick up sound from the edge.

Bidirectional mics work very well for recording two voices or instruments to one track. This is also called a figure-eight pattern.

Cardioid Pickup Pattern

A microphone with a cardioid pickup pattern hears sound best from the front and actively rejects sounds from behind. With its heart-shaped pickup pattern, you can point the mic toward the sound you want to record and away from the sound you don't want to record.

Bidirectional

Bidirectional microphones hear equally from the sides, but they don't hear from the edges. Bidirectional microphones are an excellent choice for recording two sound sources to one track with the most intimacy and the least adverse phase interaction and room sound. Position the mic between the sound sources for the best blend.

Like omnidirectional microphones, bidirectional mics tend to cause feedback problems in a live application.

Unidirectional

Most microphones demonstrate a unidirectional characteristic, often called a *directional* or *cardioid pickup pattern*. The pickup pattern is visually represented by a heart shape—rounded in front and dimpled in the back. The unidirectional mic is most sensitive (hears the best) at the part of the mic into which you sing; it's least sensitive (hears the worst) at the side opposite the part into which you sing.

The advantage to using a microphone with a cardioid pickup pattern lies in the ability to isolate sounds. You can point the mic at one instrument while you're pointing it away from another instrument. The disadvantage to a cardioid pickup pattern is that it will typically only give you a full sound from a close proximity to the sound source. Once you're a foot or two away from the sound source, a cardioid pickup pattern produces a very thin-sounding rendition of the sound you're miking.

In a live sound setting, directional mics are almost always best because they produce far less feedback than mics with omnidirectional or bidirectional pickup patterns.

There are five directional pickup patterns for consideration in normal use: cardioid, supercardioid, hypercardioid, ultracardioid, and subcardioid. Here is a comparison of their fundamental polar response patterns. Each microphone is unique in design and may exhibit its own rendition of these response patterns. Also, keep in mind that results may vary throughout the frequency spectrum.

Cardioid Pickup Pattern

A microphone with a cardioid pickup pattern hears sound best from the front and actively rejects sounds from behind. With its heart-shaped pickup pattern, you can point the mic toward the sound you want to record and away from the sound you don't want to record.

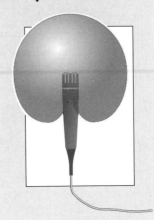

Cardioid

The cardioid pickup pattern demonstrates full response at the front of the microphone and a decrease in sensitivity of up to 25 or 30 dB at 180 degrees off-axis. In relation to a cardioid pickup pattern, the supercardioid and hypercardioid pickup patterns each become progressively narrower on the sides, with an increased area of off-axis sensitivity.

Supercardioid

A microphone with a supercardioid polar response is more directional at the front than a microphone exhibiting a cardioid pattern, with a decreased sensitivity on the sides and an area of sensitivity about 170 degrees off-axis.

Hypercardioid

A microphone with a hypercardioid polar response exhibits a high degree of directionality at the front, with a decrease of about 12 dB on the sides and an area of least sensitivity at about 110 degrees off-axis.

Ultracardioid

A microphone with an ultracardioid polar response is very focused and directional in front with a small area of sensitivity at 90 degrees and 180 degrees. A focused polar pattern, such as the ultracardioid, offers greater off-axis rejection.

The subcardioid polar response is wider and extends further than the cardioid pattern, approaching the non-directionality of an omnidirectional microphone.

Practical Applications

In a live sound application, base your choice of microphones with varying directional characteristics on a few considerations:

- Does the mic sound great?
- Where are the monitors positioned?
- How close together are the sound sources?

The live sound operator must consider several aspects of his or her unique situation in order to choose the best mic. The sound of the mic is a consideration, but it doesn't matter how great the mic sounds in a controlled environment if it's a feedback nightmare through the sound system. Most large-diaphragm mics provide the most accurate and pleasing results in a recording environment, but they are seldom the best choice for live use. The large diaphragm and increased sensitivity typically provide a boomy sound with heightened susceptibility to feedback at moderate to high volumes.

A true cardioid polar pattern provides minimal sensitivity 180 degrees off-axis. These mics are typically best for use when the monitors are directly in front of the instrumentalist or singer, because directly off-axis is the point that will provide minimal feedback.

Hyper-. super-, and ultracardioid mics provide a narrower pickup pattern than the traditional cardioid mic, and their areas of least sensitivity are typically with 10 to 70 degrees of 180 degrees off-axis. As an example, because of their decreased width in effective pickup area, these polar patterned microphones would provide better separation between backing vocalists that are grouped closely together. In addition, they will provide less

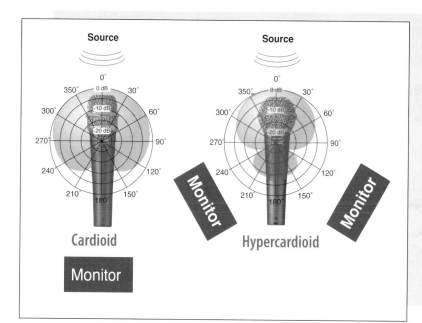

Cardioid

Hypercardioid

Choosing Monitor Positions

Position the stage monitors as closely as possible to the microphone's area of maximum off-axis discrimination. When using a cardioid microphone the least amount of feedback is likely to be created with the monitor directly behind the mic. When using a hypercardioid microphone, position the monitors between 120 and 150 degrees off-axis for the least amount of feedback.

feedback when the monitors are aimed from the sides, ideally from the point of least off-axis sensitivity.

Operating Principle

Although there are hundreds of different microphones available from a lot of manufacturers, they essentially all fit into three basic categories: condenser, moving-coil, and ribbon. Condenser and moving-coil mics are the most common of these three, although they may all be used in recording, as well as live, situations.

There are other types of microphones with operating principles that differ from what we will cover in this course, and each type of microphone has its own individual personality. Mic types other than condenser, moving-coil, and ribbon are usually selected for a special effect in a situation in which the music needs a unique sound that enhances the emotional impact of the song.

Transducer Types

A transducer is any device that transforms one type of energy into another type of energy. For instance, a speaker converts electrical energy into acoustic energy. The amplifier sends an electronic signal (a continuously varying flow of electrons) to the speaker, which responds to the electronic signal by moving air, which is, at that point, acoustical energy. Your ear is another example of a transducer because it converts the acoustic energy into electrical energy, which is then sent to the audio perception portion of your brain.

Whether in a recording or live sound application, the microphone is the first transducer, past the source, in our signal path. The microphone converts acoustic energy into electrical energy. There are other transducers that might have been involved prior to the actual sound source. For our purposes in audio, we'll consider the microphone as the first transducer within our control.

Of the three types of microphone we'll study, there are only two types of transducers: *magnetic induction* and *variable capacitance*. The transducer is the actual microphone capsule—the point at which the acoustic energy from the sound source reaches the mic and begins the flow of electrons. The microphone might contain amplifying circuitry, which is insignificant in our understanding of transducer functionality.

Magnetic Induction Transducers

A magnetic induction transducer utilizes a process in which metal, which has magnetic properties (in other words, it can be magnetized), is stimulated into motion around, or is attached to, a magnet. Of the three mic types that we'll study, two use magnetic transducers: moving-coil and ribbon.

Mics that use magnetic induction, whether moving-coil or ribbon, are *dynamic microphones*. Often, moving-coil mics are generically referred to as dynamic microphones, and ribbon microphones are differentiated as…ribbon microphones. It is more accurate to differentiate them as moving-coil and ribbon mics.

Variable capacitance transducers operate on an electrostatic, rather than a magnetic, principle. A variable capacitance microphone capsule utilizes a fixed, solid conductive plate adjacent to flexible piece of plastic that's been coated with a conductive alloy, often containing gold.

A *capacitor* is a device that stores an electrical charge. When the movable plate is electrically charged, an electrostatic charge is stored between the two surfaces. As sound waves vibrate the alloy-coated plastic diaphragm, the area of stored electrical charge emits a continuously varying flow of electrons that accurately portray the waveform. Variable capacitance transducers require power to charge the plates and to power an amplifying circuit within the microphone.

A microphone that use a variable capacitance transducer is sometimes called a *capacitance* microphone, although more often it's referred to as a *condenser* microphone. The old-school name for a capacitor was *condenser*—same device, different name.

Operating Principle of the Moving-Coil Mic

A moving-coil microphone operates on a magnetic principle. When an object that can be magnetized is moved around a magnet, there is a change in the energy within the magnet. There is also a continual variation in the magnetic status of the object moving in relation to the magnet. The moving-coil microphone uses this fact to transfer the changing air pressure, produced by an audio waveform, into a continually varying flow of electrons that can be received by the mic preamp.

In a moving-coil mic, a coil of thin copper wire is suspended over a fixed magnet, enabling the coil to move up and down around the magnet. A thin Mylar plastic diaphragm closes the top of the coil and serves to receive the audio waves. As the crests and troughs of the continually varying audio waveform reach the diaphragm, the coil is forced to move

Moving-Coil Microphones

Copper wire is wrapped into a cylinder. This cylinder is then suspended around a magnet. The copper coil moves up and down in response to pressure changes caused by sound waves.

The crest of the audio wave moves the coil down, causing a change in the coil magnetism. The trough of the audio wave moves the coil up, again causing a change in the coil magnetism.

As the coil moves around the magnet, it receives a continually varying magnetic image. The continually varying magnetism will ideally mirror the changing air pressure from the sound wave. This continually varying magnetism is the origin of the signal that arrives at the mixer's mic input.

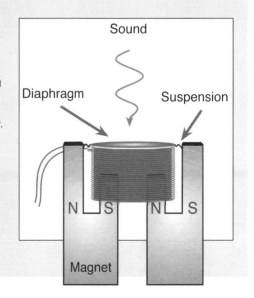

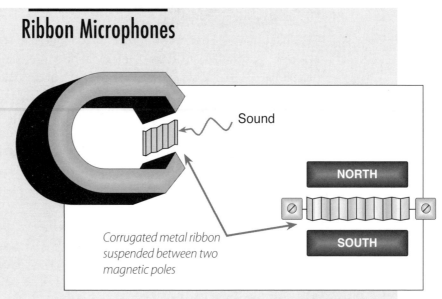

Ribbon Microphones

A thin metal ribbon suspended between two poles of a magnet vibrates in response to each crest and trough of a sound wave. As the ribbon moves in the magnetic field, it continually varies in its magnetism. These changes of magnetism are the origin of the signal that is sent to the mic input of your mixer.

The signal produced by the ribbon is typically weaker than the signal produced by the moving-coil. In practical terms, that means you'll usually need more preamplification at the mic input to achieve a satisfactory line-level signal.

Sound

NORTH

SOUTH

Corrugated metal ribbon suspended between two magnetic poles

around the magnet. The movement of the copper coil around the magnet is what causes the changing flow of electrons that represents the sound wave.

Operating Principle of the Ribbon Mic

A ribbon microphone operates on a magnetic principle like the moving-coil. A metallic ribbon is suspended between two poles of a magnet. As the sound wave vibrates the thin ribbon, the magnetic flow changes in response, causing a continually varying flow of electrons. As the ribbon moves between the poles of the magnet, it is being magnetized in varying degrees of north and south magnetism, in direct proportion to the changes in amplitude produced by the sound wave. This continually varying flow of electrons is the origin of the signal that reaches the microphone input of your mixer.

Historically, ribbon microphones have been very fragile; the ribbons needed to be a certain length to generate enough signal strength, and they needed to be thin enough to respond accurately to sonic nuance. Therefore, vintage ribbon mics, such as the RCA 77 DX, constantly need maintenance, although they sound great.

Modern ribbon microphones are capable of using smaller, stronger magnets, which enables the use of shorter ribbons. This has resulted in the production of more durable ribbon mics, although they still are the most fragile of the three types we use.

Because ribbon mics operate on a magnetic principle, they don't require a power source to operate, although there are some new ribbon mics that use phantom power to drive internal amplification circuitry—phantom power can be damaging to ribbon mics other than these. A vintage microphone ribbon tends to act like a fuse when it receives phantom power—it blows. (Pop, kaput, gone, send in for replacement…get my point?)

Operating Principle of the Condenser Mic

A condenser mic operates on a fairly simple premise, although it is physically based on a different principle. Whereas the moving-coil and ribbon microphones operate on a magnetic inductance principle, the condenser mic is based on a variable capacitance principle.

Condenser Microphone Capsule

The diaphragm of the condenser mic is very thin and vibrates in response to sounds. It is lightly coated with a metallic alloy so that it can conduct electricity. The crest of a sound wave moves the metal-coated plastic membrane inward. The trough moves it outward. The moveable plate is continually responding to the varying air pressure caused by the sounds around it.

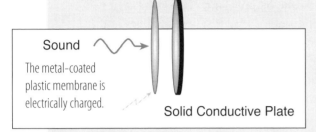

Sound

The metal-coated plastic membrane is electrically charged.

Solid Conductive Plate

The moveable plate responds to the crest and trough of a sound wave by moving inward and outward, creating a variance in the capacitance.

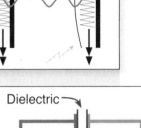

When positive and negative terminals (left) are shorted together, current flows virtually unimpeded, heating up and burning the power supply out.

Battery or Power Supply

The dielectric interrupts the short circuit. The diagram to the right represents a simplified capacitor (condenser). The electrons stored in the system are influenced by changes in the distance between the plates (vibrations of the diaphragm).

Dielectric

Battery or Power Supply

From the power supply, the positive terminal is connected to the moveable plate (diaphragm), and the negative terminal is connected to the backplate. If the diaphragm and backplate of the condenser capsule were connected together, the battery or power supply would short out (heat up, explode, catch on fire, and so on). The distance between the plates determines the amount of electrical charge that is stored on each plate. When sound waves vibrate the diaphragm, the motion causes a variation in the distance between the plates. This, in turn, influences the electrical charges to move on and off of the plates, ideally in direct proportion to the air-pressure variations in the sound wave. The electric current caused by the varying capacitance is the audio signal.

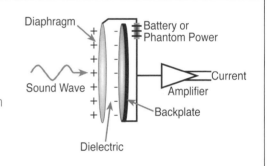

Diaphragm · Battery or Phantom Power · Sound Wave · Current · Amplifier · Backplate · Dielectric

A charged (positive or negative) electrical current is applied to a metal-coated piece of plastic. The plastic is a little like the plastic wrap you keep on your leftover food. The metallic coating is thin enough to vibrate in response to sound waves; in fact, the technique used to apply the coating to the membrane is called *sputtering* because the alloy is so lightly applied. Its function is to provide conductivity for the electrical charge while not inhibiting the flexibility of the plastic membrane. The ingredients of the alloy vary from manufacturer to manufacturer, but the key factor is conductivity—it must be able to carry an electrical charge.

The metal-coated plastic will vibrate when it's subjected to an audio wave because of the physical reality called *sympathetic vibration*. The principle of sympathetic vibration says that if it is possible for a surface to vibrate at a specific frequency, it will vibrate when it is in the presence of a sound wave containing that frequency. The metal-coated plastic membrane, the *diaphragm*, in a condenser microphone must be able to sympathetically vibrate when in the presence of any audio wave in our audible frequency spectrum.

This metal-coated piece of plastic is positioned close to a solid piece of metallic alloy called a *backplate*. As the moveable plate is electrically charged, or *polarized*, electrical energy begins to accumulate between the two metallic surfaces. The area of electrically charged air between the diaphragm and the backplate is called the *dielectric*.

As the crest and trough of a sound wave meet the thinly coated plastic, the plastic vibrates sympathetically with the sound wave. As the diaphragm vibrates, the area between the solid metal surface and the moveable metal surface changes. These changes in the dielectric create a discharge of electrical current. This electrical discharge exactly represents the changing energy in the sound wave. In other words, you have an electrical version of the acoustic energy you started with at the sound source.

Because there is very little mass in the condenser microphone's metal-coated membrane, it responds very quickly and accurately when in the presence of sound. Therefore, the condenser capsule is very efficient at capturing sounds with high transient content, as well as sounds with interesting complexities.

The signal that comes from the capsule is very weak and must be amplified to mic level. Most condenser mics use transistors in the internal amplifying circuitry; transistors provide a very clean and accurate amplification. Some condenser microphones utilize a vacuum tube instead of the transistor because of the smooth and warm sound it produces. Many of these tube microphones are well respected and highly acclaimed.

Once the signal from the mic reaches the mixer, it's boosted to line level at the input preamp.

Phantom Power

The capsule of a condenser microphone requires power to charge the metal-coated membrane. Power is also required to amplify the signal from the capsule up to microphone level.

Some condenser microphones will house a battery to power the capsule and the amplifying circuitry. However, *phantom power* provides a more efficient way to get power to the condenser mic because it's efficient, constant, and reliable. The phantom power supply is typically in the mixer. The power is sent to the mic through the balanced mic cable.

If you use batteries to power a condenser mic, always be sure the batteries are fresh and that they're supplying sufficient voltage to optimally run the microphone's circuitry.

If your mixer doesn't provide phantom power, use a commercially manufactured external phantom power supply. It receives 120-volt AC current and transforms it to the proper DC voltage and amperage. External mic preamplifiers also supply phantom power.

Phantom power does not pose an electrical danger to the user because it is low voltage and very low amperage DC current. Phantom power voltage is typically 48 volts, although it can range from around 11 volts to 48 volts. Each condenser mic draws current from the phantom power supply based on its electrical needs.

Amperage is the force behind current measured in a unit called an *amp*. A milliamp is one thousandth of an amp. Condenser mics draw a very low-amperage DC current, ranging from less than 1 mA (less than one one-thousandth of an amp) to about 12 mA. By comparison, a typical household circuit carries 15 to 50 amps of 120-volt AC current.

Phantom power has no adverse effect on the audio signal being carried by the mic cable. The DC voltage is applied equally to pins 2 and 3 of the XLR connection relative to pin 1, which is at ground potential. The fact that it functions undetected in the background on the same cable the mic signal travels on explains the term "phantom" power.

Vintage tube mics often don't require phantom power from the mixer because the power supply is external to the mic. The external power supply receives 120-volt AC current, which provides power to the external amplifying circuitry; the charging voltage for the capsule element is provided by the external power supply. The mic connects to the power supply, and then the power supply connects to the mixer input.

Electret Condenser Microphones

An electret condenser microphone utilizes a permanently charged capsule, which doesn't require phantom power. However, power is still required to operate the internal preamp. Phantom power can still be used to power the microphone, but the decreased electrical requirements make this condenser mic efficient while receiving battery power. Consequently, electret condensers are an excellent choice for application in the field. They possess all the sonic benefits of the condenser design with a realistic expectation that the battery power will provide sufficient longevity.

Comparison Between Moving-Coil, Ribbon,and Condenser Microphones

If you possess the basic understanding of each mic type, and if you have a grasp of how each type works, you'll be able to make very good microphone selections. The microphone you select for your specific recording situation makes a big difference in the sound of the final recording. It's almost pathetic how easy it is to get great sounds when you've selected the right mic for the job and you've run the mic through a high-quality preamp.

Whereas the diaphragm of the microphone is the vibrating membrane that responds to sound waves, its makeup plays a key role in the inherent ability of the microphone to provide an accurate version of the sound it receives. Because we know that the moving-coil capsule utilizes a membrane attached to the top of a coil of copper wire, and because the sound wave must move the entire assembly around a magnet, we can draw the simple deduction that, by nature of its mass, it is physically less responsive than either the ribbon or the condenser capsule. In fact, this deduction is true. Condenser mics are the most accurate and responsive of the three mic types, and ribbon mics are typically more accurate than moving-coil mics. Though there might be anomalies to this comparison, it is generally accepted.

Moving-Coil Mics

Though moving-coil mics don't excel in capturing transients and subtleties, you can still take advantage of their tendencies and characteristics.

Moving-coil mics are the standard choice for most live situations, but they are also very useful in the studio. Here are some examples of popular and trustworthy moving-coil microphones:

- Shure SM57, SM58, SM7, Beta 57, Beta 58, Beta 52, Beta 56
- Electro-Voice RE20
- Sennheiser 421, 441
- Audio-Technica ATM25, Pro-25
- AKG D12, D112, D3500, D1000E
- Beyer M88

Moving-coil mics are the most durable of all the mic types. They also withstand the most volume before they distort within their own circuitry.

A moving-coil mic typically colors a sound more than a condenser mic. This coloration usually falls in the frequency range between about 5 kHz and 10 kHz. As long as we realize that this coloration is present, we can use it to our advantage. In our studies on EQ, we've found that this frequency range can add clarity, presence, and understandability to many vocal and instrumental sounds.

Moving-coil mics have a thin sound when they are more than about a foot from the sound source. They're usually used in close-mic applications, with the mic placed anywhere from less than an inch from the sound source up to about 12 inches from the sound source.

Because moving-coil mics can withstand a lot of sound pressure, they sound the best in close-mic applications. And because they add high-frequency edge, they're good choices for miking electric guitar speaker cabinets, bass drum, snare drum, toms, or any loud instrument that benefits from a close-mic technique. Use them when you want to capture lots of sound with lots of edge from a close distance and you aren't as concerned about subtle nuance and literal accuracy of the original waveform.

Moving-coils are also used in live performances for vocals. They work well in close-miking situations, add high-frequency clarity, and are very durable.

Condenser Microphones

Condenser microphones are the most accurate. They respond to fast attacks and transients more precisely than other types, and they typically add the least amount of tonal coloration. The large vocal mics used in professional recording studios are usually examples of condenser mics. Condenser mics also come in much smaller sizes and interesting shapes. Some popular condenser mics are:
- Shure KSM 44, KSM 32, KSM 27, KSM 141, KSM 137, KSM 109, SM 82, Beta 87
- Neumann U87, U89, U47, U67, TLM170, KM83, KM84, KM184, TLM193
- AKG 414, 451, 391, 535, C1000, 460, C3000, C-12, the Tube
- Electro-Voice BK-1
- Sennheiser MKH 40, MKH 80
- B&K 4011
- Blue Microphones Bluebird, Bottle, Cactus, Kiwi, Mouse, Dragon Fly, Blueberry, Baby Bottle
- Audio-Technica 4033, 4050, 4047, 4060, 4041
- Milab DC96B
- Schoeps CMC 5U
- Groove Tube MD-2, MD-3

- Crown PZM-30D

Use a condenser microphone whenever you want to accurately capture the true sound of a voice or instrument. Condensers are almost always preferred when recording:

- Acoustic guitar
- Acoustic piano
- Vocals
- Real brass
- Real strings
- Woodwinds
- Percussion
- Acoustic room ambience

Condenser microphones (especially in omni configuration) typically capture a broader range of frequencies from a greater distance than the other mic types. In other words, you don't need to be as close to the sound source to get a full sound. This trait of condenser microphones is a great advantage in the recording studio because it enables us to record a full sound while still including some of the natural ambience in a room. The further the mic is from the sound source, the more influential the ambience is on the recorded sound. This advantage in the recording studio is precisely the cause for problems in a live application, in which the farther the source is from the mic, the greater the chance for feedback.

Condenser microphones that work wonderfully in the studio often provide poor results in a live sound reinforcement situation. Because they have a flat frequency response, these condenser mics tend to feed back more quickly than microphones designed specifically for live sound applications (especially in the low-frequency range). There are many condenser mics designed for sound reinforcement, and there are many condenser mics that work very well in either setting. Condenser mics often have a low-frequency roll-off switch that lets you decrease low-frequency sensitivity. In a live audio situation, the low-frequency roll-off is effective in reducing low-frequency feedback.

Ribbon Mics

Ribbon mics are the most fragile of all the mic types. This one factor makes them less useful in a live sound reinforcement application, even though ribbon mics produced within the last 10 or 15 years are much more durable than the older classic ribbon mics.

The ribbon capsule is inherently bidirectional. Both the front and back of the ribbon are equally sensitive, and sound from 90 degrees off-axis cancels. Many manufacturers take advantage of this natural characteristic and produce bidirectional ribbon mics. On the other hand, there are several ribbon mics that exhibit a unidirectional characteristic; these mics utilize a ribbon with the back (180 degrees off-axis) enclosed. Once the back of the ribbon is enclosed, the capsule is inherently omnidirectional, like the moving-coil and condenser capsules.

Ribbon mics, exhibiting a unidirectional polar characteristic, are like moving-coil mics in that they color the sound source by adding a high-frequency edge, and they generally have a thin sound when used in a distant miking setup. When used as a close mic, ribbon microphones can have a full sound that is often described as being warmer and smoother than a moving-coil.

There are some great-sounding ribbon microphones available. Some of the commonly used ribbon microphones are:

- Beyer M160, M500
- RCA 77-DX, 44 BY, 10001
- AEA R84, R44C, R88
- Royer SF-12, SF 24

Ribbon mics are fragile and need to be used in situations in which they won't be dropped or jostled.

The Shaping of the Pickup Pattern

The inherent polar response characteristic of the moving-coil and basic condenser capsules is omnidirectional. Set in space with no physical housing, they are equally sensitive to sound from all directions—they don't reject sound from any direction. The inherent polar response characteristic of the ribbon capsule is bidirectional. Set in space, with no physical housing, it is equally sensitive to sound from the front and back, but it is not sensitive to sounds coming from the sides—it rejects sounds 90 and 270 degrees off-axis.

Whereas the moving-coil and condenser capsule are inherently omnidirectional, and the ribbon capsule is inherently bidirectional, in design and application the majority of microphones ever manufactured exhibit a cardioid polar response. Polar response characteristics are designed and created in two fundamental ways: physical housing design and electrical combinations of multiple capsules.

Creating the Cardioid Pickup Pattern

A microphone with selectable pickup patterns typically uses a variance in relative polarizing voltage between the front and back sides of the condenser capsule to shape the pattern between omnidirectional, bidirectional, and variations of the cardioid patterns. Other mic designs utilize the physical design of the mic housing to influence the polar pattern.

The ports on the sides of the housing provide an alternate pathway to the mic capsule for off-axis sounds. As sounds travel around the mic, though the ports, and arrive at the front and back of the diaphragm, they reduce in level because of phase cancellation. The position and quantity of the ports determines the specific frequencies that are rejected most.

The microphone below is a Shure KSM141. It switches from cardioid to omnidirectional characteristic by sliding an internal cylinder up to cover the ports.

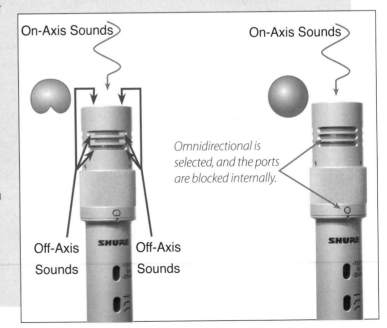

Physical Housing Design

Most microphone designs use the physical housing around the capsule to shape directional characteristics. The concept is simple once you understand phase interactions between sound waves—amplitudes in phase sum, and amplitudes out of phase cancel. For physical shaping of directional characteristics, the ribbon capsule is omnidirectional because the back of the ribbon is enclosed.

Microphones that contain a capsule at the top of a barrel housing with slots (we could also say openings or ports) around the capsule end exhibit a unidirectional polar response. These ports shape the directional characteristic. On-axis sound waves stimulate the diaphragm in the normal way; however, off-axis sound waves are allowed to reach the front and/or back of the diaphragm via multiple routes provided by the ports around the housing. As the off-axis waveforms combine at the diaphragm, the fact that they've arrived at the same point (the diaphragm) through various pathways (around the mic and through the network of ports) indicates that the length of their journey varies with the pathway. Because the same off-axis waveform has been split, and because each pathway is a different distance from the origination of the waveform, phase interaction is built into the design. The key in the mic design is the positioning and quantity of the ports. The intent of the designer is to allow negative phase interaction of off-axis waveforms through the series of ports in the physical housing as they travel multiple off-axis pathways.

If the ports are covered up, these microphones become omnidirectional rather than cardioid. If fact, some manufacturers offer multiple capsules for use with the same

Creating Polar Patterns Electrically

Large-diaphragm studio condenser mics with selectable patterns utilize a double-sided capsule with two moveable plates. The plates are charged with positive or negative polarity, in varying amounts, to shape virtually any polar response pattern. Some of these mics actually offer external control over polar response, which infinitely varies between patterns. The engineer, in the control room, shapes the microphone response to match the room and the source.

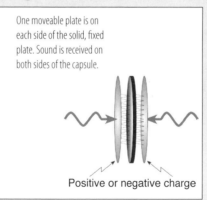

One moveable plate is on each side of the solid, fixed plate. Sound is received on both sides of the capsule.

Positive or negative charge

Selectable
Polar Patterns

Polarity requirements for each polar response pattern:

- Applying a positive charge to both moveable plates produces an omnidirectional response characteristic for the capsule.
- Applying a positive charge to one plate and a negative charge to the other produces a bidirectional pattern.
- Varying the relative intensity of the charge between the two plates, as well as changing the backside plate from positive to negative, produces any variation or permutation of cardioid, bidirectional, and omnidirectional characteristics.

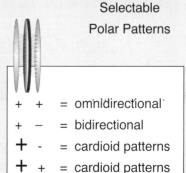

+	+	= omnidirectional
+	−	= bidirectional
+	-	= cardioid patterns
+	+	= cardioid patterns

microphone body. The capsules typically screw on and off, and their only physical difference is that the cardioid capsule has ports and the omni capsule doesn't.

The large-format condenser microphone controls directional characteristic electrically rather than through the design of the physical housing. These microphones are designed for and used in the studio and are rarely seen in a live setting; they're large, expensive, fragile, and adversely affected by environmental conditions such as humidity, smoke, wind, and so on.

Many of these microphones provide multiple pickup patterns, selectable by a switch on the mic or, externally, on a remote control. Selecting different pickup patterns causes a change in electrical polarization; nothing is altered as far as the physical housing is concerned. These microphones achieve multiple directional characteristics by incorporating a double-sided capsule. This is merely an extension of the condenser capsule we discussed earlier; however, in this design there are two moveable plates, one on either side of the fixed backplate.

In this design, the microphone's directional characteristic is controlled through the application of varying amounts of positive and negative electrical charges to the two moveable plates:

- When both plates receive a positive charge, the mic exhibits omnidirectional characteristics.

- When the front plate receives a positive charge and the back receives a negative charge, the mic becomes bidirectional.

- As the intensity of the charging voltage varies between the two plates, the pickup pattern can be shaped at will, ranging from omnidirectional to bidirectional to cardioid, and on many mics multiple patterns in between. Most of these mics let the user select between preset pickup patterns through the use of a switch on the front of the mic. A few manufacturers offer a remote control for pattern selection, with some even offering a continuously variable balance of polarizing voltage. With this control, you can evaluate the sound in the control room, capturing the balance of the direct and ambient sound that makes the most musical impact.

Some microphone designs actually incorporate two capsules to create directionality. An omnidirectional pickup pattern combined with a bidirectional pickup pattern produces a cardioid response. A mic like the Shure KSM9 uses a condenser capsule with patteren selection between the two common live polar configurations: cardioid and hypersardioid.

The Proximity Effect

It's important to understand the *proximity effect*. The result of the proximity effect is this: As the microphone gets closer to its intended source, the low-frequency range increases in relation to the high frequencies. It's not uncommon to see a rise of 20 dB at 100 Hz as the source gets close to the mic. Anyone who has used a handheld mic has probably recognized that if he or she gets closer to the microphone, the sound becomes bigger, louder, and more bass-heavy. This happens as a result of the proximity effect.

The proximity effect is most pronounced when using a microphone with a cardioid or bidirectional pickup pattern; it is least pronounced when using a microphone with an omnidirectional pickup pattern.

In effect, as a result of the proximity effect, the low-frequency range increases in relation to the high-frequency range. In reality, as the microphone moves closer to the source—a face, for instance—reflections from the source reflect back to the capsule out of phase, canceling more and more of the upper frequencies. The reflections not only cancel at the diaphragm, but they also enter through the ports, from behind the capsule, and cancel. This explains why there is less of a problem with proximity effect when using an omnidirectional microphone—there's less cancellation because there are no ports.

Most mics designed for live application make use of the proximity effect—they're designed with deficiencies in the low frequencies until they're used in a close-miking setting. Once the mics are close to the source, the sound becomes full and warm; however, when used from a distance they provide a sound that is thin and edgy.

Compensating for the Proximity Effect

To help compensate for the proximity effect, many microphones have a built-in user-selectable high-pass filter. When you're close-miking anything in which the sound becomes to bass-heavy, simply apply the high-pass filter. A typically high-pass filter sets the cutoff frequency at 75 or 80 Hz. Some mics even let the user determine the cutoff frequency, typically offering various choices between 60 and 250 Hz.

Using the Proximity Effect to Our Advantage

The proximity effect isn't necessarily a bad characteristic of a microphone design, especially when we consider the effect it has on the close-miked sound in relation to the microphone's frequency response characteristic. Often, microphones with a unidirectional pickup characteristic exhibit a decreased sensitivity in the low frequencies, especially in a distant-miking application, and they're likely to exhibit an increased sensitivity in the high frequencies between 4 and 8 kHz.

Once we understand these tendencies, we realize that many microphones used in a close-miking application benefit from close proximity positioning. The increased low frequencies fill out an otherwise thin sound, and the extra sensitivity in the high-frequency range helps clean up the sound, resulting in greater understandability, presence, and clarity.

Response Characteristic

Almost any microphone responds to all frequencies we can hear, plus frequencies above and below what we can hear. The human ear has a typical frequency response range of about 20 Hz to 20 kHz. Some folks have high-frequency hearing loss, so they might not hear sound waves all the way up to 20 kHz, and some small children might be able to hear sounds well above 20 kHz.

Frequency Response Curve

For a manufacturer to say that their microphone has a frequency range of 20 Hz to 20 kHz tells us absolutely nothing until they tell us how the mic responds throughout that frequency range. A mic might respond very well to 500 Hz, yet it might not respond very well at all to frequencies above about 10 kHz. If that were the case, the sound we captured to tape with that mic would be severely colored.

We use a *frequency response curve* to indicate exactly how a specific microphone responds to the frequencies across the audible spectrum. The frequency response curve is the line on the graph that indicates the microphone's ability to reproduce frequencies across the audible spectrum. As the sensitivity to a frequency increases, the curve ascends; as the sensitivity decreases, the curve descends.

A microphone that is equally sensitive to all frequencies across the audible spectrum is represented by a flat-line curve on the 0 dB line—this is called a *flat* frequency response.

If a frequency response curve shows a peak at 5 kHz, we can expect that the mic will color the sound in the highs, likely producing a sound that has a little more aggressive sound than if a mic with a flat response was used. If the frequency response curve shows the low frequencies dropping off sharply below 300 Hz, we can expect the mic to sound thin in the low end unless we move it close to the sound source to proportionally increase the lows.

A frequency response graph often contains more than one curve. The main, and typically more solid and bold curve, represents the on-axis frequency response. Other curves on the graph represent off-axis frequency response and sometimes comparisons between the responses in multiple sound fields.

Transient Response

The frequency response curve is one of the most valuable tools to help us predict how a mic will sound. What the frequency response curve doesn't tell us is how the mic responds to transients. We can predict the transient response of a mic based on what we already know about the basic operating principles of the different mic types. Therefore, condenser mics are expected to more accurately capture the transient, with moving-coil and ribbon mics lagging behind.

There is no real specification that quantifies a microphone's transient response. Ah! I guess there still is room for listening in this game of microphone choice!

Audio Example 12-3

Demonstration of Microphone Transient Response Characteristics

Output Characteristic

The output characteristic of a microphone quantifies factors such as noise, sensitivity, overload limits, and impedance.

Equivalent Noise Rating/Self-Noise

A microphone's equivalent noise level, also referred to as *self-noise*, indicates the sound pressure level that will create the same voltage as the noise from the microphone.

Frequency Response Curve

A mic with a flat frequency response adds very little coloration to the sound it picks up. Many condenser microphones have a flat, or nearly flat, frequency response. This characteristic, combined with the fact that they respond very well to transients, makes condenser mics very accurate.

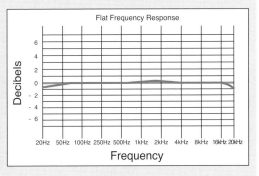

The mic represented by the curve at the right isn't very good at recording low frequencies and it produces an abundance of signal at about 4 kHz. Though this mic wouldn't be very accurate, we could intelligently use a mic like this if we wanted to record a sound with a brutal presence. Many moving-coil microphones have this kind of frequency response curve. Moving closer to the mic helps fill out the low frequencies.

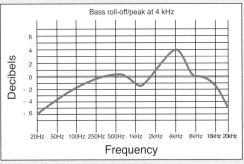

Moving-coil and ribbon microphones are very quiet. They contain passive circuitry, which poses very little noise potential, and they have a very low self-noise.

A condenser microphone produces more self-noise simply because it contains amplifying circuitry necessary to boost the capsule signal up to mic level. The amplifying circuitry, though adding to the mic's self-noise, provides a hotter signal to the mixer; therefore, it needs less preamplification at the mixer input. So, there's a bit of a tradeoff in terms of cumulative noise at the mixer output. There's slightly more mic noise and slightly less mixer noise.

Sensitivity

A microphone's sensitivity rating provides a way to compare microphone output levels. Sensitivity specifications are quantified as the microphone is given a specified frequency (typically 1 kHz) at a specified voltage. If two microphones receive the same acoustic signal, and one puts out more signal, it has a higher sensitivity rating.

Sensitivity ratings are sometimes expressed in dB, with the rating stated as a negative number—for example, –57 dB. This number quantifies the amount of boost required to amplify the signal from the mic to line level—with line level at 0 dB.

When we compare microphone output in this way, we can identify mics with stronger output signals because their sensitivity is closer to zero. A microphone with a sensitivity rating of –35 dB has a much stronger output relative to a microphone with a sensitivity rating of –60 dB.

Sensitivity ratings are also expressed as a voltage comparison, in Volts/Pascal. Air pressure at the mic capsule is specified in Pascals, and the resulting voltage from the microphone defines its sensitivity rating. If two microphones receive identical pressure (amplitude) and one puts out a higher voltage, that mic is said to have higher sensitivity. A Pascal is a standard pressure unit in the metric system, equal to one Newton per square meter—about 0.000145 pounds per square inch.

The only problem with sensitivity ratings is that manufacturers don't all use the same reference frequency or power rating. However, this specification is still useful in fundamental mic comparisons.

It is useful for us to compare the sensitivity ratings between moving-coil, ribbon, and condenser mics. When we compare apples to apples, we find these comparisons to be true:

*   Condenser microphones are typically the most sensitive, with ratings in the range of –30 dB to –40 dB or so.
*   Moving-coil microphones are next in line, with normal ratings in the –50 to –60 dB range.
*   Ribbon microphones are often the least sensitive, with ratings between –58 to –60 or so.

These comparisons are general and historically correct, considering classic microphone designs. Technology provides for extension of sensitivity in all the mic types. In fact, some of the modern ribbon mics actually receive phantom power in order to increase efficiency and to power an internal preamp much like the condenser mic.

Maximum SPL Rating

Most microphones can handle a lot of level (dB SPL) before they induce distortion. However, we sometimes need to mic loud instruments at close range, so distortion at the microphone can be an issue. It's important to be aware of the dB SPL where a specified percentage of distortion occurs (Total Harmonic Distortion). In addition, we need to realize the dB SPL, where the signal from the mic will clip, is the maximum SPL rating referring to peak SPL. Keep in mind that the peak SPL rating is typically 20 dB greater than the average, or RMS, rating.

Moving-coil mics are capable of handling a bunch of levels. They don't contain much other than the capsule, and they're usually capable of handling peak SPL well in excess of 140 dB SPL peak at acceptable distortion levels.

Condenser capsules are also capable of handling substantial level; however, the amplifying circuitry is likely to distort when subjected to loud sounds. Because of this, most condenser mics contain a pad, which decreases the signal strength from the mic capsule to its internal amplifying circuit (typically in 10-dB increments). Keep in mind that applying the pad diminishes the inherent signal-to-noise ratio by the amount of the pad, so implement it only when necessary.

Maximum SPL ratings always must be quantified at a specified total harmonic distortion (THD) and must be measured for the complete microphone (capsule and internal preamp). Most specifications relate maximum SPL to .05% THD, though some reference 1% THD.

The distortion of a circular capsule doubles with each 6-dB increase in level, so it's a simple matter to calculate distortion ratings in relation to the published specification. If a microphone specifies 140 dB SPL peak at 0.5% THD, it's implied that the same microphone will exhibit 1% THD at 146 dB SPL peak, or .25% THD at 134 DB SPL peak.

Impedance

In the modern recording world, microphones are low-impedance devices. Most low-impedance mics fall in the impedance range between 50 and 250 ohms. High-impedance microphones are not common today; they were designed and optimized for use with vacuum tube amplification. High-impedance mics fall in the impedance range between 20,000 and 50,000 ohms.

Mics with lower impedances, around 50 ohms, are more sensitive to electromagnetic hum and less susceptible to electrostatic interference compared to mics with impedances around 250 ohms. On the other hand, mics with impedances around 250 ohms are less sensitive to electromagnetic hum and more sensitive to electrostatic interference.

No matter what inherent noises mics with various impedance ratings are sensitive to, balanced low-impedance microphones are able to take advantage of the noise-canceling aspects of balanced circuitry; therefore, they benefit from the ability to run long cable lengths devoid of serious noise issues.

Stereo Mic Techniques

There are instances in which groups of instruments or voices must run through the FOH system, but using close-mic technique is either impractical or physically cumbersome. A choir, an orchestra, or a drum kit each provides an opportunity to use stereo miking. Unlike close-mic techniques, in which the microphone is within a foot of the source, most stereo miking is performed from a distance of a few feet or more.

There are some concerns whenever multiple microphones are configured in a distant-miking setup.

- Gain before feedback becomes more of an issue. It's best if the stereo mic setup is as close to the source as possible, while still allowing each mic to pick up a blend of the group rather than focusing on individuals.

- Destructive phase interactions are always a possibility when multiple microphones are used within earshot of a sound source.

- Sometimes, when miking large groups with combinations of multiple stereo mic configurations, the physical presence of mic stands and cables is awkward, if not dangerous. Permanent installations typically use condenser microphones, hung from the ceiling, to eliminate stage clutter and visual obstructions.

- Distant stereo mic techniques are effective at capturing a blend of several musicians at once; however, the mics are equally likely to capture any other sound on stage. Acoustic drums, percussion, guitar amplifiers, grand piano, and brass instruments are all very likely to bleed into any distant stereo mic setup—high frequencies and transients provide the biggest problem. Baffles around any of these instruments will radically decrease leakage while providing greater control over the stage volume—this will also help tighten the sound in the FOH system.

If your system is running in mono, stereo mic technique simply provides tighter control over the source than a single mic would. If your system is running in stereo, pan the mics apart for a wider sound. Depending on the size of the source and the physical arrangement of the instruments or voices, the FOH pan positions should be varied from hard-panned left

X-Y Configuration

On each of the following diagrams, 0 degrees represents the position of the sound source.

This is the most common stereo miking configuration. The fact that the microphone capsules are as close to the same horizontal and vertical axes as possible gives this configuration good stereo separation and imaging while also providing reliable summing to mono.

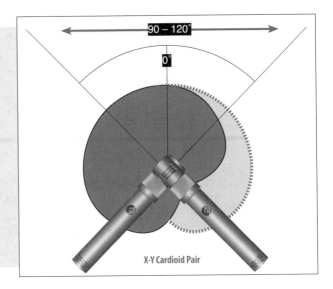

X-Y Cardioid Pair

and right to slightly panned apart. The sound operator should walk the room to evaluate the blend at several locations in the audience.

Often, stereo mic setups are integrated into the recoding of the event but not actually assigned to the main FOH system. Either way, it's important that the sound operator understands each of these techniques—they can provide excellent results in several different live and recorded sound scenarios.

Let's examine a few of the standard stereo miking configurations. Each one of these techniques is field-tested—they have proven functional and effective. Listen very carefully and analytically to these examples. Listen for left/right positioning and for the perception of distance. Are the instruments close or far away? Can you hear a change in the tonal character as the different sounds change position? Do you perceive certain instruments as being above or below other instruments? Can you hear the room sound? In other words, pick these recordings apart bit by bit.

The X-Y technique is one of the most commonly used stereo mic techniques. It works well in both live and studio applications. This technique uses two mics in a coincident physical relationship. Their capsules are as close together as possible, aimed across the same physical plane, and aimed at a 90-degree angle to each other. Listen very closely to the

Spaced Omni Pair

This configuration uses two omnidirectional mics. The ambience of the recording environment will color the sound of the recording. This setup is capable of capturing beautiful performances with great life—especially if the recording environment has an inherently good sound.

"D" on the diagram represents the distance from the center of the sound source to its outer edge. Notice that the distance from the center of the sound source to each microphone is one-third to one-half of D.

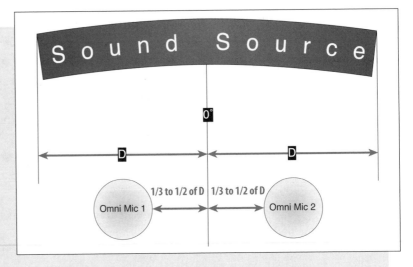

Spaced Omni Pair with Baffle

This technique retains much of the openness of the regularly spaced omni pair; however, the addition of a baffle between the microphones increases stereo separation. When miking a blended acoustical group or a stereo send from the multitrack of specific mix ingredients, this configuration provides a striking stereo image.

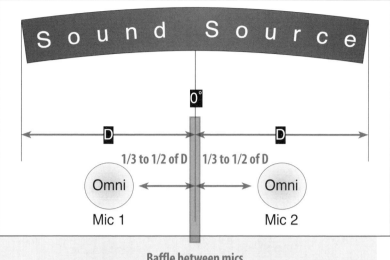

Baffle between mics

sound of each ingredient in the stereo recording. Listen to the changes in the sounds as they move around the room.

Audio Example 12-4

X-Y Configuration

The X-Y technique can be extended by separating the two microphones. They can be moved so that the capsules are a few inches or a few feet apart, depending on the width of the sound source. Always keep the mics angled about 90 degrees away from each other to maintain the stereo relationship. Never aim the two mics toward each other at the same source because the stereo image will be lost—two mics aimed at the same source functionally become a single mono mic with destructive phase interactions. Even though the X-Y pattern can be extended, the capsules should always remain along the same horizontal or vertical plane.

Crossed Bidirectional Blumlein Configuration

The crossed bidirectional configuration (also called the "Blumlein" configuration) has the advantage of being a coincident technique in that the overall sound isn't significantly degraded when the stereo pair is combined to mono. The sound produced by this technique is similar in separation to the X-Y configuration, but with a little more acoustical life.

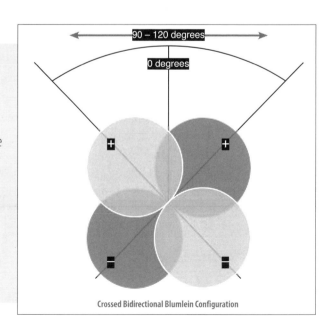

Crossed Bidirectional Blumlein Configuration

MS (Mid-Side) Configuration

Position the mid mic and the side mic in the closest proximity to each other possible. Both mikes should be along the identical vertical axis and as close as physically possible to the same horizontal axis—without touching. The MS (Mid-Side) technique is the most flexible of the stereo miking configurations. Its drawback is that it isn't simple to hook up. You must use a combining matrix that'll facilitate sending the sum of the mid and side mics to one channel and the difference of the mid and side mics to the opposite channel. In other words, you must be able to:

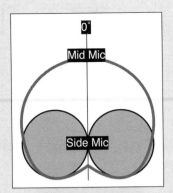

1. Split or Y the output of the mid mic and send it to both channels (or simply pan it to the center position).
2. Split or Y the output of the side (bidirectional) mic and send it to both channels.
3. Invert the phase of one leg of the side mic split. A leg is simply one side of the Y from the side mic.
4. Leave the other leg of the side mic split in its normal phase.
5. Adjust the balance between the mid mic and the side mics to shape the stereo image to your taste and needs.

High-quality, double-capsule stereo mics typically use this configuration. They demonstrate the advantages of coincident technique—minimal phase confusion between the two microphones. Also, and possibly more important, since the side mic signal is split to left and right—and left and right are made to be 180° out of phase with each other—when the stereo signal is sent through a mono playback system, the side mic information totally cancels. This leaves the mid mic signal as simple and pure as if it were the only mic used.

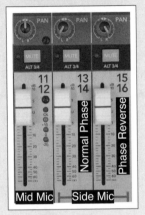

Two omnidirectional mics spaced between three and ten feet apart can produce a very good stereo image with good natural acoustic involvement. This is primarily a recording technique because the omnidirectional microphones are prone to causing feedback. When you are recording a small group, such as a vocal quartet, keep the mics about three feet apart; for larger groups increase the distance between the microphones. Use this technique only if the room has a good sound. In Audio Example 12-5, listen closely for the panning placement and perceived distance for each instrument. There's a definite difference in the apparent closeness of these percussion instruments.

Audio Example 12-5

Spaced Omni Pair

A variation of the spaced omni pair of mics involves positioning a baffle between the two mics, which increases the stereo separation and widens the image. This is primarily a recording technique because of the omnidirectional mics and the awkwardness of positioning a baffle on stage. Notice, in Audio Example 12-6, how clearly defined the changes are as the percussion instruments move closer to and farther away from the mics.

Audio Example 12-6

Spaced Omni Pair with a Baffle

The crossed bidirectional configuration uses two bidirectional mics positioned along the same vertical axis and aimed 90 degrees apart along the horizontal axis. This is similar

ORTF Configuration

The stereo mic technique developed by the Office de Radiodiffusion Télévision Française (ORTF) at Radio France provides a realistic stereo sound that offers a slightly wider image in comparison to the standard X-Y configuration. This technique produces excellent results when using two high-quality cardioid condenser microphones. The mics should be aimed 90–110 degrees apart and the distance between the two capsules should be 17 centimeters. Special mounting brackets help facilitate accurate and repeatable positioning of the mics for the ORTF configuration.

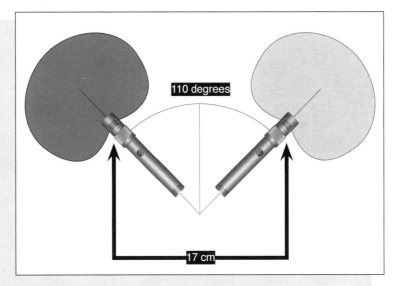

to the X-Y configuration in that it transfers well to mono, but the room plays a bigger part in the tonal character of the recording. This is primarily a recording technique.

Audio Example 12-7

The Crossed Bidirectional Configuration

The MS technique is the most involved of the techniques, but it's the best in terms of combining stereo to mono, and it also gives a very true and reliable stereo image. Most stereo mics contain two condenser capsules that are positioned in an MS configuration.

Audio Example 12-8

Mid-Side Configuration

The ORTF technique was devised around 1960 at the Office de Radiodiffusion Télévision Française (ORTF) at Radio France. It is similar in a way to the X-Y technique and it looks like a spaced X-Y with the stereo pair of mics pointing away from each other. However, the angle between the two microphones is specified as 110 degrees and the distance between them is 17 centimeters. As with nearly all stereo mic techniques, the ORTF technique works best when the stereo pair is identical and matched for frequency consistency.

chapter 13

WIRELESS SYSTEMS

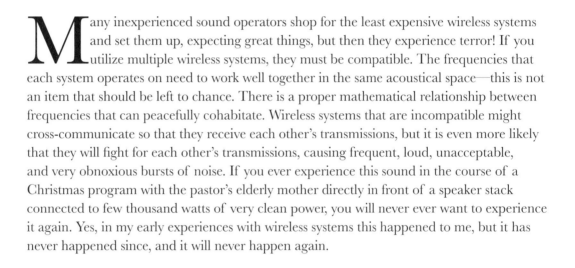

Many inexperienced sound operators shop for the least expensive wireless systems and set them up, expecting great things, but then they experience terror! If you utilize multiple wireless systems, they must be compatible. The frequencies that each system operates on need to work well together in the same acoustical space—this is not an item that should be left to chance. There is a proper mathematical relationship between frequencies that can peacefully cohabitate. Wireless systems that are incompatible might cross-communicate so that they receive each other's transmissions, but it is even more likely that they will fight for each other's transmissions, causing frequent, loud, unacceptable, and very obnoxious bursts of noise. If you ever experience this sound in the course of a Christmas program with the pastor's elderly mother directly in front of a speaker stack connected to few thousand watts of very clean power, you will never ever want to experience it again. Yes, in my early experiences with wireless systems this happened to me, but it has never happened since, and it will never happen again.

Components of the System

Wireless systems are simple to implement—they consist of three fundamental ingredients: the input source, the transmitter, and the receiver.

Input Device

Whether the input device is a lavaliere, a headset, or a handheld mic—or a guitar bass or other instrument—the wireless system simply receives the input signal, transmits it via radio frequencies, and accepts it at the receiver. Ideally, the receiver provides a signal at its output jack that is very close to the same signal that came from the input device.

Transmitter

When using a lavaliere, a headset, or an instrument, the wireless transmitter is located in a body pack that conveniently hangs on the user's belt, pants, skirt, or other clothing item. When using a handheld wireless mic, the transmitter is located in the microphone casing, typically at the end of the microphone body.

Complete Wireless System

Every wireless system must include these three ingredients:

- Input device
- Transmitter
- Receiver

The handheld wireless microphone includes the transmitter inside the mic body. The lavaliere and headset microphones must be connected to a body pack/transmitter combination.

In most systems it is very important that the transmitter has an unobstructed line of sight to the receiver. In fact, placing the body pack in someone's back pocket or holding the microphone at the bottom of the mic, around the transmitter, can easily result in unreliable signal transmission.

Receiver

The receiver listens solely to the frequency that is being transmitted from the body pack or microphone. When there is a good match between transmitted signal strength and the receiver's ability to accept the signal, everything functions seamlessly. Keep in mind that most receivers have no way of determining what the proper transmitter is, so if there is any device in the same proximity generating the frequency that the receiver is set to hear, it will hear it. These are radio frequencies. Any television or radio signal that's being transmitted at the target frequency will be picked up loudly by the receiver. This also applies to citizens' bands, walkie talkies, and such.

AM and FM Systems

Wireless systems utilize the same principle as any modern radio station transmission: radio waves. A radio wave is essentially a varying magnetic force in space. It is like a sound wave in that it is realized through a continual variation in energy; however, because the radio frequency is magnetic rather than acoustic, it doesn't require air to transmit sound pressure variations.

There are two types of radio transmission systems: amplitude modulation (AM) and frequency modulation (FM). The actual radio frequency isn't the audio information—it is really just the carrier for the audio information.

Amplitude Modulation

Amplitude modulation systems are less common than frequency modulated systems. In this scheme the carrier frequency is constant, but the amplitude changes in direct correlation to the amplitude changes in the audio waveform. For example, a 1-kHz tone causes the carrier frequency to change in amplitude 1,000 times per second. The amount of amplitude change correlates to the amplitude changes in the original tone.

This system is fundamentally simple—the audio signal is simply matched by the amplitude changes throughout the waveform. The actual specifications of the AM system are limited. Considering the maximum legal modulation, the usable bandwidth is between 50 and 9,000 Hertz and the maximum dynamic range is about 50 dB.

Frequency Modulation

A frequency-modulated radio signal uses a constant amplitude carrier at a specified frequency. The amount of amplitude determines the signal strength. The audio waveform causes a deviation (variation) in the frequency of the carrier, while the amplitude remains constant. For example, a 1-kHz tone modulates this carrier by speeding up and slowing down the carrier frequency 1,000 times per second. The amplitude of the audio source determines the amount of deviation from the original carrier frequency.

Almost all modern wireless systems utilize this FM principle—its potential for high-quality audio transmission is much greater than the AM principle. The maximum legal deviation from the carrier frequency provides for an audio frequency range between 50 and 15,000 Hz and a dynamic range of more than 90 dB.

VHF and UHF Systems

There are two primary bands that contain the accessible transmission and reception frequencies: the VHF (*Very High Frequency*) range and the UHF (*Ultra High Frequency*) range. Each of the bands is divided into a low and a high range.

- Low-band VHF is between 49 and 108 MHz. This band is not recommended for serious audio applications and is shared by cordless phones, walkie talkies, and radio-controlled toys. The band between 54 and 72 MHz is reserved for television channels 2–4.

- High-band VHF is between 169 and 216 MHz. Eight frequencies between 169 and 172 MHz have been specifically designated by the FCC for wireless microphone use. These are often called the "traveling frequencies" because they are used in most areas of the country. This band can become very crowded. It typically consists of forestry, power station, and Coast Guard communications; digital paging services; and other business and government operations. In the case of interference, the FCC has established primary and secondary radio frequency users. For example, television stations are primary users and all wireless mic systems are secondary. It is up to the secondary user to avoid interference from the primary user and not to propagate interference into their transmissions.

- Low-band UHF is between 450 and 806 MHz. This band is shared by land mobile radios and pagers, UHF television channels 14–69, and other business services.
- High-band UHF is between 900 and 952 MHz. This band is shared by cordless telephones, STL (studio-to-transmitter) links, and audio/video repeaters. The 900-MHz range has become a very common cordless phone band, which has rendered this range essentially useless for audio wireless systems.

VHF radio waves are longer than UHF waves and tend to be less affected by physical surroundings. UHF radio waves are shorter and more easily contained in physical areas. This fact makes the UHF systems more appropriate when multiple wireless systems cohabitate in a building, complex, or other acoustical space. In addition, the higher the frequency range, the shorter the receiver antennas. The low-band VHF antenna might be a meter long, whereas the upper frequency bands in the UHF range might use antennas that are about nine centimeters long.

Capture Effect

The wireless signal capture effect is similar to the concept of survival of the fittest—the strongest signal wins. Ideally, your intended signal from the transmitter to the receiver wins the struggle for success, but this is not always the case. Environmental, electrical, and physical considerations determine the effectiveness and accuracy of any wireless system. In some cases RF interference can dominate the frequency, and the intentionally transmitted signal still loses. It's important to provide the strongest signal possible from the transmitter to the receiver without overloading the audio circuit input.

Diversity

Wireless systems are susceptible to problematic interactions between the direct (line of sight) signal and reflections from the surrounding surfaces. Metal buildings provide the most difficulty because the surface reflections are very intense. When the direct and reflected signals combine at the antenna, there is a strong likelihood that the signal will not arrive intact. There might be a decrease in signal strength, an increase in interference, or even a complete dropout.

Diversity, also called *full diversity* or *true diversity*, utilizes two antennas to compensate for the problematic results of unwanted artifacts from the combination of direct and reflected transmissions. Both antennas feed the same frequency and silently switch back and forth to receive a constant signal from the transmitting device. Ideally, any time one antenna receives a corrupt or weak signal, the other antenna will receive a strong and accurate signal—the stronger signal wins, and the reception is constant.

Non-diversity systems are not completely uncommon, although they're subject to dropouts and RF noises. Lower-frequency VHF systems, which are inexpensive, can function acceptably in a fixed application where there aren't other wireless systems in close proximity and where the enclosing structure isn't metal.

Range

Most wireless systems cover a range of about 300 feet from the transmitter to the antennas. The range specification typically assumes ideal circumstances, but in most applications transmission and reception ranges are very acceptable from all wireless systems.

Most wireless systems prefer an unobstructed line of sight between the transmitter and antennas. Wood and plaster don't cause as much of a problem as metal structures, but, in any case, a clear line of sight is ideal.

Typical systems utilize antennas mounted on the receiver. It is also possible to use high-quality coaxial cable, attached to the receiver at one end and the antennas at the other, to extend the range of the system. Many installations position the antennas near the stage with the receivers in the room at the soundboard. This scenario provides a good line of sight and close proximity between the transmitter and antennas while positioning the receiver near the sound operator for ultimate control and signal analysis.

The determining factor of the range of any wireless system is signal strength. As long as the transmitter signal is stronger than the inherent radio noise at the receiver, you should experience wireless connectivity. Because radio signals decrease over distance and because physical barriers also diminish them, actual range might be far better or worse than stated specifications.

Antenna Distribution Systems

Any time you're using a system that incorporates more than a couple wireless systems, consider implementing an antenna distribution system. These systems utilize one set of antennas, typically located near the stage, that pick up the signals from multiple transmitters. Most distribution systems use a rack-mounted unit connected to remote antennas. The individual receiver antenna inputs connect to the distribution system rack mount antenna outputs.

There are two types of antenna distribution systems: passive and active.

- Passive distribution simply splits the signal between two antennas. Even though impedances are matched, there is still a 3-dB signal loss as a result of the split. Multiple passive splits are impractical because of the excessive signal loss.

- Active distribution systems are designed for larger systems in which three or more wireless systems are used at once. Many distribution systems provide antenna feeds for four systems. They also allow the user to daisy-chain two systems together for distribution to eight wireless receivers. Active systems utilize an amplifier circuit that provides unity gain (no boost or cut in the original signal strength) to all connected receivers. Chaining more than two active distribution systems together is typically impractical due potential interference and RF noise.

Antenna Distribution

When there are several wireless systems working together, the most practical way to implement the antennas is through an antenna distribution system. With this kind of system, several wireless receivers can be rack mounted together, with all antenna posts connected to the distribution hub. The hub is connected to two antennas. The antennas can either be mounted locally to the hub or they can be mounted remotely and inconspicuously out of sight. The diagram at the right illustrates a typical wireless antenna distribution setup.

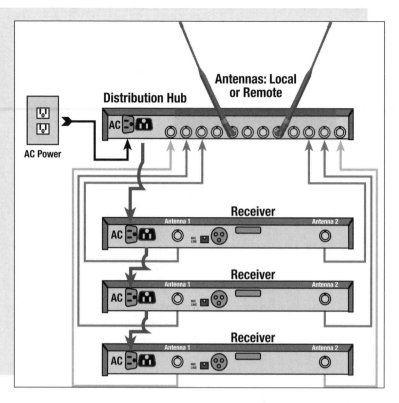

Batteries

It makes fundamental sense that rechargeable batteries would be the best choice for the constant use of a wireless system. However, there is a problem with most rechargeable batteries that makes them very risky in most live sound reinforcement applications—they are not trustworthy. It is usually best to completely discharge a battery before recharging because many battery types develop a memory, which causes them to artificially recharge. They might indicate that they're okay on the charger but, in reality, they only have a partial charge. Often, the battery even registers as charged on a meter, but within a short time after the system is in action, the battery level plummets—and there's not much time before complete failure.

Once you experience the unreliability of rechargeable batteries, you'll come to the conclusion that the importance of the show far supersedes the potential financial savings rechargeable batteries offer. Most seasoned sound operators use fresh new throwaway alkaline batteries at the beginning of every event. There are few occurrences more distracting to the flow of a service or show than dead batteries in the wireless system.

Research the cost of high-quality industrial batteries. In the case of 9-volt alkaline batteries, the manufacturer's recommended list price could be around $4; however, the same battery could be purchased in bulk for around a dollar. Some battery suppliers offer the industrial versions of the batteries found at the local grocery store. These batteries are touted to last longer and to have a longer shelf life than their consumer-grade counterparts. Wholesale suppliers such as Costco or Sam's Club often have special prices on batteries, but I've also had excellent results with battery suppliers on the Internet, such as www. jirehsupplies.com. These suppliers provide excellent batteries and typically offer programs in which automatic shipments arrive on schedule, quantity-adjusted to fit your pattern of usage.

Keep close tabs on the level of each battery. Many older wireless systems provide an LED indicator on the microphone or body pack to indicate battery power level. Unfortunately, most of the time it's too late when these indicators display that the end of the battery's life is near. Most newer wireless handheld mics and body packs provide an actual battery level gauge at both the transmitter and receiver to accurately indicate the remaining power level. These systems are efficient, and they help the operators get the most life from each battery while avoiding much-dreaded battery failure.

Diversity Indicator

Almost all diversity wireless systems provide an indicator of changes in the diversity receiving antennas. When the A indicator is on, the receiver is hearing antenna A; when the B indicator is on, the receiver is hearing the signal from antenna B. These changes are sonically transparent in normal situations. Sometimes the transmitted signal is so weak that the diversity struggles to get any signal.

If your wireless system is dropping out or if it is noisy, watch the diversity indicators. If they are continually bouncing quickly back and forth, or if there are momentary instances in which neither indicator is on, you either have a problem with the transmitter or the receiver, or there is something in between the two devices causing interference. In such a case, verify that there is an unobstructed line of sight between the transmitter and receiver and that the transmitter output level is adjusted for maximum output without signal path distortion.

Squelch

Each wireless system contains noise that constantly resides in the transmission and reception scheme. A squelch control is provided to mute the resident noise any time the intended signal is absent. In many, but not all, cases, the squelch acts like a traditional noise gate in which the control is set so that the system noise is audible, and then it is slowly turned until the noise is muted. In these systems it is important that the squelch be set properly so that the onset of the intentional audio signal is heard. If the squelch is adjusted improperly there will either be noise in the absence of signal or silence at the beginning of the signal.

A refinement of the squelch system is called the *noise squelch*. This type of squelch takes advantage of the excessive high-frequency content of radio noise and actually differentiates between the radio signal and a normal audio signal. This system lets the user set the threshold lower so that the intended audio signal can be more reliably transmitted. Also, the fact that the threshold is lower allows for a wider range of operation. Because transmitted signals decrease in strength over distance, a squelch that is set with the threshold too high (more extreme) can decrease the effective range as the transmitter moves farther from the receiver.

A further refinement of the squelch system is the *tone-key* or *tone-code* squelch. This system includes an inaudible tone with the audio signal. The squelch circuit will only open up to receive the audio signal when there is sufficient signal strength and the expected tone

Transmitter-to-Receiver Communication

Wireless systems such as the Shure UR4S provide amazingly simple use and incredibly powerful features. This system locates available frequencies and then synchronizes the transmitter and receiver at the touch of a button. Its front-panel screen provides all the information necessary to successfully complete almost any event in almost any venue. Communication frequency, power level, battery level, sensitivity, gain settings, and so on are instantly displayed. In addition, systems such as this work together with all other similarly capable wireless systems, selecting compatible frequencies for seamless integration. Further, the UR4S is IP addressable for complete computer interface and control of large and complex wireless networks. The diversity meter indicates the varying status of both antennas.

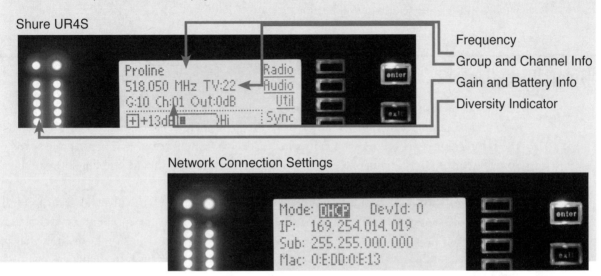

code present. This system provides excellent and reliable squelching of unwanted noise, and it provides a way to eliminate the standby switch. Because the code triggers the squelch to open, there is no chance of extraneous noises and bursts in the absence of signal. The only drawback is a momentary pause each time the transmitter is turned on. It takes a second or so for the tone code to trigger the squelch circuitry after powering up.

Microphones for Wireless Systems

There are three types of microphones that are commonly used for narrator, pastor, or lecturer: lavaliere, lectern, and headset. Each of these mic types is capable of producing excellent results when used correctly. Each of these mic types can be used in either a wired or a wireless configuration. Lavaliere and headset mics are typically wireless, so the lecturer can roam freely, unhindered by cables.

Lavaliere

Both lavaliere and headset mics let the lecturer be expressive with both hands at once. They tend to blend in, and most of the time neither the person speaking nor the audience is aware that a microphone is involved in the communication chain.

The lavaliere, also called a *lapel* or *tie clip* mic, has been the most common mic used in lecturer/audience applications for years. Lavaliere mics are fairly efficient, but whenever a significant amount of volume is required, they can be frustrating. They tend to feed back easily if the sound system isn't perfectly tuned or if the acoustical environment is very active.

Attach the lavaliere mic clip to the lecturer's tie, collar, or any other garment piece close to his or her mouth. The biggest mistake any sound operator can make when using a lavaliere is to allow the lecturer to position the mic far away from his or her mouth. Sometimes the mic ends up pointing away from the speaker's mouth. This is especially harmful if the lavaliere is directional and, in any case, it shouldn't be tolerated.

The ideal position for a lavaliere mic is a couple inches below the tie knot, facing up toward the lecturer's mouth. This position lets the mic take advantage of the resonance of the speaker's chest tones and is still close enough to the mouth to pick up sibilance and other high- and mid-frequency content.

Lavaliere mics are available in two difference types of directional configurations: directional (cardioid) and omnidirectional. Each directional characteristic offers advantages; the sound operator must weight the advantages and disadvantages for each and choose wisely.

Directional lavalieres tend to produce less feedback than omnidirectional lavalieres. The fact that they reject sounds from the back of the mic helps minimize the recirculating

Lavaliere Mic Position

Relatively small changes in the lavaliere mic position can make a huge difference in the sound quality, gain before feedback, and intelligibility. The mic must be aimed up at the presenter's mouth. As soon as it slips off-axis, pointing to the side or downward, it becomes nearly useless. Omnidirectional lavaliere capsules are more forgiving in their directional position; however, they can only be used in a sound reinforcement application where the sound system provides ample gain before feedback. Any omnidirectional mic will present an increase in likely feedback, relative to a mic with a unidirectional characteristic.

Cardioid lavaliere capsules are better at rejecting feedback, and in most instances they are preferable. All lavaliere microphones share a set of concerns of which all sound operators should be aware.

- Directionality. When the presenter turns his or her head to the side, the volume can decrease dramatically if the mic isn't aimed directly at the person's mouth when he or she is facing forward. For example, if the mic is aimed to the left, the volume will greatly diminish when the presenter looks right. The directional characteristic is increased when using a capsule with a cardioid polar pattern.
- Negative interaction with reflections. This is common with all lavalieres—it is exhibited most dramatically when the presenter steps up to a podium or holds a book or notes up to read from. The reflections off the surface combine with the direct sounds in a destructive way. The sound usually gets thicker and duller sounding.

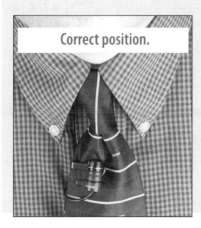

Correct position.

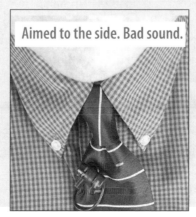

Aimed to the side. Bad sound.

Mic too low. Feedback issues.

audio feedback. The disadvantage of these cardioid mics is their tendency to exaggerate the proximity effect.

The omnidirectional lavaliere produces more feedback than its directional counterpart, but when feedback isn't a real problem, it provides a more natural sound that doesn't suffer the negative aspects of the proximity effect.

In practical terms, lecturers often look down at their notes. If the lavaliere is attached to the lecturer's tie, a couple inches below the knot (some people like to place the mic about one fist under the chin), each mic will react differently. The directional (cardioid) mic will produce a thick, bassy sound when the lecturer looks down at his or her notes; the omnidirectional mic will still sound pretty natural in this scenario because the proximity effect, caused by the chin closing in on the mic capsule, will be minimal.

The omnidirectional mic will also produce a more consistent sound as the lecturer steps behind and away from the lectern. The exaggerated proximity effect produced by the directional mic will tend to provide a boomy sound when the speaker is behind the lectern and a thinner sound when he or she steps out into the open.

In reality, the omnidirectional lavaliere is capable of providing the best sound, but in a live sound application, there are many considerations that make the cardioid lavaliere the best choice. Any time the lecturer is using stage monitors, feedback will become an issue that will be minimized by the use of a directional mic. In a very reflective room, in which the sound system is already on the verge of feedback, the omni mic will be difficult to deal with. When there are multiple lecturers on the same stage, directional lavalieres will help isolate each lecturer from the other.

Whether directional or omnidirectional, the sound that is provided by the lavaliere mic is often very active in the low mid frequencies between 200 and 400 Hz. These are the frequencies that make the voice sound small and weak. Once the lavaliere is positioned for the best sound and turned on through the system, the sound operator must evaluate the raw sound and adjust accordingly.

If the sound is hollow, containing a low-mid ring, you'll need to equalizer the signal for the best sound. This is best accomplished with a sweepable or parametric equalizer. On the mid-frequency band, set up a cut of five or six decibels, then sweep the cut between 150 and 500 Hz. Usually, there will be a position at which the sound cleans up and comes to life—this one adjustment might provide all the equalization necessary to produce a full and excellent sound. Each situation is unique, but this initial step will get the sound operator off to a good start. An inexperienced sound tech will often start boosting highs and lows to fill out the sound, and typically ends up with so many frequencies boosted that the audio quality is adversely affected and, in addition, signal path might be overdriven. The initial subtractive EQ that I suggested will provide a signal with more sonic integrity. If the sound is a little dull, even after the mid-frequency cut, try boosting the high around 5 or 6 kHz. In most acoustical spaces, the lavaliere mic will not require a low-frequency boost. In fact, it is typically advisable to activate the channel high-pass filter, which should effectively eliminate the frequency band below about 80 Hz or so.

Lavaliere mics are very effective and provide an incredibly natural sound in any controlled environment, such as a broadcast studio. They are perfect for television applications because they become visually transparent. In addition, the studio environment is typically dry and free from reflection and reverberations. Ambience is at a minimum, and the raw sound provided by the lavaliere is typically very usable.

Headset

Headset mics have become very popular for lecturers and vocalists. Because they are typically in front of, or very close to, the singer or speaker's mouth, they provide a much more intimate sound and provide substantially increased gain before feedback. Whether for a pastor, lecturer, or lead singer, the headset mic is an excellent choice.

Headset mics are also offered in cardioid and omnidirectional configurations. The benefits of each directional characteristic are the same in both headset and lavaliere mics. Because the headset mic is in a consistent position in relation to the lecturer's or singer's mouth, the proximity effect is less dramatic than when using a lavaliere.

The equalization techniques are similar when comparing the lavaliere and headset mics, but the raw sound provided by the headset mic is typically much more desirable prior to equalization than that provided by the lavaliere. Though the techniques are similar, the headset mic will typically require a less extreme amount of equalization to produce an excellent sound.

Lectern

The lectern mic is typically mounted on a long, flexible gooseneck mounted on top of the podium lectern. The lectern mic is usually a wired condenser mic and is often very small and inconspicuous. These mics sound fine most of the time, as long as the mic is positioned properly. The most common mistake sound operators make when they use a lectern microphone is allowing the lecturer to stand too far from the mic. These small diaphragm condenser mics typically feed back fairly easily, so it is best to position the mic within four to six inches from the speaker's mouth. It is also possible to get too close to these mics. Because they're typically flat with a characteristically strong output, allowing the lecturer to get within an inch or so usually produces a sound that is thick, boomy, and unattractive.

Headset Mic Position

It's important that the headset mic capsule is positioned directly beside and below the presenter's mouth—it is acceptable to place the mic capsule on the presenter's face. If the mic is pointed too far outward or downward, there will be a marked decrease in sonic intimacy and vocal intelligibility.

Handheld

Handheld wireless mics are commonly used for lead singers, announcers, and sometimes lecturers or pastors who are simply more comfortable using a regular mic. Handheld mics provide the fullest and most pleasing raw sound. Because the speaker's mouth is within an inch or two of the mic capsule at all times, the sound the handheld provides is very full and clean. It also provides the most gain before feedback of all these mic types. Handheld mics used in a live application are virtually always directional.

Control Which Mic Is Turned On

Often, in a church service or other setting, the pastor uses a lavaliere or headset mic, while other service participants use lectern or handheld mics. It is common for the pastor to move toward the lectern in order to control the visual flow of the service and, quite often, he or she might use both the handheld and the lavaliere/headset at the same time, or even stand at the lectern to speak for a while. The sound operator should turn off the lectern or handheld mic while the pastor speaks and use the audio feed from the lavaliere or headset only. Whenever both mics are on at the same time, significant phase cancellations between the mics will produce a hollow, unattractive, and unreliable sound.

Two Mics

In the broadcast industry it is common to see two lavaliere mics side by side, mounted on the same clip. It's also common to see two identical handheld or wired mics mounted together at a lectern. The mics are typically identical, and they're usually in place for a couple reasons:

- **Redundancy**: Equipment malfunctions can cause catastrophic losses if they happen at the wrong time and place, especially in the broadcast industry. For this reason, many broadcast engineers hook up two of everything. There is often an extra mixer set up and ready to go just in case the primary mixer malfunctions. The extra lavaliere or stand mic is usually connected to a separate mixer, channel, or transmitter. If the mics are wireless, they're connected to separate transmitters and possibly even to completely separate systems. As a side note, any time you use a wireless system, be sure to have a wired mic

Two Lavalieres

When there is no room for error, utilize two separate wireless systems on the same person. This is a common technique in broadcast applications where there is no opportunity for multiple takes or even retakes. Important speeches and live television demand the use of redundant systems, such as the use of two lavalieres.

readily available just in case the wireless system malfunctions or starts to receive spurious noises or random radio signals.

- **Auxiliary sends**: Sometimes one mic is connected to the house sound system and the other is connected to a recording or broadcast system. This technique allows for optimization of each mic and transmitter for its intended purpose.

Wireless Frequency Selection

Wireless frequency selection is an important consideration! Too many rookie sound operators lack even a rudimentary understanding of how wireless frequencies interact and react to each other. Random purchases of wireless equipment without considering other existing wireless systems can be disastrous. If you use two wireless systems that operate on the same fixed frequency, each transmitter will be heard on both receivers. If the frequencies are not correctly spaced, there is a good chance that the two systems will conflict, causing extraneous noises and outbursts to happen on either or both.

Most earlier-model wireless systems used fixed frequencies. In these cases, it was fundamentally important that each subsequently added system operated on a compatible frequency. If your system contains fixed-frequency wireless devices, list all frequencies from each wireless system and keep them in a conveniently accessible place. Any time you purchase additional wireless systems, be sure to supply your salesperson with all onsite wireless frequencies in use. It is especially important to choose other fixed wireless devices that are capable of peaceful and blissful cohabitation with your existing systems. Often different areas of the country use specific bands selected to complement local television and radio frequencies.

Modern wireless systems typically utilize frequency-agile technology. Most of these systems scan for any other wireless systems and automatically select unused frequencies— they can even continue to scan just in case a conflict develops. This technology has been revolutionary in the simple and pain-free implementation of wireless systems.

As digital television stations continue to activate, there are more and more possibilities for frequency conflicts. These conflicts could ruin the transmission potential of some fixed-frequency systems; however, frequency-agile systems should be able to simply rescan, selecting an available and compatible transmission frequency. It is strongly advisable that all new wireless system purchases utilize frequency-agile technology.

In recent years there has been quite a scare regarding the available frequencies and the protocol for their use by wireless microphones. The FCC was determined to allocate the spaces between television transmission bands, called "white spaces," to everything but wireless microphones. That possibility created a bit of a panic in the live sound industry, threatening touring companies, live acts, and fixed venues with the overwhelming task and expense of buying new wireless equipment and rethinking their uses of wireless systems. Through the efforts of many interested manufacturers, professionals, and wireless users, the FCC was persuaded to take a less dramatic approach than they had originally specified. The solution and its implementation are somewhat defined as of this writing; however, the details, though apparently known, are at the mercy of the bureaucratic process, which promises to be long-running. New wireless devices are likely to need some form of geolocation system, along with a national database of wireless frequency users, to assist in the recognition of nearby systems and determinations regarding available frequency bands. The efforts of the Shure microphone company and the Recording Academy (National Academy of Recording Arts and Sciences) were instrumental in the solutions agreed to by the FCC. For up-to-date information regarding this important topic, refer to www.shure.com /whitespaces or use your smart device QR-Code reader to scan this code.

Mute Switch Versus Power Switch

Considering most wireless systems, any time the transmitter power is switched off, the corresponding receiver's mixer channel must be muted to avoid spurious and often dramatic bursts of noise. If the receiver isn't locked on to the transmitter's signal, it will scan the airwaves in all directions, trying to locate it. These loud outbursts of irritating noise are the results of this futile quest for signal. If the transmitter body pack or handheld mic casing provides a mute switch, train the users to mute the transmission instead of turning off the power. These mute switches maintain communication with the receiver so that it won't search continually for its expected signal.

New-technology wireless systems often provide protection against these noises. Many system transmitters don't include a mute switch, simply providing one power switch. Usually, these systems have built-in safeguards against unwanted bursts of noise.

A Clear Line of Sight

Whether the presenter is using a handheld wireless or a lavaliere or headset with a body pack transmitter, it is very important that the line of sight between the transmitter and receiver is unobstructed. Therefore, the presenter's hand should not be around the end of the wireless handheld mic, and a body transmitter should not be stuffed in a pocket with the antenna bunched up or hidden.

Mic held correctly.

Mic held incorrectly. Hand is covering transmitter.

Caring for the Antenna

Whether you're using a body pack or handheld transmitter, you should strive to:

- Maintain an unobstructed line of sight with the receiver.

- Be certain that the body pack antenna hangs freely from, and points directly out of, the body pack. Verify that the transmitter is hanging freely and that it isn't crammed in the user's pocket with the antenna bunched up or wrapped around the body pack.

- Be certain that the handheld user holds the mic near the middle of the mic casing rather than at the end of the mic, where the transmitter usually resides.

Adjusting Levels

Most transmitter packs include a sensitivity control, which acts as a pad for excessively strong signals. Many musical instrument pickups and preamplifiers provide a signal that is at (or close to) line level. The body pack input must be sufficiently adjustable to receive these hot signals. At the same time, the body pack must be adjustable to optimally accept weak signals, such as those coming from a lavaliere or other low-level source.

Each device has a unique set of controls and ways to access them; however, it is very important that the sound operator verify the proper adjustment of the input level. A transmitter that is being overdriven at the input because of a strong source signal will tend to distort, even at low volumes. Sources that are too weak for the input setting will tend to be noisy and might even intermittently trigger the gate-type of squelch to trigger. In this case, the sound might be glitchy, and the range of the wireless coverage could be drastically decreased.

Some transmitters provide physical switches and knobs to adjust the input level. Many new wireless systems provide a user interface that has a graphical interface in which the sound operator scrolls through windows to find the level settings page. No matter how your wireless system is laid out, be sure to look for the input level sensitivity control to get the best possible sound quality.

Connecting to the Sound System

Connecting a wireless system to the house sound system is very simple. Connect the output from the receiver to the mixer channel input. Various receivers provide different types of outputs. The sound operator must confirm that the proper receiver outputs are connected to the appropriate mixer inputs. Typically, the transmitter input type will be mirrored at the receiver. A system designed for use with a microphone should offer a mic-level output at the receiver. A system designed for instrument use should use a receiver with instrument-level outputs that are typically medium-impedance, unbalanced, and quarter-inch phone jacks.

Many systems include multiple types of receiver outputs, including mic-, instrument-, and line-level, as well as balanced and unbalanced. Each type of output should be labeled clearly; however, if they not labeled, refer to the manual for specifications. Although, XLR connections are typically mic-level, and quarter-inch connections are often line- or instrument-level, there are many different reasons that they could be different than one would expect. Find the documentation that comes with your wireless system and verify that mic outputs are connected to mic inputs, and that line outputs are connected to line inputs, and so on.

In the unlikely event that your wireless system doesn't have the appropriate type of output to connect to your mixer, you might need to incorporate an adapter or a line-matching transformer.

- If your receiver only has instrument-level outputs, use a direct box (line-matching transformer) between the receiver and the mixer mic input.

- To connect a balanced XLR output to an unbalanced quarter-inch input, use an adapter that connects the XLR pin 2 to the tip of the phone plug. In addition, the XLR pins 1 and 3 should be connected to the sleeve of the phone plug.

- Other adapters can be used when needed; however, even though the adapter provides the right type of connection, the sound operator must verify that the levels and impedances are satisfactorily matched.

Most modern mixers supply phantom power through each mic input. It is best to turn the phantom power off when connecting the wireless receiver. If the receiver output is balanced, it's unlikely that the phantom power will cause a problem, but if the receiver output is unbalanced and it's connected to a mixer input that supplies phantom power, noise could be induced into the signal. This problem can be remedied through the use of a transformer or adapter with the appropriate capacitors in line, selected to absorb the flow of phantom power into the audio signal.

The Backup Plan

It can't be overstated that a wired mic should always be available as a backup in case the wireless system becomes problematic. If there isn't a backup mic ready and waiting, there's an excellent chance that the sound operator will experience a painfully uncomfortable, show-stopping, ego-damaging, and possibly job-losing pause in the flow of the show.

Using the Wireless System

Once the wireless system is functional, there are a few ground rules that need to be established between the musician or lecturer and the sound operator. It must be established and agreed upon as to who is in control of the mute switch. Many users prefer to simply leave the transmitter on at all times so that the sound operator is in complete control. This is a good system as long as the sound operator is sufficiently attentive that he or she never misses a word. In a fixed show, in which there are minimal variables in transitions and show order, this approach works particularly well.

Often, in the case of a lecture or church service in which the lecturer or preacher takes great liberties as to when and where he or she interjects comments and clarifications, whether during a music service or a multimedia presentation, it is advisable to put the control in the hands of the speaker and out of the control of the sound operator. This approach lets the speaker use the mute switch to cover coughs and sneezes and such, and it gives the speaker ultimate control over when and where he or she is heard.

Either system is satisfactory in the best-case scenario. For specific situations, it's important to develop a routine that is consistent and efficient. When the control resides with the sound operator, great focus is required to catch every important word or note while still muting the signal for privacy and optimum sound quality. When the control resides with the musician or speaker, that person simply needs to understand that if he or she forgets to unmute the signal, his or her voice won't be heard.

I've seen both of these approaches work very well, and there are good reasons for each. Base your procedures on the dependability and preferences of those using the system.

Always strive to maintain a clear, unobstructed line of sight between the transmitter and the antennas. Often, with the soundboard near the back of the room, the audience is in the way of the line of sight. Depending on the system, the distance to the antennas, and the existing battery power, the audience might cause dropouts and other unwanted noises. In such cases, elevate the receivers to regain an unobstructed line of sight. It is also advisable to remotely locate the antennas closer to the transmitter. To accomplish this, use high-quality coaxial cable to connect the antennas to the receiver. There are several mounting systems for wireless system antennas, from mic stand mounts to wall-mount brackets.

chapter 14

LOUDSPEAKERS

When you visit the pro audio department at your favorite music store and look at all the pretty speaker cabinets, it's very difficult to know what the best choice is for your situation. They all look great and they all look like they should sound good. However, they don't all sound the same, and they definitely don't all sound good. In addition, some of them won't last very long under any intense usage. It is your job as an astute sound operator to learn enough about loudspeakers to develop an intelligent opinion about how they might fit into your system.

In this chapter we'll look at a few different aspects of loudspeakers: commercially built cabinets (enclosures), crossover configurations, speaker impedance, and components. In addition, we'll consider the advantages and disadvantages of powered speaker cabinets.

Commercially Built Cabinets

For the majority of applications, commercially built speaker enclosures are capable of providing excellent audio. There are several manufacturers that offer excellent products. In this industry the brand name means something. A product with an excellent reputation and critical acclaim is worth considering.

An off-brand—one that you haven't heard of before and that has an unrealistically low price—is rarely a good deal in the long run. There's a reason these products are inexpensive, and it's not that the manufacturers just aren't interested in making a profit—it's usually because the components are poorly constructed, lightweight, and unreliable.

A reputable manufacturer with an excellent track record in the industry will almost always provide a superior product, excellent technical support, and a respectable warranty. In system development, it is tempting to save money on speaker cabinets or components, so that there's extra money left for other less critical devices—that strategy is ill-advised.

Each cabinet offers a list of features and selling points, such as:

- Size
- Cosmetic appeal
- Ability to handle power
- Type of components

- Built-in electronics
- Acoustical design
- Rigging points
- Manufacturer reputation
- Durability
- Warranty
- Impedance
- And so on

These are all important considerations, but there are two considerations that must be met before any others.

- Does it sound good?
- Will it last a long time?

Even if you're designing a system from reputable manufacturers, you must listen to several cabinets to find the ones that fit your taste. We recently replaced the sound system in

The Cone Speaker

It is instructive to understand the cone speaker and its components. A moving-coil microphone and a cone speaker operate according to the same principle. At the core of both devices is a coil of copper wire suspended around a magnet. In a microphone, acoustic audio waveforms vibrate the diaphragm, connected to the copper coil, which cause a variance in the status of the magnet, which moves the speaker cone —this is the source of the audio signal.

In a speaker, the electrical signals from the power amplifier continuously vary the electromagnet at the core of the speaker, causing movement of the coil around the magnet and the attached speaker cone—this is the source of the variations in air pressure that we perceive as sound.

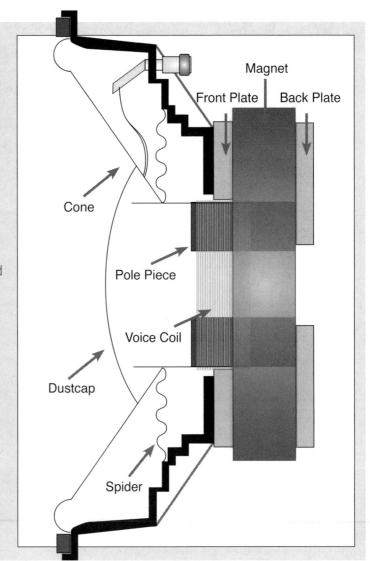

our church. The existing system had been in place for about 10 years, but it just didn't sound good in the sanctuary. It had excellent components throughout the system, but the cabinets didn't suit the room.

In redesigning the system, we brought in several different cabinets to hear how they sounded in the room. They were representative of various price ranges and feature lists, but they were all from reputable manufacturers. Each of the cabinets sounded good, but there were definitely some that sounded better. There was one that sounded better than all the rest—it offered a much more pleasing sound when given the same source. In addition, it was a cabinet that worked together with other identical boxes to form a line array.

Line arrays offer several acoustical advantages and a tighter, cleaner sound that suits the acoustical design in our sanctuary. These cabinets were a little more expensive than the others, but in the long run, they will be very cost-efficient. Rather than building a system that we could outgrow in a few years, we built a system that will still sound excellent in 10 or 15 years, or more. In addition, if we ever want to add to the system for greater coverage, we can simply add more cabinets.

Manufacturers

There is a relatively short list of audio equipment manufacturers that have been around for more than 10 years or so, and who have always offered products that consistently perform at or beyond expectations. These companies are constantly developing new technologies and designing products that address specific consumer needs.

Most of the time a system is designed around a couple factors: budget and need. Established manufacturers realize that they need to fill several needs—they want you to spend your money on their product, whether you're spending a thousand dollars or a million. Therefore, they offers products in all price ranges. Companies such as JBL, Mackie, Apogee, Bag End, and Yamaha have developed trust in the industry because they answer needs with products that, although they're not always the least expensive, last a long time and provide consistently high-quality audio.

There are a few manufacturers who take advantage of inexpensive foreign labor, reverse engineer their products (in other words, they copy the design of an already successful product), and sell their version for half the cost of the original. These companies, who make their money from stealing the designs of others, provide a product that looks like the original and offers many of the same specifications; however, they don't typically offer good technical support, and their products are inherently inferior—they're not manufactured to the same tolerances and standards as the original. In my experience, the copies have never sounded as good and they've been unreliable in the heat of battle. If you're serious about audio quality and if you're in a situation where you depend on your equipment to function properly day after day, week after week, and year after year, purchase reputable equipment from a reputable audio professional.

Pro Audio Suppliers

An important ingredient in the design of any system is a relationship with a professional audio equipment supplier. Without someone at hand to help weed through the available equipment—a real person you can rely on to set up demos and provide current information—the quest for excellent audio system components can be very overwhelming.

Make an effort to find a pro audio supplier who is willing and capable—one with whom you can develop a longstanding relationship. It is his or her job to stay in touch with the offerings from all of the manufacturers and to help guide you through the maze.

It is also very important that you do your own research and become an informed consumer. The retail audio industry is very competitive. In addition, prices are similar whether you are purchasing on the Internet or in a large music store; however, even if you pay a little extra through an established professional, it will be a worthwhile investment in the long run.

Passive Crossover

The passive crossover receives a powered, full-bandwidth signal from the power amplifier. This signal is split into high- and low-frequency content through filters. Crossover frequencies and the severity of the filter slopes are carefully selected depending on the speaker components and intended application.

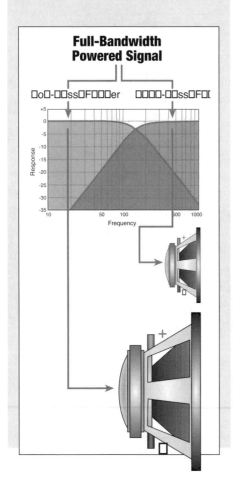

Crossover Configurations

Speaker cabinets typically contain components that reproduce specific frequency bands, ranging from the lowest frequencies to the highest. In addition, many systems are designed to include multiple speaker cabinets that each cover a specified frequency band from lowest to highest.

The full-range audio signal is split into frequency bands by an electronic circuit called a *crossover*. Typical crossovers divide the audio into two, three, or four bands. Each band is sent to a component (or components) in a cabinet or, in large systems, to separate cabinets optimized for a certain band. A separate power amplifier is typically used for each band. Therefore, each frequency band can be powered according to its requirements—low frequencies require substantially more power to accurately reproduce than high frequencies.

Using a crossover typically produces the best sound quality, but it must be adjusted correctly for the system components. It's important that the sound operator understands the concepts and considerations involved in the utilization of crossovers in audio system design.

Passive Crossover

There are two types of crossovers: passive and active. A passive crossover is typically built into the cabinet and receives a full-range signal from the power amplifier. High-pass, low-pass, and band-pass filters divide the frequency spectrum for delivery to the target components.

Passive crossovers don't require electrical power to operate—they are merely circuits that eliminate (filter out) specified frequency bands. These filters ramp up and down at different rates specified in decibels per octave—most commonly at a rate of 6 dB per octave (first order), 12 dB per octave (second order), 18 dB per octave (third order), and 24 dB per octave (fourth order).

Passive crossovers are typically less expensive than active crossovers and they don't require electrical power to operate. The fact that they receive their signal from a power amplifier is important. In order to withstand the sheer energy of a modern, high-wattage amplifier, the passive crossover circuitry must be substantial. Even if it is capable of withstanding a constant barrage of power for over long periods of time—like from the beginning of the performance to the end—the passive crossover's electronic components will usually heat up. As electronic circuits heat up, their response characteristics typically change, causing them to produce an inaccurate and uncharacteristic sound.

Active Crossover

Whereas a passive crossover, typically located inside the actual speaker enclosure, receives a signal from the output of a power amplifier, an active crossover receives a line-level signal, either from the mixer or in-line processor. Active crossovers require electrical current to operate. The filters included in an active crossover are typically adjustable and accurate. These valuable tools let the system designer adjust the size and strength of each band.

The point where one band transitions to the next is called the *crossover point*. Systems comprised of custom-built or specifically selected components often require carefully chosen crossover points in order to realize the full system potential. Active crossovers provide ample control over the band selection and distribution.

Because the active crossover splits the frequency spectrum into multiple bands, keep in mind that each band will then require at least one power amplifier to operate its target cabinet or component. Many stereo systems incorporate a separate mono power amplifier for each cabinet in the system. A stereo or left-center-right system could easily utilize 8 to 12 power amplifiers—a

Active Crossover

Whereas the passive crossover receives a powered signal from the output of the power amplifier, the active crossover receives a line-level signal from the output of the mixer or signal processor. The line-level signal is divided into multiple frequency bands and then sent to the line input of separate power amplifiers.

Because the active crossover receives line-level signals rather than powered signals, there is less demand on its circuitry; therefore, the active crossover output is much more stable over the course of extended use than the passive crossover output.

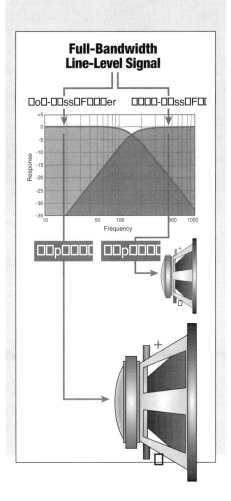

separate amplifier for high, mid, and low frequencies at each pan position, plus a separate subwoofer for each.

Analog Versus Digital Crossovers

Many great systems have been built around high-quality analog crossovers. Modern systems often take advantage of digital crossovers and the additional flexibility they offer. The digital filters are very precise, plus the A/D converter takes the analog signal into the digital domain. Once the signal has been digitized, it is very easy to include access to digital signal processing, such as compression, limiting, and time alignment. The processing can be applied separately to each band before returning to the analog domain through each channel's D/A converter.

Most of the digital crossover systems, such as the dbx DriveRack and the BSS Soundweb, provide the ultimate in system control, feeding multiple bands and channels simultaneously with high-quality and precisely processed audio.

Full-Range Cabinets

Full-range cabinets use passive crossovers to distribute the incoming, powered signal to the individual components. Some speaker components act as full-range devices, but they are rarely used in high-quality professional sound systems.

Many cabinets accept a full-range powered signal and pass it through a passive crossover, while at the same time offering access to the individual components in the cabinet directly, bypassing the passive crossover. This affords the sound operator the opportunity to divide the frequency spectrum through an outboard active crossover, then route the specific bands through separate power amplifiers, which are then connected to the individual components. This typically creates a two- or three-way system from a full-range cabinet.

Two-Way Cabinets

Two-way cabinets, traditionally referred to as *bi-amplified*, contain components that are optimized for two specific frequency bands: highs and lows. The components determine the exact crossover frequencies. A two-way system is better than a full-range system, but it isn't the optimum in most circumstances. The high-frequency drivers require less amplification than the low-frequency drivers—a two-way system can adequately separate the highs, but the remaining band is so broad that it becomes inefficient as it tries to simultaneously push the mids and lows through the same component.

Two-way systems work fairly well in a studio monitor application where the speakers are within about one meter of the listener's ears, but in a live application, a three- or four-way system will provide superior sound quality.

Three-Way Cabinets

Three-way systems are capable of providing excellent sound. High-quality components combined with excellent power amplifiers should faithfully reproduce a good audio source. Three-way cabinets provide access to components that are designed to reproduce highs, mids, and lows. Sometimes the low-frequency components are unduly stressed as they attempt to reproduce the low-mid, bass, and sub-bass frequencies at the same time.

Depending on the style of music and the desired sound pressure level, a three-way system could provide perfectly acceptable or woefully marginal results—it's really all about the sub-bass frequencies below about 100 Hz. These frequencies require the most power to accurately reproduce and they are the most adversely affected by the three-way design.

Four-Way Systems

Four-way systems typically add one or two subwoofers to a three-way system. The addition of components capable of receiving the sub-bass frequencies along with sufficiently powerful amplifiers provides a noticeable increase in sub-bass frequency content. At the same time, removing the sub-bass frequencies from the components that handle the low-mid and low frequencies radically increases their efficiency.

Most large professional systems use this configuration. Some very large systems split the frequency spectrum into more than four bands, but not always. Four bands with intelligently chosen crossover frequencies and ample power can adequately represent a well-mixed, full-range audio source.

One-, Two-, Three-, and Four-Way Cabinets

Two-, three-, and four-way cabinet designs typically include a crossover for distributing the appropriate powered frequency bands to each speaker component. In addition, there are usually separate input jacks that provide access to each speaker if the sound operator chooses to use an active crossover to separate the line-level audio signal into separate bands.

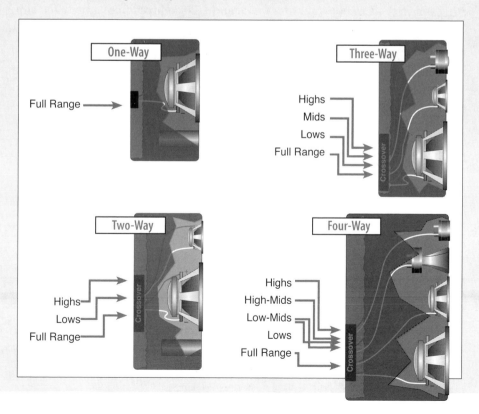

Powered Cabinets Pros and Cons

Several years ago, JBL introduced the EON series. This powered speaker cabinet, designed specifically for use in sound reinforcement, combined an excellent high-frequency driver and a 15-inch cone speaker in a lightweight enclosure, along with a power amplifier designed specifically for its components. These cabinets were immediately put to work around the world—they sounded great, looked great, and were easy to set up. The built-in amplifier provided the correct amount of power to the two-way speaker design. This feature alone removed a key variable in comparison to previous systems that relied on the user or salesperson to pair the cabinet with the amplifier.

The instant success of the EON system has continued since its introduction. Mackie has also introduced a substantial line of powered cabinets and, like JBL, now offers powered cabinets in a variety of sizes, from enclosures designed for the smallest of venues to cabinets containing 18-inch subwoofers. Other manufacturers have also joined in with their renditions.

Powered speaker cabinets address some very important consumer needs:

- Portability
- Ease of setup
- Lightweight construction from durable molded synthetic materials
- Elimination of cumbersome and heavy power amplifiers
- Amplification that is perfectly matched to the speaker components, resulting in improved sound quality
- Molded construction, which typically includes a built-in pole mount receptacle for use atop lightweight tripod stands, along with rigging points for hanging installations
- In most models, line inputs for quick and easy connection to any mixer, along with a single microphone input for public address applications in which a mixer is unnecessary
- Complete connection using microphone cables
- Enclosure design that accommodates quick setup either on end for front-of-house use or laying on its side for use as a floor monitor

There are a couple drawbacks to the powered cabinet design:

- There needs to be AC power available for each cabinet.
- In large installations, they don't always offer the flexibility required for complex zone coverage, pattern control, and tonal balance.

Powered cabinets provide an excellent solution for portable systems as well as for mid-sized permanent installations. They sound good and are easy to cart around. They fill a huge need in the live sound industry and they've proven to be dependable and rugged.

Speaker Components

There are only a couple different types of components found in the majority of sound systems for live reinforcement. They're commonly referred to as *horns* and *cone speakers*, but more specifically, they are *direct radiator drivers* and *horn compression drivers*. Most speaker enclosures utilize some combination of these designs.

Drivers

The driver is the actual device that moves the air in a speaker. A speaker is a form of transducer—transducers simply convert one form of energy into another. The driver converts electrical energy from the power amplifier into acoustical airwaves.

Interestingly enough, the fundamental mechanical process that we use today to convert electrical impulses into airwaves was patented by Ernst W. Siemens in 1874. This system utilizes a copper coil of wire suspended around an electromagnet. The magnet polarizes in response to the amplified audio signal, causing the suspended coil to move axially in response to the continually varying north and south magnetism. Axial movement involves a motion in two directions, such as in and out, up and down, or side to side.

Though there have been refinements of the process and technical advances that have provided exceptional materials, this is still the way most speakers work.

The home theater and hi-fi industry offers some very nice new-technology speakers that sound incredible; however, the live sound industry still requires the brute strength and power provided by these relatively massive drivers.

Direct Radiating Cone Speakers

The cone speaker typically provides the low- and mid-frequency components of the audio signal, although the actual effective frequency range is dependent on the materials used and the diameter of the speaker. Commonly used direct radiating cone speakers vary in size from as small as a couple inches to as large as 18 inches—small speakers are more capable of reproducing high frequencies, and large speakers are more capable of producing low frequencies. Professional sound reinforcement systems rarely use cone speakers smaller than 10 inches because, in a designed enclosure, they primarily reproduce low and mid frequencies. Horn drivers are more efficient at handling high-frequency content of the audio signal.

Cone speakers need an enclosure to realize their low-frequency output potential. In space, without an enclosure, their response characteristic suffers from back-to-front cancellation. The energy created by the front of the speaker cone is, in reality, 180 degrees out of phase with the energy created by the back of the speaker—while the front of the speaker pushes the air in front of the cone, the back of the speaker pulls from the air behind the cone.

The principles of diffraction indicate that sound bends around obstacles, especially low-frequency sounds. The air movement from the back side of the speaker cone diffracts around the speaker and combines with the air movement from the front side of the cone in a destructive phase relationship. In fact, low-frequency wave forms are so much longer than the length of the path from behind the speaker cone to the front of it that phase cancellation approaches 180 degrees.

Horn Compression Drivers

Horn compression drivers utilize the same core functionality as the cone speaker. They both use a voice coil around a magnet to instigate the movement of air; however, the compression driver is loaded down by air pressure created by a horn lens. A phase plug is used to help dissipate heat and lower distortion.

Compression drivers are more efficient than cone speakers and they are typically used to reproduce upper midrange and high frequencies. These drivers require less power to achieve the same SPL as cone drivers. Whereas a low-frequency cone driver might need to receive 1,000 watts for optimal performance, a high-frequency compression driver might only require 200 watts.

Impedance ratings for compression drivers are in the same range as the large cone drivers—they tend to fall in the range of 4 to 8 ohms. Therefore, the previously mentioned considerations regarding series and parallel connections still apply.

Horns can be used both on cone speakers and on compression drivers. The opening of the driver influences the potential frequency response character—larger openings are more able to reproduce lower frequencies.

The Importance of Enclosures

The capabilities of the drivers combine with the physical design of the cabinet to create a device with sonic characteristics. High frequencies rely very little on the enclosure dimensions and physical construction; however, low frequencies are radically colored, molded, and shaped during the design process. The cabinet size and construction often provide exaggerated pathways within the enclosure, which allow the low-frequency information to develop more fully. These pathways channel the low-frequency energy, focusing and enhancing it to the desired balance in relation to the mid and high frequencies.

Sealed Enclosures

Sealed enclosures eliminate diffraction by closing off the path by which the rear-generated sound wave can combine with the front-generated sound wave. Absorption within the

The Cone Driver Outside the Enclosure

The enclosure plays a key role in realizing the deep tone that the cone driver is capable of producing. Air moves in response to in and out motion of the cone, creating the acoustic sound. As the cone moves outward, the air in the front of the speaker compresses—as the cone moves inward, the air in front of the speaker expands (compression and rarefaction). With the speaker in open air, there is one important problem—each time the cone pushes outward in front, it also pulls on the air behind the cone. This results in the simultaneous creation of two waveforms that are 180 degrees out-of-phase.

Because the low frequencies are very long, the distance from the back of the speaker to the front is insignificant; therefore, frequencies longer than about one foot tend to cancel, resulting in a thin and unreliable sound. Placing the speaker in an enclosure effectively removes the rear waveform information from the equation, allowing the low frequencies to reproduce accurately.

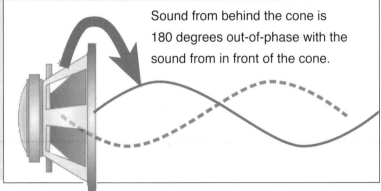

Sound from behind the cone is 180 degrees out-of-phase with the sound from in front of the cone.

enclosure helps minimize destructive interactions between sound waves within the enclosure and the movement of the cone creating the sound being projected from the enclosure.

Infinite Baffle

The infinite baffle design utilizes a very large sealed enclosure in an attempt to take advantage of the resonance within the enclosure. These cabinets provide a very deep tone when properly designed, but they aren't typically known for providing a very flat or controlled frequency response characteristic.

Acoustic Suspension

Acoustic suspension enclosures utilize a relatively small enclosure, which is sealed to eliminate diffraction. The inside of the cabinet is lined with absorptive materials to minimize the destructive influence of mid-frequency reflections within the cabinet and help decrease resonance. This design is much more linear in its response in comparison to the infinite baffle design; however, it suffers a decrease in efficiency due to the small enclosure size.

Bass Reflex

A bass reflex cabinet attempts to release the audio energy generated inside the enclosure in a way that augments its low-frequency response rather than decreasing it. In this design, an opening in the enclosure, called a *bass reflex port* or just a *reflex port*, lets the energy from inside combine with the energy being projected from the front of the speaker. If the enclosure dimensions are just right, and if the internal sound has been routed and released in the correct way, the bass reflex cabinet offers an excellent low-frequency response characteristic while providing a smooth midrange response.

Absorption is used within the enclosure to minimize destructive interactions from midrange reflections within the enclosure. The actual response impact created by the port is a result of the size of the port and the length on the pathway that the waveform must travel to reach its exit.

The port, a tube extending into the enclosure from the opening in the front of the cabinet, acts much like a flute. When air is blown across the hole, a tone is produced. A small flute (a piccolo) produces higher frequencies than a large flute (a bass flute). The same concept applies to the response of the reflex port in a speaker enclosure—the larger the port and the longer the tube, the lower the frequency produced.

The resonance of the port can make a reflex cabinet sound boomy at a specific low frequency. Ideally, the port is tuned so that it supports and reinforces the frequency-response deficiencies created by the enclosure's physical size.

The Passive Radiator

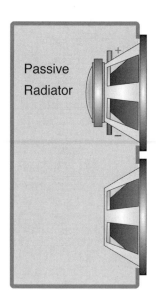

Passive Radiator

The passive radiator design utilizes a "dummy" (passive) speaker in addition to the active speaker. This passive speaker has no magnet or coil and it doesn't receive a powered audio signal. Its purpose is to support the motion of the active speaker by moving additional air in response to the activity from the air movement created by the active speaker. This design tends to result in a dip in frequency output at the enclosure's resonant frequency.

In relation to the bass reflex cabinet design, the passive radiator can potentially provide an excellent low-frequency response in a smaller enclosure.

Transmission Line

Transmission Line

The transmission line concept utilizes an extended-length port in an attempt to shift the phase of the sound waves generated by the rear of the speaker cone by at least 90 degrees. This is the type of system used by some manufacturers of small devices with unnaturally excellent bass response.

A physical pathway is constructed within the enclosure—its purpose is to delay the escape of the sound waves inside the enclosure. Ideally, it will be delayed by an amount sufficient to allow it to combine in a constructive way with the sound wave emanating from the front of the speaker cone.

Component Alignment

Speaker enclosures that contain separate high-, mid-, and low mid-frequency components must accommodate the size difference between the individual drivers. If all drivers are mounted on a flat surface—and if they are fed the same audio signal—there is a phasing discrepancy between the component outputs.

The physical size variation between tweeters, midrange, and low-frequency drivers results in a staggered point of origin for the audio that projects from each device. Because the sound doesn't originate from the same vertical plane, any shared frequencies, such as those that reach into the crossover point, will be out of phase at the frequency that equals the length of the physical displacement.

In addition, timing incoherence between the frequency bands allocated to each driver blurs the audio image. Sounds that contain bandwidth that takes advantage of multiple drivers won't be reconstructed accurately if their content isn't reconstructed in a perfect phase relationship. The goal in any speaker design is to faithfully reproduce a broad bandwidth transient. A design has audio integrity if the transient audio signal is divided into frequency bands, through the crossover network, and then reproduced by the components in the same phase relationship as the original source transient.

The process of positioning speakers so that the sound they produce is identical in phase to the source audio signal is often referred to as *time alignment*, *phase alignment*, *phase coherence*, or

Component Alignment

Often, the speaker components are simply mounted on the flat surface on front of the enclosure—this is not always optimum. The physical alignment of the voice coils is sometimes adjusted by changing the shape of the enclosure. This is done so that there can be a more cohesive phase relationship through the overlap frequencies at the crossover point. Aside from physical positioning of the components, alignment adjustments can also be created electronically through delay circuits. Sometimes, speakers are shifted to compensate for inherent delays in the crossover circuitry.

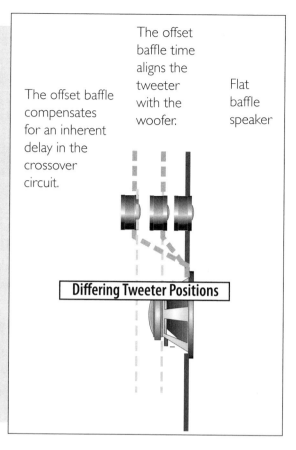

The offset baffle compensates for an inherent delay in the crossover circuit.

The offset baffle time aligns the tweeter with the woofer.

Flat baffle speaker

Differing Tweeter Positions

transient alignment. Shifting the audio output from each individual driver can be accomplished through physical design of the enclosure or through electronic delays.

Enclosures in which the tweeter sets back farther than the midrange driver are simply positioning the tweeter driver in line with the midrange driver. This design is effective and keeps the electronic circuitry simple.

Rather than creating an enclosure with the tweeter driver physically lined up with the midrange driver, some manufacturers simply mount the drivers on a flat surface, delaying the output of the tweeter so that it starts just as the signal from the midrange driver is on the same plane as its driver. Some manufacturers offset the physical position of a driver to compensate for a delay that is inherent in a particular crossover circuit.

Precise alignment is very important in the development of near-field reference monitors for use in the recording studio because the focal distance is very specific, typically within a meter or two. In these designs the sweet spot is along the same plane as the recording engineer's ear—standing up or changing position can radically alter the perceived speaker response. Although live sound reinforcement systems are designed for a much larger coverage area than near-field reference monitors for studio use, it is still very important to understand the concept that displacement of components, whether they're in the same enclosure or just part of a large system, affects the combined phase coherency and transient accuracy.

Speaker Impedance

Speakers exhibit impedance characteristics, which must be matched to the amplifier outputs. Amplifiers are rated according to the load range of the source. Most amplifiers specify minimum operational impedance. The components of a speaker box are wired together in a way that produces the desired load (the impedance at the input enclosure's connector).

Impedance math is pretty simple. Each amplifier is rated at minimum impedance, which is typically 4 or 8 Ω—many modern amplifiers are rated down to 2 Ω. You simply need to be aware of the impedance load you are putting on the amplifier to optimize the power output and rating. Never connect speakers in such a way that the impedance load is less than the specified minimum. If the speaker offers too little resistance to the amplifier signal, it's just like trying to push a little water through a big pipe—it can't sustain the push and eventually will overheat and fail. Proper impedance matching provides a balance, which results in the efficient transfer of power.

Calculating Speaker Impedance Loads

There are two ways to wire two speakers together: in *parallel* and in *series*.

- Multiple speakers wired in parallel present a lower impedance load to the amplifier output. To calculate the resulting impedance, divide each speaker's impedance by the number of speakers. Two 8-Ω speakers wired together in parallel result in a 4-Ω load. Parallel connections are the most common between multiple speaker cabinets. This is the kind of connection you're making when you use a speaker cable to daisy-chain boxes, or when you stack dual banana connectors. In this scenario the positive post on each speaker input is connected to the positive post on the amplifier output. Also, both negative speaker inputs are connected to the amplifier's negative post.

- Series wiring is more common in internal speaker design using multiple drivers in a single enclosure, or whenever a large array is designed for certain applications. Multiple speakers wired in series present a higher impedance load to the amplifier output. To calculate the resulting impedance of two speakers wired in series, the load presented to the amplifier output is double each speaker's individual impedance. Two 4-Ω speakers wired in series present an 8-Ω load to the amplifier. Series wiring completes a circuit from the positive post-amplifier output, through two speakers and back to the negative post on the amplifier. The positive post on the amplifier connects to the positive side of speaker 1; the negative side of speaker 1 connects to the positive side of speaker 2; then, the negative side of speaker 2 connects to the negative post on the amplifier.

Parallel and Series Wiring

Two speakers, equal in load and wired in parallel, cut the impedance load in half. Two speakers, equal in load and wired in series, double the impedance load. Where "S" equals speaker impedance, the equation used to calculate the resulting impedance from a parallel connection is Load=S1 * S2÷(S1+S2). The equation used to calculate the impedance resulting from a series connection is Load=S1+S2+S3...

When you connect enclosures together in a daisy-chain from the power amp, parallel wiring is the result.

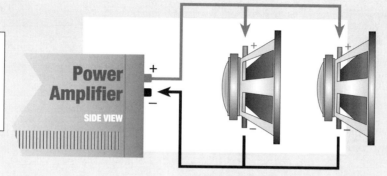

Parallel Wiring: Load=S1 * S2÷(S1+S2)

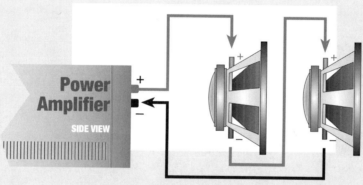

Series wiring is typical within the multi-driver enclosure as a design technique to facilitate an overall enclosure impedance.

Series Wiring: Load=S1+S2+S3...

chapter 15

THE RACKS

Depending on the size of the sound system, there might be several racks of outboard equipment. In a professional system these racks will be very organized and logically laid out, typically consisting of the drive rack, amp rack, and effects rack.

Drive Rack

The drive rack includes all of the system processors—processors that connect between the mixer output and the power amplifier inputs. These include equalizers, dynamic processors, and system time-alignment delays. Typically, the drive rack also includes a real-time analyzer or other diagnostic tools.

It is very important that the processors in the drive rack are top-quality because the entire program passes through them on the way to the amplifiers and speakers. Any substandard devices at this point permanently degrade the resulting audio quality. It doesn't matter how much power you use to run the best enclosures ever manufactured; if you provide a two-dollar sound at the drive rack outputs, you'll get a very accurate rendition of that two-dollar sound at the front of house.

In a large system the drive rack components are often separate rack-mounted devices that connect together with cables in the back of the rack; however, modern devices contain so much DSP (*digital signal processing*) power that many manufacturers offer products that fulfill all of the drive rack functions in one rack-mounted unit. Such devices are manufactured by DBX (the DriveRack) and BSS (the Soundweb).

These powerful multi-effects devices offer all of the tools necessary to finely tune almost any system, including processing, such as:

- 48- and 96-kHz operation with Wordclock input
- Full-color QVGA display
- Four analog and AES/EBU inputs
- Eight analog and AES/EBU outputs
- Optional CobraNet I/O
- Optional Jensen® I/O transformers

- Full band-pass filter, crossover, and routing configurations with Bessel, Butterworth, and Linkwitz-Riley filters
- 31-band graphic and 9-band parametric EQ on every input
- 6-band parametric EQ on every output
- Loudspeaker cluster and driver alignment delays
- Selectable DSP inserts on all inputs and outputs, including compression, limiting, and advanced feedback suppression, among others
- Ethernet networking and control
- Wall panel control

Dynamics

Dynamic control is important in any system. In overall system control, compressors and limiters help with a couple important considerations:
- Overall system level control
- Protection against damaging peaks and extraneous volume surges

Compressor/Limiters

Compressors and limiters are really the same device with the exception of the ratio setting. Compressors use ratios between 1:1 and 10:1. Limiters use ratios above 10:1 up to infinity:1.

The use of dynamics on an entire system differs conceptually from their use on individual channels even though both applications utilize the same processing concept. Applying dynamics to an entire system serves to put a lid on the loud and out-of-control portions of the program, rather than constantly commanding the overall mix level. Extreme compression that constantly turns the system down and then back up again as the signal crosses the user-set threshold typically results in feedback problems during the quiet portions of the show.

The resulting effect of compression is that the portions of the program that would be loudest—that generate the most voltage—are turned down. In the recording world this process allows the engineer to turn the entire track up according to the specific amount of maximum gain reduction in order to regain the original peak signal level. In the live sound world, the listener shouldn't recognize that anything changed during the loud parts; however, the quiet parts of the program should be more apparent and discernable (louder) in direct correlation to the gain adjustments made to regain the original signal level. If the processor reduces the maximum level by 6 dB, then the track or mix can be turned up by 6 dB to regain the original level, resulting in an increase in the minimum track levels of 6 dB and no change in the maximum level.

Limiters are simply a more extreme degree of compression. The controls on the limiter are typically set to the maximum ratio, the fastest attack time, a fast release time, and the output adjusted to make up the lost gain. Limiters are frequently used on the final output of the mixer. They don't typically ride the overall mix level in the same way as a normal compressor. They do, however, control the level of transient peaks that can damage equipment and irritate listeners. A limiter, used correctly, shouldn't cause feedback problems,

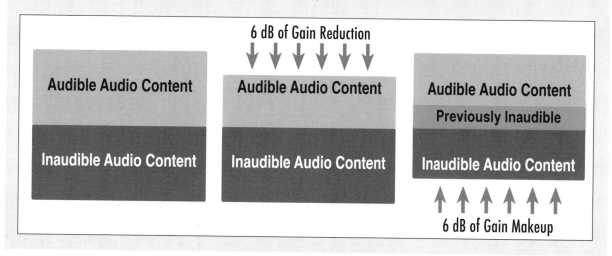

Compressing and Limiting the House Mix

Notice in this very simplified drawing that the end result of compressing and limiting is realized when the gain is made up. Decreasing the loudest portions of the mix decreases the distance between the softest sounds and the loudest sounds. Every decibel that the mix is turned up to re-achieve its original level results in increased audibility and understandability .

although it can help shape a mix that is more aggressive and professional-sounding. Part of the reason a record sounds like a record is the use of a peak limiter during the mastering process. This same process can help your mix sound a little more polished, full, and well-rounded.

Equalizers

Nearly every live sound system uses an equalizer between the mixer and the power amplifiers. The settings and adjustments have less to do with tonal shaping of the mix and more to do with compensating for room acoustics. Each room colors the sounds that happen within its walls. These acoustic colorations can adversely affect the live sound in several ways. They hold the potential to:

- Make low frequencies sound boomy and out of control
- Make the system sound dull and lifeless
- Cause feedback problems across the audible spectrum, depending on acoustic tendencies
- Make some frequencies very loud at some locations in the room and nearly nonexistent at others

The proper use an excellent 31-band graphic equalizer, used in conjunction with a real-time analyzer, can solve many system equalization problems.

Real-Time Analyzers (RTA)

A real-time analyzer utilizes a calibrated microphone, capable of picking up evenly across the frequency spectrum, connected to the RTA input. When pink noise (all frequencies at an equal level) is generated through the sound system, the RTA displays the frequency content as heard through the calibrated microphone. The RTA displays graphically onscreen or

The RTA and the 31-Band Equalizer

Typically, when a real-time analyzer is used to set up the frequency response characteristic of a sound system, the mixer outputs are each routed through a 31-band graphic equalizer.

A calibrated, flat-response microphone is plugged into the RTA input and, while pink noise is generated through the speakers, the RTA display indicates the energy reading of 31 specific frequency bands. These bands correspond to the frequency center points of each control on the 31-band graphic equalizer.

To create a flat frequency response, simply adjust the graphic equalizer faders in response to the amplitude readings on the RTA display.

Although the ultimate goal for any system is a flat frequency response across the audible spectrum, it is common for the actual response to decrease below about 50 Hz and above 15 kHz or so.

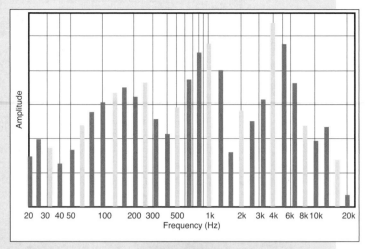

RTA Display

through a series of LED-segmented meters, typically on 31 separate meters that correspond to the bands on a 31-band graphic EQ.

With the pink noise generating over the front-of-house speakers, simply adjust the 31 graphic sliders to achieve the most flat frequency response reading on the RTA.

Video Example 15-1

Using the RTA

Keep in mind that the adjustments to the system equalizers represent the frequency response at the calibrated microphone position, which might or might not be representative of the coloration at other locations in the room. Once the equalization curve is set at the mix position, move the mic to other locations in the room to verify the overall pertinence of the curve. If radically different readings show up as the mic is moved, try averaging the settings so they represent a compromise between locations.

In large rooms, low-frequency waveforms, which contain the greatest amplitude, can create acoustic patterns. As a sound wave reflects back along the same path, there is an increase in amplitude. In frequencies with waveforms that fit evenly between opposing surfaces, these reflections tend to accumulate, developing substantial energy. This energy is heard as an increase at the standing frequency.

Low frequencies are typically between 20 and 40 feet long. In small rooms these frequencies aren't able to develop a standing pattern—this explains why small rooms don't have an abundance of low-frequency coloration. On the other hand, large rooms, such as gymnasiums, coliseums, or theaters, offer ample space for sound waves to develop standing patterns. The low, droning sound of a band playing in a gymnasium or domed arena is caused by standing frequencies in the acoustic space.

Any time there is enough space for sound waves to form standing patterns, doubling in amplitude, there is an equal likelihood that there are frequencies that stand and partially or almost completely cancel. Try generating low-frequency tones over the sound system in a large acoustical space and then walking from the back of the room toward the speakers. You'll find that there are tones that sound very loud at one point in your walk, then there will probably be a spot within 10 or 20 feet or so, toward the speakers, where the same frequency seems to disappear.

This phenomenon is due to the acoustic design of the room, and short of redesigning the room, there's not a lot the sound operator can do other than averaging the mix so that it covers the room as well as possible. Because the problem frequencies are low and omnidirectional, repositioning the speakers doesn't have much effect on the coloration. Adjusting the system equalizer so that the curve is an average of multiple locations can help the operator provide a workable mix.

Crossovers

As described previously, crossovers divide the audible spectrum into segments so they can be sent to separate speaker components. Many systems, especially software-based drive rack systems, allow for separate limiting of each crossover band. The adjustment of limiters on each crossover band should be done carefully in a live setup because extreme settings offer great potential for dramatic feedback problems and erratic sonic tendencies across the volume spectrum.

The exact choice of crossover frequencies depends on the specific components and their reproduction potential. A characteristic of a well-designed system is the smooth transition in frequency response across each crossover point. It is not uncommon to find

The Slope

Notice the shared regions in the graphs below. High-, low-, and band-pass filters ramp up and down at a rate specified in decibels per octave. This rate of energy variation is called the slope—the most common filters ramp in or out at 6, 12, or 24 dB per octave. More extreme slopes help minimize the sonic confusion throughout the shared frequency bands. Less extreme slopes help soften the transition from one component to the next.

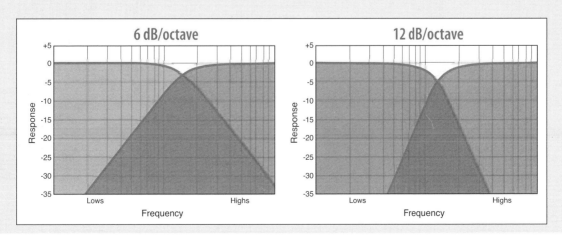

a 3- to 6-dB boost at the crossover point. This is the result of audio accumulation as one frequency range ramps up while the adjacent frequency range ramps down.

A typical two-way system will divide the spectrum somewhere around 250 Hz to 1.5 kHz. The exact crossover point is deteremined by the size and potential of the low-frequency driver and the efficiency of the mid- and high-frequency component.

A typical three-way system will divide the low frequencies from the mids around 200 to 500 Hz. It will also split the mid and high frequencies in the range of 2 to 5 kHz.

A typical four-way system often matches the crossover point of a three-way system, except the sub-bass frequencies below about 100 Hz are sent to the subwoofer.

Some complex systems include components optimized for the subsonic (infrasonic) frequencies below 50, 40, 30, or 20 Hz. The exact crossover point selection is dependent on the components and their response potential.

In addition, some systems include components optimized for the untrasonic frequencies above 16 or 20 kHz. The exact crossover point selection is dependent on the components and their response potential.

Delays

A system with multiple components often requires certain speaker sends to be delayed so that all of the frequency components arrive at the audience in phase and working together rather than against each other. These delay times are often very short, sometimes in fractions of a millisecond; however, proper time alignment results in a sound that is full and powerful throughout the room, rather than weak and hollow-sounding at some locations.

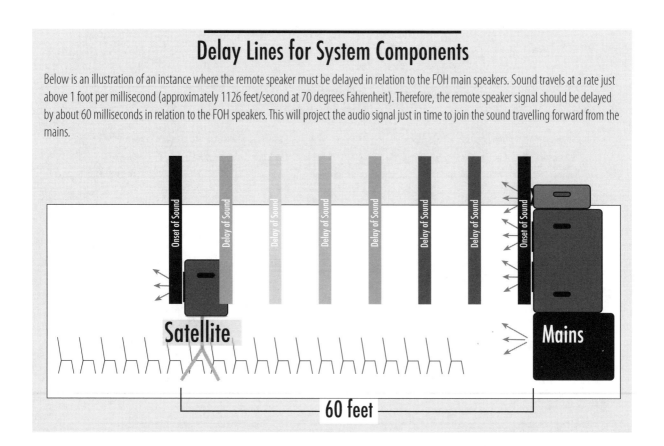

Delay Lines for System Components

Below is an illustration of an instance where the remote speaker must be delayed in relation to the FOH main speakers. Sound travels at a rate just above 1 foot per millisecond (approximately 1126 feet/second at 70 degrees Fahrenheit). Therefore, the remote speaker signal should be delayed by about 60 milliseconds in relation to the FOH speakers. This will project the audio signal just in time to join the sound travelling forward from the mains.

Satellite

Mains

60 feet

The Amp Rack

The proper selection of power amplifiers for any system is not random. First of all, in this industry reputation is everything, whether you are considering who to work with or what equipment to purchase. If you find a power amplifier that seems too good to be true—its specifications are incredible, it's lightweight, and it's less expensive than anything else you've seen, but you've never heard of the manufacturer—be careful. There are a few good names in the professional power amplifier field, and they are very worthy of consumer trust. Any brand that hasn't been proven in the field and favorably reviewed is likely to add drama to your life right in the middle of the most important performance. I don't know why things fall apart in the middle of the biggest show of the year, but somehow that's just how it often works out.

It's important that each amplifier is judged on a level playing field. Power ratings are sometimes confusing to the inexperienced buyer. Terms such as *RMS*, *peak*, and *continuous power* aren't equal, but they can all be useful when considered thoughtfully and when they're assessed fairly. Always combine the implications of multiple specifications. For example, two amplifiers might claim an RMS rating of 500 watts; however, one is rated at .05 percent distortion (THD) and the other at 5 percent distortion (THD). In this case, the amplifiers sound like they're very similar at first glance; however, in reality, they're quite different.

Peak Power

Peak power is the highest instantaneous power potential in an amplifying circuit. It is essentially the absolute maximum amount of power that a circuit can muster up in a mere instant in time. It far surpasses the sustainable power level. Sustained peak power levels result in circuit failure.

Peak power ratings are numerically impressive, but offer little to no value in the assessment of a power amplifier.

RMS Power

RMS stands for *root mean square*, which is technically the square root of the arithmetic mean of the square's set of values. There is a fair amount of math involved in the actual calculation, but for our purposes we need to remember just a few simple conceptual principles. If you are mathematically inclined, visit www.wikipedia.org and look up "root mean square."

Amplifiers are tested using sine waves as audio sources. The RMS value is calculated as .707 times the zero to peak rating of the sine wave. Conversely, the peak rating is calculated from the RMS value by dividing by .707.

Practically speaking, the RMS rating resembles an average power rating in that it closely resembles the average metering of the typical VU meter. RMS power closely equates to perceived volume because, like the human ear, it tends to ignore instantaneous peak levels. In reference to a square wave rather than a sine wave, the RMS value is the equal to the peak value.

Continuous Power

Continuous power represents an amplifier's continuous reliable power output potential and is often equated to the RMS power rating. This rating indicates the power level for continuous duty, typically in reference to a constant load.

THD (Total Harmonic Distortion)

As we've seen previously, it's the harmonic content of the waveform that shapes the timbre (tonal character) of the sound. An accurate amplification of a signal is linear, meaning that the amplified waveform is directly proportional to the source waveform. When plotted on an X-Y graph, where X=time and Y=amplitude, all points on the source waveform are multiplied by the same number to achieve the amplified rendition.

THD is essentially the addition of harmonic content during the amplifying process. It is expressed as a percentage comparing the energy of the harmonic additions to the energy of the original waveform (power of the harmonics [P1] ÷ power of the original waveform [P2]). Imagine delivering a pure sine wave to the power amplifier input, then comparing the actual amplified waveform with the theoretically linearly amplified waveform. From our previous look at phase relationships, we know that combining a waveform with an identical waveform that is 180 degrees out of phase results in silence. When the original waveform is combined with the amplified waveform at the same fundamental amplitude, then combined with the original waveform 180 degrees out of phase, the resulting waveform should contain any distortions that occurred within the circuit, but none of the original signal. Simplistically, it is this resulting power that is compared to the original sine wave to achieve the THD rating.

The lower the THD, the better. It is commonly believed that the human ear can't detect distortions less than about two percent, but taking into account the cumulative nature of audio signals and their anomalies, a low THD rating is important.

Total harmonic distortion that is less than one percent is considered acceptable. Bear in mind that the THD rating should be specified in reference to operating sensitivity, impedance, and frequency. A typical THD rating might be displayed as THD < 0.05% @ 4 ohms, +4 dBm,
20 Hz – 20 kHz.

IMD (Intermodulation Distortion)

Intermodulation distortion (IM or IMD), usually the result of poor amplifier design, is typically less subtle than harmonic distortion. Whereas harmonic distortion is essentially a rebalancing of the naturally existing harmonic content, IMD adds previously nonexistent content. These distortions are typically the sum and difference of frequencies that are present in the original audio waveform.

Because it is not harmonically related to the original waveform, intermodulation distortion is typically more noticeable—and generally perceived as more harsh and grating—than THD. IMD produces higher levels of ear fatigue in comparison to a similar amount of THD.

Speaker Sensitivity

The sensitivity rating quantifies the speaker's ability to output energy when a specified voltage and frequency band are supplied at its input. Typical sensitivity ratings assume the source, delivered to the speaker or cabinet input, is a 1-kHz tone at one watt (2.83 volts @ 8 ohms). It also assumes that a calibrated microphone is placed one meter from—and directly in front of—the speaker and that the calibrated microphone is connected to an accurate audio meter reading dB SPL.

Considering these quantifications:

- Normal loudspeakers have a sensitivity of 85 to 95 dB.
- Nightclub speakers have a sensitivity of 95 to 102 dB.
- Rock concert, stadium speakers have a sensitivity of 103 to 110 dB.

Although sensitivity rating provides a way of comparing the efficiency of speakers and enclosures, it doesn't provide the complete picture. Directional characteristics and the physical ability of components to move large amounts of air also play into the equation. For example, cone speakers can be compared as to the maximum amount of travel distance of the voice coil around the magnet—this movement is called the *cone excursion* and is specified as the X_{max}. Though this is more commonly specified in car audio system subwoofers than in professional audio speakers, the concept provides an instructive visual image.

Power, SPL, and Sensitivity

Bear in mind that there is a fixed relationship between power and SPL. Doubling the power, given the identical input source, results in a 3-dB increase in SPL. Therefore, a 100-watt speaker with a sensitivity of 95 dB will produce 98 dB when given 200 watts. Conversely, when sensitivity is considered, a speaker with a 3-dB increase in sensitivity will require half the power to achieve a given SPL. Seemingly small decreases in sensitivity imply substantial increase in wattage demands.

The assessment of power-handling capabilities, sensitivity, and desired SPL go hand in hand. As with other speaker assessments, specifications only provide useful information when considered altogether and from a quantifiable vantage point.

Correct Power Ratings for Each Cabinet

The cabinet must be capable of handling the power supplied by the amplifier at the specified impedance. For example, a cabinet rated to handle 600 watts RMS at 8 ohms should be paired with an amplifier that can deliver 600 watts RMS to an 8-ohm load. A more powerful amplifier could cause damage to the speaker components. Conversely, though an amplifier with less power than the specified cabinet rating might not damage the components, it also might not produce optimal sound quality. Sometimes, an underpowered amplifier works too hard to achieve the desired SPL, delivering a distorted waveform to the cabinet that is irritating to the listener.

The ideal combination delivers a very accurate powered signal with minimal distortion characteristics to sufficiently rugged and accurate speaker components.

Bridging the Power Amp Outputs

Be sure your power amplifier is designed to be bridged before you attempt this procedure. If your amp isn't specifically designed for bridging, it will be damaged.

Bridging the stereo outputs results in a mono output with increased power. This power amp offers two sets of powered outputs: dual banana and Speakon. Using the banana connection in normal stereo mode, the channel one and two red and black terminals connect to the corresponding speaker enclosure terminals. Notice that in mono the single banana connector is flipped horizontally to connect to the positive terminals of channels one and two.

To bridge to mono, simply connect a banana plug to the red posts or select BRIDGE mode and connect to the Speakon MONO BRIDGE output.

Bridging a Stereo Amplifier

The bridging connection turns a stereo power amplifier into a more powerful mono amplifier—it combines the left and right outputs from the power amplifier into a mono output with increased power.

The bridged connection is very simple to accomplish. From the power amplifier, connect the positive post from output 1 to the positive post on the enclosure, then connect the positive post from output 2 to the negative post on the enclosure.

Bridging is common in large systems in which multiple components are powered by several amplifiers. A theoretical four-way stereo system might include two high-frequency horns, two mid-frequency cabinets, two enclosures for the low-mid frequencies, and a subwoofer for the left channel, along with an identical complement of speakers on the right. That adds up to seven cabinets per side, for a total of 14.

In a large system, each cabinet will typically be fed from its own bridged power amplifier. Bridging is an effective way to provide more power to each enclosure while increasing the fidelity and flexibility of the system. Modern speaker components are efficient and capable of handling impressive amounts of power—power that is effectively delivered by bridging.

Be sure that the power amplifier is designed to be bridged. Any amplifier is easily bridged by simple connection modification, but if the amp isn't designed to be bridged, it could be damaged.

Some amplifiers provide a switch to select bridge mode with a diagram showing proper speaker connection. Amplifiers using banana connections typically position the outputs so that bridging is easily accomplished by connecting the plug across the left-right outputs instead of between the positive and negative posts on just one of the outputs.

Power Amplifier Specs for Bridging

Notice the specified power ratings on the following two amplifiers: the Mackie M-Series and the QSC Powerlight series.

M-Series Specifications

	M•2000	M•3000	M•4000
Continuous Average Output Power in watts, both channels driven			
20 Hz-20 kHz into 8 ohms per channel	325	475	650
20 Hz-20 kHz into 4 ohms per channel	525	800	1050
40 Hz-20 kHz into 2 ohms per channel	800	1200	1600
Bridge Mono: 20 Hz-20 kHz into 8 ohms	1050	1600	2100
Bridge Mono: 40 Hz-20 kHz into 4 ohms	1600	2400	3200
Maximum Output Power in watts, both channels driven			
1 kHz @ 1% THD into 8 ohms per channel	400	600	800
1 kHz @ 1% THD into 4 ohms per channel	650	1000	1300
1 kHz @ 1% THD into 2 ohms per channel	1000	1500	2000
Bridge Mono: 1 kHz @ 1% THD into 8 ohms	1300	2000	2600
Bridge Mono: 1 kHz @ 1% THD into 4 ohms	2000	3000	4000

Mackie M-Series

PowerLight Series

Specifications

	PL3.4			PL3.8^x	PL4.0	PL6.0II
Stereo Mode (both channels driven)				Continuous average output power per channel		
		Ch 1	Ch 2			
8Ω / FTC 20 Hz - 20 kHz / 0.1% THD	725 W	900 W	450 W	900 W	1150 W	
8Ω / EIA 1 kHz / 0.1% THD	750 W	–	–	1000 W	1300 W	
4Ω / FTC 20 Hz - 20 kHz / 0.1% THD	1150 W	1400 W	800 W	1400 W	2050 W	
4Ω / EIA 1 kHz / 0.1% THD	1225 W	–	–	1600 W	2200 W	
2Ω / EIA 1 kHz / 1% THD	1700 W	2400 W	1400 W	2000 W	3500 W	
Bridge Mono Mode						
8Ω / FTC 20 Hz - 20 kHz / 0.1% THD	2300 W	–		2800 W	4400 W	
4Ω / EIA 1 kHz / 1% THD	3400 W	–		4000 W	7000 W	

QSC Powerlight Series

When the amp is bridged, the resulting power output more than doubles. Because two identical amplifiers work together in bridged mode, it's easy to see how the power output might be doubled. Additionally, the two sides working together create a decreased impedance load, likewise increasing the power output potential. The perfect power amplifier in bridged mode, with ideal output devices and a sufficient power supply, would provide four times the single left or right channel potential. In the real world, the bridged power potential is typically about three times the single left or right channel rating—this is a result of inadequate power supplies and output devices.

The following specifications represent the power specifications of a couple professional power amplifiers offered by reputable manufacturers.

Effects Rack

The modern live sound setup uses most of the same tools used in the recording studio, so an effects rack could potentially be somewhat massive and, to the novice, intimidating. Most effects racks simply include multiple versions of just a few conceptual designs, including:

- Mic preamplifiers
- Dynamics processors
- Effects processors
- Equalizers
- Music playback devices
- Recording devices

Mic Preamplifiers

If there is one tool that shapes the fundamental essence of the miked sound, it's the microphone preamplifier. Dynamics and effects processors indeed shape the sound provided from the mic, but the essence of the sound comes from microphone choice, placement, and the quality and accuracy of the preamplifier.

It's the preamp that brings the mic-level signal up to line level. The signal that is produced by the microphone is typically between 30 and 60 dB below line level—the level that the mixer circuitry expects to receive. The quality of the preamplifier circuit is crucial.

A poorly designed preamp delivers a substandard rendition of the miked sound to the beginning of the signal path. If the sound isn't excellent at the beginning of the chain, the sound operator's job becomes much more difficult than it should be. A well-designed preamplifier delivers the best possible rendition of the miked sound to the mixer signal path, giving the sound operator the best chance of creating a great-sounding mix.

Every mixer includes a mic preamplifier with each mic input. Some of these preamplifiers sound excellent and provide world-class audio to the beginning of the signal path. Even in such a case, the preamp so dramatically affects the sound quality that an experienced engineer might choose to bypass the mixer preamplifier, instead inserting an external preamp that adds personality or character to the sound.

Yamaha, Mackie, Soundcraft, Midas, and so on include excellent preamplifiers on all their mixers. Modern circuitry is clean, transparent, and extremely accurate. It's not really necessary to use outboard mic preamps; however, on certain tracks such as the lead vocal, the unique personality provided by a classic tube—or other high-quality—preamplifier can help accentuate the singer's tonal character and musical style.

The microphone plugs into the preamplifier instead of the mixer. The preamp output is then connected to the mixer line input. Phantom power is either provided by the preamplifier or, in case of some tube mics, from the external microphone power supply.

Some of the favored outboard preamplifiers are manufactured by:

- Avalon VT-700
- Presonus ADL 600
- Focusrite Liquid Channel

- Demeter VTMP-2C
- Summit TPA200B
- Mackie ONYX
- Neve 1073
- DBX 786
- GML 2020 and 8302
- Universal Audio 610
- Groove Tubes Vipre
- Manley TNT

These outboard mic preamplifiers are all pretty expensive, but they offer tonal character that is unique and clearly superior.

Dynamics

Dynamics proqcessors include:
- Compressors
- Limiters
- Gates
- Expanders
- Duckers

Each of these devices adjusts signal level in relation to an observed voltage, either at the main input or the external input. Bear in mind that most modern dynamics processors are capable of performing multiple dynamic-control functions—many devices control dynamics in any conceivable way.

Dynamics processors are often inserted on individual channels, so a well-equipped effects rack might contain several. A good mix is strategically molded so that the individual mix ingredients occupy a very intentional portion of the mix space. The proper and musical use of dynamics processors and equalizers lets the sound operator shape each channel so that it works together with the rest of the channels to form a complete mix.

Visually consider each mix you build. Mixes are multidimensional, so your mix image must include:
- **Frequency range (EQ)**: High to low, top to bottom
- **Dynamics (volume)**: Size
- **Depth**: Reverberation and delays
- **Pan position**: Left to right, LCR, surround

Compressor/Limiter

Compressor/limiters help the sound operator define the precise dynamic range of selected channels so that they essentially stay where they're put in the mix. For example, an acoustic guitar typically provides a wide range of volume. In order for it to be heard and appreciated during quiet passages, its level must be turned up; however, during the louder passages the guitarist will most certainly play a more aggressive musical part and, in addition, he or she will strum harder.

In the mix, this variation in dynamics could result in an appropriate sound for the quiet passages and an obnoxiously loud sound during the loud passages. Conversely, the acoustic guitar would probably be inaudible during the quiet passages if its level were set for the loud passages. This concept applies to all mix ingredients with a wide dynamic range, including:

- Acoustic guitar

- Vocals

- Bass guitar

- Drums

- Percussion

WARNING: Considerations as to the use of compressor/limiters are different between recording studios and live sound applications! The recording studio is a controlled environment in which feedback is not an issue. Extreme dynamics control in a live sound environment usually results in feedback as the volume-control circuit releases its control of a loud passage. The really obnoxious aspect of compressor-induced feedback is that there is no feedback during the loud passage where everyone is playing full blast; however, once the texture thins out and it's time for the quiet and gentle portion of the song, the compressors release their control, the mics turn back up, and feedback abounds. Compression and limiting should be used carefully and minimally in live sound as a tool to help provide a controlled and professional-sounding mix. However, the sound operator still must actually mix the music, always paying attention and adjusting the settings for the best and most musically intriguing sound.

HINT: Distance matters! Mics that are more distant from the sound source will typically result in substantial feedback problems if they are over-compressed. Mics that are very close to loud sound sources won't usually cause much problem during the compression process.

CONSIDERATION: Especially in an active acoustical environment, the sound operator often struggles to achieve a tight, intimate sound. Keep in mind that several compressors active at once can detract from an intimate sound. When the miked source (vocal, acoustic guitar, percussion, and so on) is inactive, the level-control circuits return the mic to its uncompressed level, increasing their ambient leakage into the overall mix. The choice to include dynamics processors and to what degree should be based on the type of music and the density of the orchestration. It's most important that the sound operator be aware of the positive and negative ramifications when he or she chooses which tools to use to create a great-sounding mix, and the operator should adjust accordingly.

Expander/Gates

As we've seen previously, gates and expanders are functionally opposite in relation to compressors and limiters. Whereas the compressor/limiter decreases the signal level when it exceeds the user-set threshold, the expander/gate decreases the signal level when it is below the user-set threshold. For example, most guitarists include their own complement of processor, which they use to build the perfect guitar sound for each song. The problem is that guitar sounds often consist of extreme compression use along with one or two types of overdrive—another type of compression. This typically results in a signal containing substantial noise, which is then routed through other potentially noisy processors (chorus, delay, EQ, and so on).

All this results in a great sound, as long as the guitarist is playing; however, whenever the guitarist is idle, there is a very audible hissing, grinding noise coming from his or her equipment. In this scenario, it's the expander/gate that comes to the rescue. The sound operator must simply adjust the threshold so that it is just above noise level—when the guitarist plays, the gate opens up, allowing the full sound to be heard, but when the guitarist isn't playing, the gate turns the channel off, eliminating the noise.

WARNING: If the threshold is not set correctly, portions of the musical ingredient could be completely lost. The threshold must be set so that it opens anytime the musician plays or sings; however, if the threshold is set too high, some of the more quiet and subtle passages might not be strong enough to open the gate.

HINT: There needs to be a clearly defined dynamic difference between signal level required to exceed the threshold and the desired musical source. A distant mic that's used for overall pickup is usually ineffective when gated. The closer the mic is to the sound source, and the louder the sound source is, the better the chance that gating will be effective. Gating and expanding typically work well on loud electric guitar sounds, close-miked drums, and loud brass instruments. Gating and expanding don't usually work well when miking drum overheads, lead vocals, or large groups of instruments or voices.

CONSIDERATION: When miking loud instruments, compression is often useful in creating a polished, professional sound; however, the constant gain increase when the source is inactive can increase the ambient component of the mix in the most exposed musical moments. Using a gate in conjunction with a compressor can help minimize this problem. Patch the channel through the gate first so that the original dynamic range triggers the gate to open—be sure the gate is set to react as quickly as possible and that its release time allows for a natural sound and musical decay. Patch the output of the gate into the compressor/limiter and shape the sound perfectly for the mix. When this technique is applied carefully and the settings are correct, the sound operator can take advantage of the dynamic shaping provided by the compressor and the isolating characteristics of the gate. Many dynamics processors include compression and gating functions.

Here's a list of commonly used and well-respected dynamics processors in the low-to-moderate price range:

- dbx 266, 1046, 576, 166
- Aphex 240 Dual Logic Gated Compressor
- Drawmer DL251, Tube Station, DL241, Three-Sum
- TC Electronic C300, Triple C
- Summit Audio TLA-50
- ART TCS
- Aphex 320A, 720, 661 Tube Expressor
- Presonus ACP-88, BlueMax
- Mackie Quad Comp
- BBE MaxCom
- Alesis 3630

Here's a list of commonly used and well-respected dynamics processors in the high-priced and esoterically cool price range:

- Universal Audio LA-2A, LA-3A, 1176
- Empirical Labs Distressor
- Drawmer 1968
- Manley Variable Mu Dual-Channel Tube Limiter / Compressor, ElectroOptical
- Tube Tech CL2A
- Groove Tube Glory Comp
- Avalon AD2044
- Summit Audio DCL-200, MPC-100
- Tube Tech MEC 1A
- Focusrite Red3
- dbx 160
- GML 8900. 2030
- Solid State Logic XRack
- Neve

Effects

In many live sound venues, the room acoustics provide more reverberation than you can use. In these types of acoustical environments, the use of reverberation and delay effects might be ill-advised; however, even in these reverberant rooms, if the mix is loud—in a stylistically correct way, of course—the sound system could be so dominant in the space that certain effects are manageable.

If the room acoustics provide ample reverberation, delay effects such as slapback and doubling can help define the size of the mix without adding to the accumulation of a wall of decaying echoes.

On the other hand, many venues are extremely dead, providing very little reverberant character. In these types of acoustical spaces, the sound operator must shape the sound and power of the mix in ways that are similar to the recording studio engineer. Reverberation, delays, and chorus effects help trick the audience into feeling as if the musicians and singers are in a large and impressive space.

The number of effects processors included in the effects rack depends almost completely on the size and capacity of the front-of-house mixer. Effects processors are almost always connected to an auxiliary bus so that varying degrees of the effect can be selected for the mix ingredients. If the auxiliary buses are all being used for monitors, recording devices, and overflow feeds, the sound operator might need to devise a compromise to simply free up a bus to connect to an effects device. On the other hand, most large-format live sound mixers contain several aux buses. In addition, they often deal with monitor sends a little differently than small systems, so there might be several auxes available for effects sends.

Whether the sound operator has one or ten auxiliary sends available for effects, it is important that these devices are used musically, tastefully, and effectively—they shouldn't be overused, but the operator also shouldn't be afraid to use them. In the chapter that addresses mixing techniques, more information will be presented to provide a list of valuable and viable considerations when utilizing effects during the creation of a professional-sounding mix.

Here's a list of commonly used and well-respected effects processors in the low-to-moderate price range:

- Lexicon MXP500, MX200, MPX1, PCM Series, MX400,
- TC Electronic G-Force, M-One, M2000, M350, FireworX, M300
- Yamaha SPX900, SPX90, SPX2000, REV7, SPX1000, REV 100
- Alesis Quadraverb, MidiVerb4, NonoVerb
- Antares Vocal Producer
- Behringer V-Verb Pro
- Focusrite Platinum TwinTack Pro
- DigiTech S200, S100

Here's a list of commonly used and well-respected effects processors in the high-priced and esoterically cool price range:

- Eventide Eclipse, H3000, H8000, DSP7000, H7600
- Lexicon 224, 480, 200
- Kurzweil KSP8

Equalizers

Most modern mixers provide ample equalization control. However, each device offers a sonic ingredient that adds character and personality. Outboard equalizers can help the sound operator shape the mix into something that's just a little more exciting and unique than it might be without them.

Certain equalizers, such as the Focusrite Red series, Tube Tech PE1-C, or virtually any equalizer made by Solid State Logic, Neve, Pultec, or Universal Audio has helped shape the sound of modern music recordings. To build a mix that resembles the original recordings, it is helpful to have access to these tools.

Practically speaking, most sound operators can't justify spending the amount of money it would take to have several of these classic devices available in the rack. However, a single splurge or a prudent perusal of eBay and of online merchants could provide one processor that could help add a world-class character and personality to the single most apparent sound in the mix, which is almost always the lead vocal. Aside from system-wide applications, equalizers are usually patched into the insert of one channel at a time.

Here's a list of commonly used and well-respected equalizers in the low-to-moderate price range:

- Klark Teknik DN360
- Drawmer Model 1961
- Apogee CRQ-12
- dbx IEQ31, 2031
- Rane MQ-302
- Mackie Quad EQ

- Yamaha Q2031B
- BSS DPR-901

Here's a list of commonly used and well-respected equalizers in the high-priced and esoterically cool price range:
- Never 1073
- GML 8200
- Manley Mic EQ 500
- Empirical Labs Lil FrEQ
- Summit EQP-200, MPE-200, EQ-200
- Tube Tech PE1-C, EQ 1A
- Urei 545
- Great River EQ-2NV

Music Playback Devices

Virtually every live sound system requires the use of music playback devices for filling gaps before, after, or during the program. Whereas a good CD player is an acceptable tool for this task, new technology is providing some more convenient and flexible options. An iPod or laptop computer provides the type of flexibility that allows the sound operator to choose and build playlists that are custom-designed for a specific performance.

Recording Devices

A well-equipped rack typically contains a recording device that is used for simple stereo recordings of the board mix during a performance. A high-quality rack-mountable CD recorder is an excellent addition to any rack.

Most mixers provide multiple stereo outputs—simply connect one of these outputs to the recorder input. Be sure that the operating levels and impedances match between the mixer output and the recorder input.

An excellent addition to the recording signal path when recording the board mix is a multi-band peak limiter. The incorporation of a multi-band limiter helps polish the rough edges off the mix, compressing it into a more commercial-sounding dynamic range. If you like to listen to board mixes, you'll like them better when they've been run through a good multi-band limiter.

Patch the mixer output to the limiter input, and then connect the limiter output to the CD recorder input. Many modern peak limiters will accept an analog input, and then provide a digital output for connection to the CD recorder. It's usually best to use digital connections where possible. When using a digital mixer, use digital connection throughout the signal chain.

Most CD recorders let the operator manually insert track markers on the fly. While recording the performance, simply press either the play or record button while recording to manually place another track marker. When it's time to listen to the CD or to transfer it into a computer, you'll appreciate the convenience of the inserted markers.

Some sound operators prefer to connect a laptop computer with an external hard drive to the mixer output, recording directly into the computer. Once the program is in the computer, it can easily be processed, trimmed, edited, and compiled into a polished playlist.

Patch Bays

A patch bay is nothing more than a panel with jacks in the front and jacks on the back. Jack #1 on the front is connected to Jack #1 on the back, #2 on the front to #2 on the back, and so on.

If all available ins and outs for all of your equipment are patched into the back of a patch bay, and the corresponding points in the front of the patch bay are clearly labeled, you'll never need to search laboriously behind equipment again just to connect two pieces of gear together. All patching can be done with short, easy-to-patch cables on the front of the patch bay.

Patch bays are used for line-level patches such as channel line ins, direct channel outputs, signal processor inputs and outputs, and any recording device inputs and outputs. Any line-level connection can be routed to a patch bay. Microphones inputs should all be patched directly to the mixer or to an external mic preamp. Line-level outputs from outboard mic preamps are often routed to the patch bay.

Warning: Don't use the patch bay for powered outputs, such as the speaker outputs of your power amplifiers.

The concept of easy and efficient patching through the use of a patch bay becomes obvious when it's explained and even more so once it is experienced. The routing flexibility provided by a patch bay is very convenient, especially when the setup changes frequently. However, there are also some disadvantages, including:

- Intermittent connections
- Loss of signal integrity
- Confusion for inexperienced operators

Intermittent Connections

Any time a connection point is added in a signal path, there is an increased likelihood that a problem will occur. A patch bay introduces four potential pitfalls in the signal path

Keep the Patch Bay Clean

The patch bay provides an easy way to connect and reconfigure the connections between the components in your system. The primary drawback in the use of a patch bay is the potential for intermittent connections. Any metallic material holds the potential for corrosion. Depending on the humidity in your region and other environmental considerations, you might find that your patch bays quickly develop crackles, pops, and noises. The patch bay cleaning tool looks like the other jacks that connect to the front of the bay; however, its surface is rough. It is designed to rough up the contacts in the jack in order to scrape away corrosion from the metallic surface.

connectivity: the front of the patch bay, the back of the patch bay, and both ends of the patch cable. These are aside from the original output and mixer input. In addition, the patch bay is stagnant most of the time, with some patches rarely, if ever, used. An unused patch point tends to corrode in the course of time, so a patch point that works well today might stop working tomorrow.

Patch bay cleaning tools are readily available through your local pro audio supply store or through www.markertek.com and www.patchbays.com. The two primary cleaning tools are the burnisher and the injector. They are both the same size as the connectors on the patch cables, and they're both necessary for thorough cleaning. The burnisher, which looks like a normal plug with abrasive surfaces, simply plugs into the holes in the patch bay—it cleans on its way in and out, plus when the plug is inserted, it is twisted for further cleaning. The injector acts as a nozzle that plugs into the patch bay for the delivery of cleaning solution to the internal connections.

Loss of Signal Integrity

Every time a patch point is added to a signal path, there is a chance that an otherwise strong signal will be weakened. If you're utilizing a patch bay, be sure it is of the highest quality and that you use the best connectors at each point. In addition, keep the connectors and connections clean. Even if a patch point isn't intermittent, there's ample opportunity for it to slowly lose its ability to accurately transfer signal.

Confusion for Inexperienced Operators

If several operators share duties using the same sound system, a patch bay might confuse the inexperienced operator. If the patch bay is truly necessary and adds to the creative and technical freedom in the live sound process, it is well worth the time it takes for all operators to learn to use it. Once the layout is familiar and the patching process becomes routine, all users will appreciate the ease and flexibility in using a patch bay.

Are They More Trouble Than They're Worth?

If the system is small and understated, a patch bay is probably more trouble than it's worth. The potential for intermittence and signal loss outweighs any potential benefit. If, however, your system expands to the point where there are several effects devices and dynamics processors involved, the inclusion of a patch bay might be wise. If a patch bay helps eliminate mess and speeds up the setup procedure, it is worth trying.

Selecting the Appropriate Connection Format

Choosing the best patch bay format for a specific setup is pretty simple. The primary consideration has to do with how many connections you need access to and how much space is available for the patch bays.

Quarter-Inch

The most trustworthy patch bays utilize quarter-inch tip-ring-sleeve connections, simply because they provide more surface area to make contact with the inserted plugs and on the internal jack connections.

Tiny Telephone

Tiny Telephone (TT) patch bays are very common in large studios because they are small and because they provide the ability to include more contact points within a smaller area. These patch bays are often expensive because they use excellent components and because more of them can fit in each single rack space. Typically, TT patch bays are custom-wired, and the connections in the back of the bay are soldered instead of simply plugged in.

RCA

RCA patch bays aren't suitable for professional sound systems. The connections are not trustworthy, and, more importantly, the RCA connector—with only tip and sleeve connections—doesn't have enough contacts for a balanced connection.

Types of Patch Bay Connections

Patch bays contain multiple rows of jacks. These jacks are always arranged in vertical pairs—one above and one below. Whereas the horizontally and diagonally adjacent jacks are discrete (not connected), there are a few different ways the vertical pairs are wired: normalled, mult-normalled, and open.

Normalled

The normalled patch bay connection refers to the rear patch points being normally connected within the patch bay. This is very useful for certain types of connections. For example, it is very convenient to be able to access the mixer's auxiliary outputs. In a typical setting, an auxiliary bus is connected to a power amplifier, which runs a set of stage monitors. If that connection loops through a patch bay, the aux output plugs into the top patch point, and the amplifier input connects to the bottom patch point directly below the aux output. The front jacks remain empty, but the aux output connects to the amplifier input inside the patch bay.

Half Normalled—Bottom

Most patch bays arrive from the manufacturer with the top and bottom jacks normalled; however, most patch bays also provide a means of altering the type of internal connections made. The regular patch bay connection between the vertical pairs is called "half normalled." This connection normally connects the rear jacks together, but when a patch cable is inserted in the lower corresponding front jack, the normalled connection is broken. This enables a new source to connect to the device input.

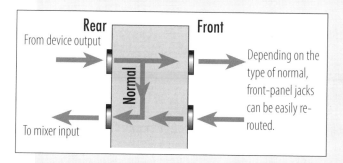

Rear — From device output — **Front**

Normal

To mixer input

Depending on the type of normal, front-panel jacks can be easily re-routed.

Normalled Connections

Notice how the send is looped through the patch bay to the return. This lets the sound operator use a patch bay in its most common functionality without using patch cables. When any reconfigurations are needed, they can be completed with simple and neat patch cables.

However, if a patch cable is connected to the top jack on the front of the patch bay, the normally connected rear jacks remain connected together. This type of connection lets the sound operator gain access to the source signal without interrupting its normal internal connection, essentially splitting the signal. This is a very convenient way to access any mixer channel, input, or output that loops through the patch bay.

Half-Normalled—Top

Most patch bays allow for simple rotation of the channel circuit board, which typically includes the jacks and all connection points, so that the rear connections are still normally connected, except that patching into the top-front panel jack breaks the normal connection, and patching into the bottom-front jack doesn't.

Full-Normalled

A full-normalled connection operates like the half-normalled connection except that the normally connected circuit is broken when a patch cable is plugged into either the top or bottom jack on the front of the patch bay.

Parallel

The parallel patch bay connection normally connects the rear top and bottom jacks; however, patching into either corresponding front jack leaves the normal connection intact.

De-Normalled, Non-Normalled, or Isolated

Patch bays can also be set up to simply connect the back jacks to the front, with no internal connections at all. De-normalled patches are convenient whenever you are preparing access for extra playback gear, such as cassettes, carts, CDs, MiniDiscs, and computers that aren't normally connected to the mixer. It takes up less space to receive a series of left and right output with the left patch above and the right below, rather than occupying a set of four jacks for each device simply to conform to the convention that equipment inputs are typically on the bottom and outputs are on the top.

Additionally, when connecting equipment that's not normally connected to the mixer to the patch bay, it is still convenient to access the inputs on the bottom and the outputs on the top, but when they're not being used there is no reason to want the output normally connected to the input.

What Goes into a Patch Bay

Each setup is unique and requires customized configuration of the patch bay. Planning out an intelligent layout is most of the work whenever you are implementing a patch bay. Quarter-inch patch bays typically offer 48 patch points in a single rack space—TT patch bays can offer 96 patch points in the same single rack space.

Be sure to use patch bays designed to connect balanced devices. These bays require three-point tip-ring-sleeve jacks on the patch cables. If your patch bay specifies that it should use patch cables with a tip and sleeve only, it isn't configured to connect balanced devices, and you should exchange it for a bay with a balanced configuration.

Keep in mind that it isn't necessary to run all connections through a patch bay. Only include the connections that provide the most flexibility and creative freedom during the setup and performance.

Mixer Outputs

There are several mixer outputs that could loop through the patch bay. Many mixers provide direct outputs on every channel. These outputs deliver the channel signal to a variety of convenient locations. In a modern system with an in-ear monitor system, such the Aviom system, these direct channel outputs provide the perfect send to the in-ear system hub. Simply connect the direct mixer channel outputs to the rear patch bay upper jacks and the in-ear system inputs to the lower rear jacks. Leave the patch points in half-normalled (bottom) position so that they're normally connected, yet alternate sources can be easily patched into the in-ear hub, and so that the channel direct outputs can be split for a send to a recorder or other device without breaking the normal connection.

All aux outputs should route through the patch bay on the way to their destinations. In many cases the auxes normally connect to the monitor system power amplifier inputs, or they might connect to an effects device, a recorder, a broadcast feed, or an extra listening area, such as an overflow room, nursing mothers' room, or specific main room zone.

Extra main mix sends should arrive at the patch bay. Often these extra sends are normalled to recording devices, broadcast feeds, or monitor systems for the video control room.

Subgroup outputs should connect to the patch bay. These outputs are typically routed internally to the main mixes, but there are several creative ways to use sub-mixes. They can be used for alternate audio feeds; they can also be used to feed a mix group, such as backing vocals or drums, to a separate dynamics or effects processor.

Mixer Inputs

It is convenient to be able to access the mixer channel line inputs at the patch bay. With the direct channel outputs and the channel line inputs all available at the patch bay, it is a simple matter to rearrange the mixing surface for easier operation. For example, it is always easier to keep track of multiple backing vocal mics when they are in a continuous group of faders on the mixer—one that corresponds left to right with the stage position, from the sound operator's perspective.

Sub-group inputs should be available in case the sub-group outputs need to be routed through a processor and returned to the mix.

Channel inserts should be available at the patch bay. The insert is typically one jack on each mixer channel, which routes the send and return through the tip and sleeve of the TRS connector. Therefore, the single mixer insert jack occupies the top and bottom jacks on the patch bay. In addition many mixer channel inserts are normalled so the circuit is broken when plugs are inserted into the send and return jacks. For this reason, the patch bay must be normalled so that the channel signal path remains intact with the plug in the mixer insert jack.

Auxes often have corresponding mono or stereo returns for use with outboard effects. These inputs should be connected to the effects outputs, looping through the patch bay. This gives the sound operator the flexibility of rerouting the effects return to channels, if that makes his or her job easier, or patching an alternate device into the effects return channels.

Processor Inputs and Outputs

All processor inputs and outputs should route through a patch bay, either normalled to sends and return channels or discretely available for quick and easier implementation. If the processor isn't normalled to a mixer or other device, set the patch points so that they're isolated. In this setting the front patch bay jacks are not connected internally no matter what's patched into the front. This provides the most flexibility and the least chance of catastrophe for the sound operator.

Playback Device Outputs

Playback devices should loop through the patch bay on their way to the mixer. These connections should be normalled (bottom) so that the output of the device can be split when necessary, and yet the mixer input channel can still receive an alternate input, breaking the playback device connection.

Tie Lines

With multiple patch panels and snakes around a stage, it is easy to have more stage channels than there are mixer channels. Often there are a few extra snake channels available at each location. These extra channels should be numbered and labeled as tie lines and connected to

Patch Bays

If all available ins and outs of your equipment are patched into the back of a patch bay, and if the corresponding points in the front of the patch bay are clearly labeled, your sessions will be more efficient. You'll free yourself from searching behind equipment in all sorts of contorted positions, just to connect two pieces of gear together. All patching can be done with short, easy-to-patch cables on the front of the patch bay. Patch bays are made using most standard types of jacks:

- RCA patch bays are the least expensive and work well when connecting gear with RCA connections. However, these do not allow for balanced ins and outs.
- TT (tiny telephone) patch bays use a small tip-ring-sleeve connector. These take the least amount of space and work very well in a professional studio where ins and outs must be balanced and massive amounts of patch points demand efficient use of space.
  - 1/4" phone patch bays are very solid and are the most dependable for ultimate plug-to-jack contact. Though they occupy a lot of space, they offer all the advantages of TT patch bays and provide better signal transfer.

the patch bay. They provide an excellent way to get balanced line-level signals from the stage to the mixer.

Many synthesizer and guitar pedal effects provide balanced professional-level outputs. These outputs can be run through the tie lines to the patch bay, and then connected to the channel line input or other mixer inputs. This procedure often produces the cleanest results, and it also opens up other channels that could be used for microphones.

Tie lines also provide a great way to get an onstage sub-mixer signal cleanly into the main mix. Sub-mixers are often used to consolidate onstage keyboard systems or for turntable mixers.

Recording Device Inputs and Outputs

Recording devices should loop through the patch bay. They are typically normalled to a set of main mix outputs and sometimes are returned to mixer channels. However, there is no real need to patch the output of a recording device back into the mix except for special occasions. It is better to leave the recording device output either disconnected from the mixer or patch bay or simply connected to two isolated jacks.

The main problem with routing the recording device output anywhere near the main mix is that it provides a very real and likely opportunity for the holy mother of all feedback—a real shock and awe experience. If the sound operator forgets to turn the recording device playback channels off in the mix, and then puts the device into record mode as the show begins, an intense feedback loop is created as the mixer hears the recorder hearing the mixer hearing the recording, and so on. This is a showstopper, and I personally don't even like to give opportunity for such an occurrence. Most of the time the recordings are listened to after the show in a car or home system anyway, so unless you have a real need for them, leave the recording device outputs disconnected—you can always use headphones or your car stereo if you need to verify the recording quality.

Planning and Laying Out an Efficient Bay

List all of the connections you want to include in patch bays. You'll probably find an astronomical amount of patches on your first draft. It could be that you actually need every patch on the list, so you might need to purchase multiple patch bays. There's a good chance that you can trim down the list to patch points that you really need to access—this could cut your list in half. You also can develop and implement your patch bay needs in phases. Start out with the patch points that you absolutely must have. This will give you time to decide what else you really need to access. Once you connect the gear to the mixer and mix a few shows, you'll soon decide what you really need to include in your next patch bay phase.

Keeping it Neat

Anything that goes into or out of the patch bay should be clearly labeled on the front and back of the bay. Don't use permanent markers when labeling the bay. Most patch bays provide a white area for each patch point that the sound operator can write on; however, keep in mind that even though you might think your system is perfect and will never need to be changed in any way, there will be a time—probably sooner rather than later—when you'll realize everything needs to be moved around. If you have used a Sharpie or another permanent marker, you might have difficulty re-marking the jacks.

Most manufacturers suggest using a dry erase marker to label the patch points. This is a very non-permanent way to label a patch bay. In fact, with relatively little use, the dry erase labels wear off pretty quickly. Try using a top-of-the-line label maker that can produce labels with very small text to generate small stick-on labels. I like to either make a template for my computer or download one from the manufacturer's website and develop a label strip that attaches all the way across the patch bay. It's pretty easy to get these to be perfectly sized and then to print them on any laser printer on waterproof inkjet printer. They're easy to read, they're easy to change and regenerate, and they're relatively simple to attach. While you're developing your system, simply attach each end of the strip to the patch bay with Scotch tape.

Once the system is pretty well locked in place, try printing out the label strip, attaching two-sided tape (carpet tape) to the back, using scissors to trim the label strip and carpet tape to the correct size, removing the carpet tape backing, and sticking the new label strip in place. These will last for a long time, and yet they're still pretty easy to remove when it's time to make a new label.

Keep your patch cables neatly organized. It is much easier to patch and re-patch if your patch cables are laid neatly in place. Hopefully, most of your connection will be normalled in the patch bay so you won't need to do a lot of patching; however, stuff happens, and you might need to make a lot of connections. Try to keep the bends in the patch cables uniform and grouped together by function—you can usually turn one end of the patch cable to get it to lay a certain way.

Locking Various Components

Each live system contains devices that just shouldn't be casually adjusted. These devices include almost anything that is adjusted to tune the entire system. System equalizers, time-aligning equalizers, system limiters and compressors, and crossover settings should be calibrated during the setup procedure and should rarely be adjusted during a performance. If they do need to be adjusted, only an experienced operator should adjust them—someone who understands the ramifications of the changes he or she makes.

These system control devices are often locked up, especially when inexperienced sound operators constantly use the system, such as in a church.

Lockout Devices

There are several types of lockout devices. There are different sizes, depending on how many rack spaces you need to secure. There are key-locked covers, which are very convenient, especially when fine-tuning the system. It works well to simply use a key to gain access to the devices rather than having to remove rack screws to get the cover off. Most equipment covers are in the form of a solid metal grill for maximum ventilation.

Usually, it is sufficient to simply attach the equipment rack covers with the same rack screws that hold the gear in place—this is usually plenty of protection to deter the average sound operator. Sometimes the overly zealous, yet unqualified, sound operator just pulls out a Phillips head screwdriver and removes the cover to make changes. In this case, the next level of security involves using non-standard rack screws, such as the kind you see on restroom doors, or even hex head screws. Anything that requires a screwdriver that isn't onsite should do the job. If that doesn't work, try involving the CIA, FBI, or CTU—that should stop them.

The Box with No Knobs

Modern system control is often performed through a DSP grid that has no knobs—the only interface might be the onscreen graphical interface provided by the system software. These systems are sure to keep intruders out. They're really pretty simple and convenient, but when they're used, there should be two or three people who know how to gain access and make adjustments.

Devices such as the BSS Soundweb provide ample system control choices and are very functional and powerful. The interface is pretty straightforward, but the system tuning is critical. The fact that there are so many options and so much flexibility makes it important that someone experienced with the device helps set the system up the first time—he or she should also help train the sound operators in the process.

The dbx DriveRack also offers loads of internal processing, routing, and configuring options; however, it provides access to parameter adjustments though knobs and windows on the device, without relying solely on a computer interface.

BASIC EQUIPMENT NEEDS

The following suggestions are designed to help the up-and-coming church sound operator build a system that is cost-appropriate and functionally efficient. The equipment recommendations are based on equipment that I have used or that is from manufacturers who I trust. In addition, each recommendation should be followed with the words "…or the equivalent."

These lists of suggestions assume that the music services use a live band, complete with guitars, bass, drums, keyboards, and vocals. If your church is more traditional or more contemporary, you'll simply need to adjust everything to fit your unique situation.

Generic Recommendations

The purpose of this book is to help you operate and understand the equipment used in a modern-day church service. It's my hope that you will:

- Gain enough insight that you'll be able to get the most out your gear
- Learn to connect the equipment in the most efficient way
- Learn to create an excellent mix
- Better understand the relational and spiritual aspects of church life and ministry

The purpose of this book is not to teach you to be a master system designer. If you know and understand everything in this book, you should be capable of putting together a system, but once the system gets to a certain level of complexity, you should consider hiring an audio professional who specializes in church sound systems.

Professional Help

Most pro audio dealers will be able to lend a credible opinion, and many can even help you set up and tune the system—after all, service is what they are really selling. If your needs are very involved and require system documentation, schematics, and complex design ideas, they'll probably ask for a consulting or design fee. It's fair for them to ask, and it's right that they get paid.

Most small systems won't require much more than a simple understanding of the audio connections; however, larger systems with complex connections and interconnections can be confusing. Even though you might be able to figure out how to connect and use the gear, you might not intuitively optimize the system's functionality, so you might as well take advantage of as much experience and knowledge as you can. Besides, it's Scriptural.

Ecclesiastes 4:12
And if one can overpower him who is alone, two can resist him. A cord of three strands is not quickly torn apart.

Microphones

Don't cut back when it's time to buy microphones. If you buy inexpensive microphones, they won't work very well and they probably won't work for very long. As your church grows, your system will grow, too. Cheap mics will end up in "the round file" within a year or two, whereas a high-quality mic will last for many years.

Microphones from Shure, AKG, Audio-Technica, Sennheiser, Beyer, or Neumann are worth considering. Cost is a primary concern for most churches, so you must choose the mic that fits your need and price range, but most of these manufacturers provide excellent mics in each price range. Buying microphones, as with making other sound equipment purchases, involves choosing from a lot of different options and deciphering information you hear and read. If you have a good relationship with a local music equipment retailer, they'll probably let you try out several different products in your price range. This is a very valuable service and is what sets a local audio dealer apart from mail-order services.

A good, new-technology wireless lavaliere or headset mic is a must. Choose a name-brand system. Wireless technology is ever-changing, and digital television stations are making wireless transmissions even more complex. Buying old technology with a fixed frequency isn't a good idea. Any wireless system should be frequency-agile, and it should choose the best available frequencies automatically.

Cables

If you purchase the least expensive cables in the "cheap bin" at the local music store, you will be sorry. That statement might sound a little harsh and black-and-white, but it is true. These inexpensive cables are prone to breakage—their contacts are unreliable, and the wire is substandard. They might seem inexpensive but, in the long run, they are the most expensive cables in the store. You'll be constantly replacing them, and they'll provide very intermittent functionality.

Most trusted cable manufacturers warranty their cables so you only need to buy them once. In the unlikely event that they fail, the manufacturer will replace them.

Direct Boxes

It's important to use high-quality, well-respected direct boxes from trusted manufacturers. Radial Engineering, Countryman, Whirlwind, and Behringer offer suitable direct boxes for most applications. Instruments that contain lots of acoustical transients, such acoustic guitar, benefit most from the exceptional DIs. Radial Engineering, Countryman, and Demeter offer exceptional DIs that sound great in situations where increased detail, transparency, and signal integrity are most crucial.

Playback Devices

CD players are currently very inexpensive, and most of them will provide adequate service for most churches. Most sound operators are finding that the convenience and flexibility provided by a laptop computer or an iPod is vastly preferable to a simple CD player for most playback needs.

Cases

All reputable manufacturers offer cases for their equipment. It is well worth the investment to purchase cases—even the padded gig bag–style cases can help protect your equipment. Heavy equipment requires heavy duty cases. Most case makers provide products that are very functional and diverse in their cost and intended application. SKB and Anvil both provide excellent products. Allcases.com offers rolling cases with retractable handles and wheels for convenient portability.

Small Church: Fewer Than 75 People

Small churches with an average weekly attendance of 75 or fewer people require a sound system that is primarily designed for vocal reproduction, along with some light instrumentation. Because these churches typically meet in smaller rooms and because they're usually portable—setting up and striking before and after each service—equipment that is compact and lightweight is fundamentally important.

Many small churches use a full complement of singers and instrumentalists; therefore, they might require nearly as many mixer channels as a medium or large church. It is fairly common for all these churches to utilize some or all of the following musicians:

- Music leader
- Four or more backing vocalists
- Keyboardist
- Acoustic guitarist
- Electric guitarist
- Bass guitarist
- Drummer

Small Sound System #1

Handheld mic and stand -
:ements and special music

Headset Mic

CD player, iPod, Computer

ht, non-powered speakers

Music Versus Voice Reproduction

Small churches are unique in nearly every way because they are so very dependent on each member. Because there are very few members, the responsibility for carrying out church duties falls on a few people. There might not be any musicians in the church, and, therefore, there won't be a lot of attention paid to a music reproduction sound system. This church requires a sound system that's capable of amplifying the pastor's sermon, some pre- and post-service music, and a worship leader with a couple accompanying musicians.

On the other hand, some very small churches end up with a disproportionate number of singers and instrumentalists. They might immediately need to step up to a larger sound system.

Sound System Recommendation #1

The minimal church system is also the easiest to set up and tear down. It is self-contained, typically including the mixer, amplifiers, equalizers, and effects in one lightweight package. Simply connect the mixer unit to a set of speakers using speaker cable, and you're ready to rock. Most these devices provide between 6 and 12 channels.

An excellent small-church setup should include some of the types of items discussed in the following sections.

Mixer/Amp/Processor

Fender manufactures the Passport series, which is completely self-contained. It has most of what a small church needs, plus it all cleverly connects together in one suitcase-sized package—mics and cables are even included. You can have all this for between $450 and about $1,000, depending on the model you choose.

Mackie and Yamaha make similar types of systems with an excellent core unit that connects to a couple speakers with speaker cable. These are excellent products. They might be a little more expensive than the Fender system, but they offer a bit more rugged design and different features.

Speakers

Speakers for this application should be non-powered, lightweight, and durable. Speakers from JBL, Mackie, and Yamaha are dependable. Products from these manufacturers or an equivalent will provide years of excellent service. Most of these speakers also contain built-in stand ports for easy placement on lightweight tripod stands.

The nice thing about starting with a small system such as this is that it will never go to waste. As the church grows and steps up to a larger system, other ministries will be glad to use a small system such as this. It should have a long and productive life.

Sound System Recommendation #2

The following systems are excellent choices for a small church with a musical congregation. Many new churches grow out of a group of people who love to worship together. Typically, many of the congregation members play guitar, and there are usually enough musicians in attendance to start out with a respectable worship team.

If your church fits this description, you'll need a little more from your sound system than the self-contained units can offer. The next step in power, flexibility, and number of

Suggestions for Inexpensive Sound Systems

Here are some specific recommendations for inexpensive system components.

Manufacturer: Peavy
Model: Sanctuary Series S-14P
MSRP: $2,000
Street Price: $1,899
I/O Configuration: 12 x 2 x 2
Feature set:

- 10 XLR mic and two stereo line inputs
- Four Automix channels with compressors
- Two-band EQ with low and high shelf
- Special mid-morph EQ circuitry with cut/boost
- Feedback Ferret seeks and eliminates feedback frequencies
- Dual 500-watt power amplifiers
- Monitor blend controls mix of vocal and music tracks in monitors
- Remote and record outputs with auto-level circuitry
- Onboard digital reverb and vocal enhancer

Manufacturer: Mackie
Model: PPM1012 Powered Mixer
MSRP: $1,449
Street Price: $1,099
I/O Configuration: 12 x 2 x 2
Feature set:

- Full-featured 12-channel, 1600W (800W mains and 800W monitors) powered desktop mixer
- 4-segment LEDs and solo PFL on every channel
- 800W + 800W Mackie-designed Class-D, Fast Recovery power amplifiers
- 32-bit dual RMFX+ processor featuring 24 "Gig Ready", usable reverbs, choruses, and tap delay for live applications
- 8 mic/line mono inputs and 2 stereo line inputs for flexible connectivity
- Mackie's 3-Band EQ with a mid sweep (100Hz–8kHz) on mono channels & 4-band active EQ on stereo line channels 7 & 8
- 2 built-in DI boxes for direct instrument connections on channels 7 & 8
- 4 aux sends—2 FX & 2 Mon
- Dual 9-band graphic EQ on the Mains and Monitor outputs for tuning rooms and monitor mixes
- Mono output with level control & low pass filter selectable from 75Hz to 100Hz

channels utilizes a non-powered mixer along with powered speaker cabinets. These systems also provide a greater capacity for monitor systems, playback devices, effects, and recording.

In a small church, portability is very important, so even if there are an ample number of musicians available, the sound equipment should be lightweight and easy to connect.

Suggestions for Inexpensive Sound Systems (cont.)

Manufacturer: Yamaha
Model: EMX5016CF
MSRP: $1,249
Street Price: $999
I/O Configuration: 16 x 2 x 2
Feature set:

- Up to 12 mics, 16 inputs total
- 84 stereo inputs
- Dual aux sends
- 500W
- Yamaha speaker processing
- Dual SPX processors
- One-knob compression
- 9-band digital graphic EQ
- Multi-band maximizer

Powered Mixers and Speakers

There are many mixers available that are portable, feature-laden, and excellent. Because we're still trying to keep the system small and easy to set up, the mixers that contain equalizers on the main outputs as well built-in effects are an excellent choice. Any of these types of mixers that is offered by a respected manufacturer is capable of providing excellent results.

Mackie, Yamaha, JBL, and Soundcraft offer some excellent components for this kind of system.

Mackie

- The CFX series mixers include system equalizers, effects, and an excellent signal path.
- The SRM350 powered speaker cabinets are small and powerful.

Yamaha

- The EMX series mixers provide EQ, effects, and an excellent signal path.
- The Yamaha MSR100 powered cabinet provides excellent audio.

JBL

- The Soundcraft E-8 mixer with the powered JBL Eon-10 cabinets is an excellent choice. The Eon-15 revolutionized the portable sound system market.

Any systems from respected manufacturers that are equivalent to these systems will perform admirably. These mixers and speakers are easily redeployed as the church grows. In addition, these powered cabinets all perform very well as stage monitors when incorporated into a large system.

Small Sound System #2

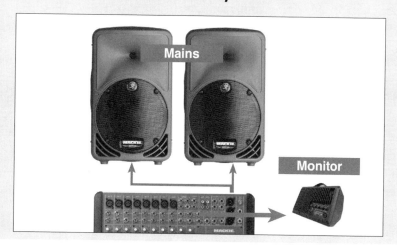

Monitors

If your church meets in a small location, and especially if you set up and strike for every service, consider using small powered monitors, such as the powered Hot Spot from Galaxy Audio. These workhorse monitors are small, they set easily on a microphone stand, and they focus the sound directly at the musician's ear. Unlike a floor monitor, which is designed to be set on the floor five or six feet away from the user, the Hot Spot is best from a distance of two or three feet. In a small room, Hot Spots provide a much more controlled monitor level compared to floor monitors.

Other manufacturers provide similar products. The TC Helicon VSM series and the Fender personal monitor are worth considering.

These monitors require AC power and accept line-level signals from a mixer aux send. They provide volume and tone controls. In addition, they include an XLR microphone input for those very small, intimate gatherings.

Medium Church: Between 76 and 300 People

As your church edges toward 100 people—and then upward to 200 or 300—your audio needs will change. Covering the space occupied by a couple hundred people involves some planning. You'll need to cover a larger area for the congregation, and yet volume will still be a critical issue. You might need to employ more speakers or larger speakers. Your worship team will probably grow, so you'll probably need a larger mixer, more microphones, and increased monitoring flexibility.

This phase of growth is one of the most exciting. There are more people, there may be a need for an additional service, and everyone is up and excited about church life. It's an exciting time in the worship and sound ministries because the demand of more people in a larger acoustical space requires more gear and fun new toys.

The medium-sized church has a new set of concerns for the sound operator. Whereas a small church is typically in a small room and some of the instrumentalists probably play through their amplifiers, a medium-sized church is typically in a large enough acoustic space that the instruments must go through the sound system just so they cover the room evenly, without excruciating volume at the amplifier.

Mixer

In a medium-sized church where the coverage area is much larger, there is much more dependence on sound reinforcement. Therefore, there's a need for a larger mixer. It's easy to use up 24 to 32 channels for a normal worship band configuration.

- **Drums**: If the drummer uses acoustic drums, it will be necessary to use at least a few mics on the kit.
- **Bass**: Whereas the bass amp alone is more than satisfactory in a small church, in a larger room the bass should be run through the sound system so it can be heard and felt evenly across the room.
- **Acoustic guitar**: The acoustic should be run through the sound system so that it can be heard clearly and evenly throughout the room
- **Electric guitar**: Again, the guitar amp can't fill the room evenly, so the guitar will need to run through the house mixer. Most modern guitar multi-effects devices, such as the Line 6 pedal board, sound great running directly into the mixer. In addition, they sound much better in stereo than in mono, so the electric will probably take up a couple channels
- **Electronic keyboards**: The keys sound better in stereo than in mono also, so if your system is stereo, they'll probably take up another couple channels. If your keyboardist uses multiple keyboards and sound modules, it is most efficient to submix all the devices. Position a mixer with the keyboardist and let him or her control the relative keyboard levels from the submixer, then send the stereo submix to the front-of-house mixer. As long as the player listens to the stereo feed from the monitor system, all the parts should blend together and fit nicely into the mix. All this should happen with little effort on the sound operator's part and with minimal frustration on keyboardist's part.
- **Horns, strings, kazoo, and beyond**: Each individual instrument will need to run through the sound system. Even if the instrument is capable of impressive volume, it's better to ask the musician to understate the volume so that the room can be covered evenly.
- **Backing vocalists**: Most churches learn to love the efficiency of placing fewer singers on individual microphones because it sounds better and is much easier to put together musically. You'll probably need between two and six vocal mics for harmony singers. I prefer two harmony singers with no one doubling the melody. Not only is it much easier to get this to sound good, but it is easier to understand lyrically and cleaner harmonically.
- **Worship leader**: You'll need one handheld mic, and it should be wireless. The worship leader might not move around much, but this mic typically gets put to use during the service for announcements, prayers, baby dedications, and so on.
- **Pastor's lavaliere or headset**: This should be wireless and rock-solid reliable. Spend the extra money to get this one right! Your pastor will be thankful. The message that's being preached should not be messed up by wireless dropouts, static, random radio stations, or an occasional drive-by trucker. Get a great microphone and new wireless

Manufacturer: Soundcraft
Model: Si Compact 32
MSRP: $7,499
I/O Configuration: 32 x 2
Feature set:

- TOTEM—The One Touch Easy Mix
- One control = one function
- Soundcraft FaderGlow
- 32 mono inputs
- 4 stereo channels
- 14 sub-group / aux busses
- 4 FX busses
- 4 Matrix busses
- LRC Mix busses
- Mix and matrix busses fitted with compression, parametric, and graphic EQ
- 4 stereo Lexicon effects engines
- 4 mute groups
- AES in and out
- 4 analog insert send-return loops

Manufacturer: Presonus
Model: StudioLive 24.4.2
MSRP: $3999.95
Street Price: $3,175
I/O Configuration: 24 x 4 x 2
Feature set:

- 24 high-headroom XMAX mic preamps
- Built-in FireWire recording and playback interface—32-in/26-out
- More than 90 signal processors with a studio-grade compressor, limiter, gate/expander, 4-band semi-parametric EQ, and highpass filter available on each channel
- Two programmable 32-bit floating point, stereo DSP effects engines, packed with 50 reverb, delay, and time-based effects presets
- 10 aux busses and 4 subgroups give you a high level of control
- QMix wireless aux mix control software for your iPhone and iPod Touch!
- Includes StudioOne Artist DAW software and streamlined Capture recording software

technology, be sure the gain structure is properly set, and, above all, be sure it's ready to go when the pastor talks.

There some excellent analog and digital mixers that are well-suited to the needs of a medium church. They are typically loaded with features, they have dynamics and effects processors available on every channel, and they provide excellent audio quality. However, a

Manufacturer: Yamaha
Model: IM8-40
MSRP: $7,198
Street Price: $5,399
I/O Configuration: 40 x 8 x 2
Feature set:
- 40 input channels
- Precision mic/line preamplifiers
- Versatile input connectivity
- One-knob compressor on all mono channels
- 4-band EQ
- 8 AUX sends
- Channel output control and routing
- Additional channel I/O
- Four stereo aux returns
- Matrix out
- Mute masters
- 2TR in/USB and rec out/USB sections

modern digital mixer typically requires a little practice and study to thoroughly understand it.

As the sound operator and someone who inherently has an aptitude for audio and logical thinking, you should expect to be up and running quickly on any digital mixer designed for live application. It is fairly simple, logical, and intuitive in most cases, and the flexibility and features the mixer provides are fun. On the other hand, if you expect the pastoral staff or volunteers to just be able to walk in and flip a switch and be ready to go, you might be disappointed. You will need to spend some time training the ministry leaders in the basic mixer functions. For a while, you should expect panicked phone calls and reports of dismay, but eventually others will start to understand the system, and church life will ramp back down to its normal hurried pace.

Speakers

Powered speaker cabinets are still very applicable to the medium-sized church. As the church grows, more powered cabinets will probably be required to evenly cover the congregation. You might need to consider upgrading to larger powered cabinets. Most of the respected manufacturers offer sizable versions in their powered series.

The addition of a subwoofer to the sound system will immediately provide better sound quality as the low frequencies below about 100 Hz are channeled to enclosures and amplifiers designed to accurately reproduce them. The proper implementation of subwoofers could be the single most significant leap in the sound of your system. Not only will the low frequencies be more powerful and distinct, but the mids and highs will dramatically increase in clarity and intelligibility.

Monitors

As the church size increases and larger acoustic spaces demand more from the sound system, more intelligent and controlled monitor systems should be considered. In an appropriate room for a medium-sized church, floor monitors are often used. This presents a problem because most singers and instrumentalists want the monitors so loud that the room quickly becomes filled with monitor volume. It is common for the front-of-house sound operator to have too much volume in the room without even turning up the main speakers.

It is difficult to get everyone in the team to function with minimal monitor levels and, in my experience, it doesn't usually last too long once a reasonable monitor volume is set. I am a big fan of in-ear monitor systems in which everyone wears headphones or ear buds to hear themselves and others. There is much more information about in-ear monitor systems in the next chapter.

Large Church: Between 301 and 1,000 People

It takes a large church a while, sometimes years, to settle into a functional routine. If the growth has happened quickly, the development of a capable team of leaders, ministers, and musicians is a challenge. Churches of this size typically hold multiple weekend services in order to accommodate more people in the existing church space.

Somewhere between medium and large, the large church moves to a permanent location. Obviously, many mainstream denominations already own church buildings no matter what the size of the congregation, but many independent and charismatic denominational churches start as a plant from a larger church—they're small, they meet in a school or other public building, and they're portable. They develop their paid and volunteer staff as the church grows, and they have to accumulate the resources to buy or build a permanent building.

At this point, instead of setting up and striking each week, the music team and tech crew get to develop a permanent system. Having helped start a portable church and having been deeply involved in a mega church in a permanent location, I can say with great confidence, "Permanent is better. Not setting up and striking every week is preferable." Now it's time to build something that serves a broad range of ministry needs—to develop a sound ministry that focuses on better communications and procedures that help the church body in exciting new ways.

A church of up to 1,000 people brings several new considerations that place new demands on the sound operator and the sound system.

- There are an increasing number of ministries that require audio and video help.
- There is a larger talent pool.
- Having more talented congregants typically increases the likelihood of special music, dramatic presentations, concerts, and music groups that serve specific ministries (men, women, youth, and so on).
- A larger congregation typically means that there are more funds available for equipment, so there are continual installations, deployments, and re-deployments of gear.

- Because there are more functions, there needs to be several sound operators, so everyone needs to learn to share their toys and peacefully cohabitate.

Professional Help

No matter how knowledgeable and skilled you are, consider taking advantage of a design specialist when you are developing a complex sound system for a large church. Even if you have substantial knowledge and experience regarding music and sound, a design specialist who has successfully implemented several systems brings a valuable depth of insight to the table. In addition to audio system issues, he or she will be able to help predict future audio needs as well as interconnectivity requirements for video, multimedia, intercom systems, recording, and so on.

Mixer

A large church requires a large front-of-house mixer. The increased demands that result from special music, dramatic presentations, concerts, and so on can quickly overwhelm a single sound operator and the mixer. A large church should have a mixer with at least 32 channels. They shouldn't all be used all the time—if you use all 32 channels during weekly services, it's time to buy a larger mixer with 48 or 56 channels.

An active church—especially an active church with strong programs in the arts—will have several services throughout the year that require more mixer channels than a regular weekly service does. In addition, there are often soloists or guest musical groups that participate in the church service. You need enough available channels so that your regular setup can remain intact—complete with all wireless mics, choir mics, and so on—while special music and media presentations connect painlessly to the system.

Try to keep at least eight extra channels available for special circumstances. An event that requires more than eight extra channels should typically run through a submixer—the stereo output from the submixer should simply be routed to a stereo pair of inputs on the main mixer. The primary disadvantages of using a submixer are:
- A separate snake will need to run from the stage to the submixer.
- Providing a good monitor mix to the musicians becomes cumbersome because the individual submixer channels aren't available in the normal aux sends, which are connected to the floor monitors or an in-ear system.

Large-format consoles typically offer excellent features, but they're not all exactly the same. The equalization sections are similar between manufacturers, but some provide fully parametric mid bands, while others only provide sweepable, fixed-bandwidth EQ. Parametric equalizers are conceptually preferable; however, all circuits are not created equal.

A large-format console is difficult to demo in your building—they're often special-order items only, plus disconnecting your old console and connecting a new one is time-consuming. This is a case in which you'll probably need to judge the purchase according to reputation. Ask your pro audio dealer to provide references—people who own the mixer you're interested in. Ask them how they decided on their mixer, how well they like the sound and features, and whether they've had any problems with manufacturer support and warranty.

The mixer in a large church absolutely must be dependable. If it breaks down, the manufacturer must find a way to keep you functional while it's repaired. Dealer service and manufacturer support are far more important than most people realize—if your mixer ever

Suggestions for High-Priced Sound Systems

Manufacturer: Avid
Model: VENUE Profile Main Console
MSRP: $18,895.00
I/O Configuration: 24 x 8 x 2
Feature set:

- A component of a larger system
- Small footprint: 45.3" x 31.1" x 6.65" (1150mm x 790mm x 169 mm)
- 24 input faders, 8 output faders, one Mains fader
- Single row of assignable encoders
- Provides up to 128 inputs when integrated with FOH rack and two stage racks
- Eight GPI inputs and eight GPI outputs allow profile to send or respond to simple switch closures
- Eight assignable function buttons
- The input channel section provides immediate visual and tactile control in a familiar channel strip layout. Each channel features an independent, assignable encoder for quick adjustment of key parameters—such as gain, pan, compression, gate threshold, and aux send levels—without having to assign each channel to the console's central control section.
- Four fader banks allow easy access to the full 96-input capacity of the 24-fader D-Show Profile console.
- Multi-function input faders offer access to all four banks of input faders as well as the dedicated effects returns. They can also be used as encoders for 1/3-octave graphic equalizers.
- View and adjust all parameters for a selected channel within the Assignable Channel Section (ACS). The ACS features dedicated controls for assigning and adjusting busses, auxiliary sends, onboard and plug-in EQs and dynamics processors, and more.
- View mode controls provide immediate access to every D-Show software page—such as the inputs, outputs, filing, snapshots, patchbay, plug-ins, and options pages—at the push of a button. Preview and recall all snapshots with the rotary control.
- The D-Show Profile master section features assignable encoders to control numerous operations, including output pans and mono and stereo (PQ) matrix mixes. The encoders are also mapped for plug-in control—all parameters appear below each encoder in the six-character LCD display.

breaks and you're left orphaned by a mail-order catalog and poor technical support, you'll instantly realize their value.

It's always advisable to purchase a FOH mixer from a reputable manufacturer, such as Yamaha, Mackie, Soundcraft, Midas, and so on. Unknown brands might be good, but they're a gamble, and I don't think the gamble is worth it. For example, there is one manufacturer who likes to copy Mackie. Their gear is pretty popular because they're very inexpensive and they've copied a good design, but I've heard several horror stories about their support. If your mixer is being used at home and you're a hobbyist, maybe it's worth the risk; however, if you're in a professional situation in which people are depending on absolute functionality at all times, spend a little more money for the right tool for the job. In the long run, buying smart is much cheaper than buying cheap.

Suggestions for High-Priced Sound Systems (cont.)

Manufacturer: Soundcraft
Model: Si3+
MSRP: $28,000
I/O Configuration: 64 x 4 x 2
Feature set:

- 64 inputs, 4 stereo inputs, 24 bus outputs
- 8 matrix outputs, 12 VCAs, 8 mute groups
- 4-band fully parametric EQ with high and low cut filters
- On-board dynamics
- Four independent Lexicon processors
- A physical output and meter for every bus

Manufacturer: Yamaha
Model: M7-CL
MSRP: $38,599
I/O Configuration: 48 x 8 x 2 or 32 x 8 x 2
Feature set:

- Centralogic™ Total Access for absolute control
- StageMix provides remote control of M7CL functions
- Versatile channel module functions
- Waves SoundGrid system
- Sends on Fader in M7CL V3 Editor
- M7CL-48ES: Onboard EtherSound and 3rd port
- M7CL-48ES: Auto-configuration for plug-and-play convenience
- M7CL-48ES: Analog insert via OMNI I/O

For a more detailed list of low- to high-priced mixers, visit:
billgibsonmusic.com/media/Church_Sound_Mixers.pdf

Speakers

A large church often requires a complex complement of speakers. There are probably multiple zones and a large primary coverage area, which could require many cabinets hung strategically for the best possible coverage. Powered cabinets are not usually the best choice for these types of systems—the fact that they require electrical current to run the built-in amplifiers is sometimes a problem in a complex install. It's typically more efficient to use power amplifiers and non-powered cabinets.

Potential Mixer Channel Count

It's amazing how quickly the demand for mixer channels increases. Even when providing sound for a medium-sized venue, purchase a mixer with the future in mind. Given the choice between purchasing a mixer that just fits your needs and buying one that has some room for growth, choose the latter. There will be several times when you'll need to add another musician, presenter, drama group, or other special presentation. Plan ahead and spend a little more money to get the right mixer—the mixer that serves your needs over the course of time. It is less expensive to purchase the right tool for the job, up front. It is more expensive to purchase a mixer that will not satisfy the long-term needs, only to end up purchasing a satisfactory mixer shortly thereafter.

The following range of track needs is very realistic.

Instrument	Tracks
Drums	4 – 8
Bass Guitar	1
Acoustic Guitar	1
Electric Guitar	1 – 2
Keyboard Sub-mix	1 – 2
Backing Vocals	2 – 6
Lead Singer	1
Narrator/Presenter	1
CD Playback	1 – 2
Video Playback	1 – 2
SUBTOTAL	**14 – 26**
Horns, Strings, and so on	1 – 4
Choir	2 – 4
Special Guest Speaker	1
Drama Presentation	1 – 4
Special Music	1 – 4
TOTAL	**20 – 43**

The large church system is typically three- or four-way, and the individual components must always be correctly matched. The speaker cabinets should receive the recommended power, and all connections should be made with high-quality connectors and wire.

Monitors

Most large churches, though they might occupy sizable rooms, are still not large enough to absorb the sheer volume of sound produced by onstage monitors. In-ear monitors are usually a better choice for large churches. A good in-ear monitor system provides several benefits when compared to floor monitors, the most dramatic of which are:
• Dramatic reduction in the monitor volume throughout the room
• Dramatic improvement in the team's ability to play and sing together as a musical unit

Portable Systems

Most large churches utilize several portable systems in addition to their main FOH system. Ministry functions throughout the church require sound reinforcement for musical worship, for pre-recorded background music, or simply for spoken word. The best-case scenario is that a large church re-deploys the portable system that served as the main FOH system during the initial phases of church growth.

If the technical crew has made wise purchasing decisions, buying quality merchandise for every need, they will eventually realize the brilliance of their ways. The purchase of a high-quality system, such as the Mackie ONYX series mixer along with powered SR350 or SR450 FOH cabinets, at the beginning of the church is a good decision—it will probably still be in use after 10 to 15 years of service or more. Even though a system like this might cost between $1,500 and $2,000, and the less-expensive options cost $400 to $500, the extra money is insignificant once you consider all of the options.

The inexpensive system will probably fail within a few years and will be relegated to life on the shelf (at a loss of $500). On the other hand, the more expensive system will function beautifully for several years—once it's time to redeploy it to another ministry use, it will perform very well and completely satisfy the need (for no additional expenditure, by the way). If this system cost $1,500 when it was purchased, but there were no additional purchases required until redeployment, then the total cost over the product's life is $2,000. On the other hand, the less-expensive system that fails represents an immediate loss of $500. In addition, once there's a legitimate need for a portable system, the price probably will have increased to $2,500, resulting in a total expenditure of $3,000. Using this logic, it is easy to see that buying the right tool in the beginning could have saved the church $1,000 or so, plus your church will have had the use of an excellent system for the duration.

Mega Church: More Than 1,000 People

Mega churches come in different forms. Some of them feel like home—more like a medium to large church. Other mega churches feel huge—like the big time, where money is no issue and everything you see is the best of the best.

The church where I was on staff for 12 years had a few hundred attendees when we first started attending. The building was a great size for that church because it could comfortably hold 400 people. The church grew very quickly, and we used that same space as the church grew to the point where more than 5,000 people called it their home church. It always felt like the same church, and I think a lot of people liked the fact that the congregation felt smaller than it was. We did end up doing five weekend services with a packed overflow room and people seated in the foyer watching the service on a row of makeshift television monitors—all very exciting, but taxing on the staff.

There are a few common denominators, no matter which type of mega church we consider.

- A church of a thousand or several thousand people is typically operating on an annual budget of at least several hundred thousand dollars—in many cases, the annual budget is in the millions of dollars.

- There's a good chance that a church of more than a thousand people will include highly-skilled professionals in many areas of business, science, and the arts.

- Whereas a small church might have one or two people who share a particular need or emotional concern, a mega church will probably have several. In a church, any time there's a large enough group of people with a common concern, need, or desire, you'll probably find an official church ministry to serve the need. For the sound operator and tech crew, this means increased support requests and staff demands.

- There is an increased need for clear and accessible organization and structure in a mega church. Even in many large churches, the "fly-by-the-seat-of-your-pants" approach can fool some of the people some of the time; however, a mega church requires organization to avoid chaos. Schedules, e-mails, phone calls, planning meetings, staff meetings, budget reviews, and meetings at Starbucks can quickly dominate the schedule of the sound operator team leader.

- As ministry demands increase, so does the likelihood of paid staff positions. Because the music and sound ministries are such an integral part of most churches, the pastor and church leaders usually reach a point of frustration—when no one is officially accountable for the quality of the music and the excellence of the sound—that results in hiring staff.,

Professional Help

In a large facility—or in the case of many mega churches, complex of facilities—there are so many considerations having to do with sound, lighting, video, drama, interconnectivity between systems, recording, networking, and so on that you should consult with experienced professionals for all major equipment choices, installations, and expansions. No book or class can provide the type of information that years of experience designing and implementing systems provides. You will be money ahead to consult with a highly qualified and well-respected system designer or other industry professional when it's time to build, upgrade, or rebuild a system.

Be sure to tap the resources in your congregation. Most technically-adept professionals are pretty busy in the secular world, but they are almost always thrilled to be able to help in church. If it comes to a point when the church needs to hire them for their services, expect that they'll offer a fair rate; however, I've never met a seasoned pro that wasn't anxious to help at almost any level. Helping in church is obviously a labor of love, but most people are pretty humble and are reluctant to step forward to even ask if you need help.

A mega church usually has the financial resources to do the job right the first time—and that is truly the only way anything should be done. Hold your standards high. When you're speaking with industry professionals, they will respect the fact that you want to do a great job and, in most cases, if they can't help you, they'll know who can.

There is one school of thought that says it is best to hire everything from outside the church because it guarantees accountability from the service or merchandise provider. Although there might be some wisdom in that concept, I don't know if it's really how things should work in a church.

When congregants are involved in a church task, they must know that there is a serious intent to achieve excellence. If you hire a church member for a job, be sure to communicate the scope of the job and the expectations for the date of completion and quality level. In my experience, there have been few problems hiring a church member when there was a clear line of communication before, during, and after the job. Usually, the church gets a good rate, and the church member is blessed by the work. When handled lovingly with excellent communication between the church and the service provider, the project typically gets done sooner and better than expected because there's ownership in doing a job for your church and for God.

Mixer

A mega church needs a large-format mixer with ample capacity to perform weekly services and special performances by soloists, choirs, and other musical and dramatic groups. It at least needs to have:
- Forty-eight or more channels
- Several auxiliary sends for monitors and other auxiliary feeds
- Multiple outputs for recording feeds and overflow rooms
- Excellent equalization
- Eight subgroups
- Talkback capability for speaking into separate auxes or the FOH speakers

- A combining matrix for providing a different blend of the subgroups and main sends
- Mute grouping capability

In an active mega-church setting, it is easy to take advantage of most of the features included in any mixer because there are so many different and ever-changing demands.

Analog

There are a lot of analog FOH mixers that are intelligently laid out and functionally efficient. Mackie, Yamaha, Midas, Soundcraft, and a few others make very nice analog consoles that are suitable to a mega church FOH application.

Digital

Digital consoles are becoming more and more reasonably priced. In any new installation, a mega church should consider a large-format digital console—especially if the arts play an important part in the church life. The fact that snapshots can be taken and recalled during rehearsals and services typically results in a more professional presentation. In addition, the fact that each channel includes dynamic processing and ample sends to onboard delay and

Digital Versus Analog Consoles

The mixers below represent some of the most popular analog and digital models. Notice the size differences and consider that the digital boards have several dynamics and effects processors available on every channel. Also, the digital mixers, although about half the physical size of their analog counterparts, typically offer multiple user-selectable layers to utilize the same control surface for additional input channels, group masters, effects masters, and so on.

In defense of the analog consoles, they are much easier to operate, especially for the inexperienced sound operator. The controls are what they are—there is much less mystery in the functionality of an analog mixer in comparison to a digital mixer. Almost anyone should be able to turn on a sound system and get immediate results from an analog board—that's not always the case when using a digital console. On the other hand, digital consoles offer amazing capabilities and, once the operator is familiar with its operation, are extremely efficient, straightforward, and powerful.

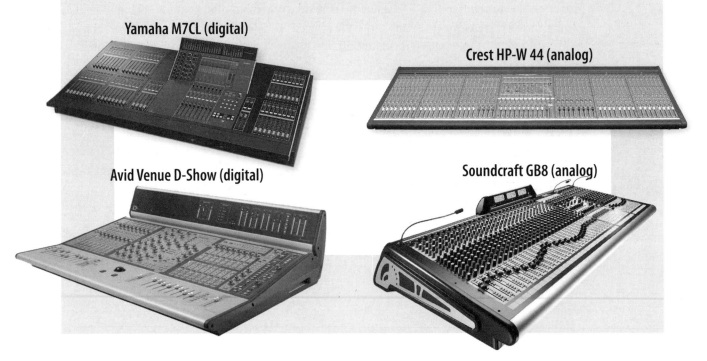

Yamaha M7CL (digital)

Crest HP-W 44 (analog)

Avid Venue D-Show (digital)

Soundcraft GB8 (analog)

reverberation effects makes a modern digital console a bargain compared to a large-format analog mixer with similar processing power.

Speakers

Any room that holds 1,000 or more people requires careful deployment of a well-chosen and expertly designed complement of speakers. Speakers designed for deployment in an array provide excellent coverage and increased intimacy.

Monitors

A large room can typically absorb the monitor volume without destroying the house mix. However, in-ear systems provide for more exact reference for the musicians and help avoid the negative influence of floor monitors on the first few rows of the sanctuary.

Gymnasiums

Many mega churches house full-sized gymnasiums. There isn't a much more difficult room than a gymnasium when it comes to sound reinforcement. Acoustically treating the walls can help a little, but the sheer size and the nature of six long, flat, hard surfaces provides a sound reinforcement challenge, to say the least.

How you proceed depends on how the gym is laid out and whether you want to cover the entire gym—as in, both sides and both ends—or just one side. You also need to determine whether you need a voice-only system, such as that used to announce a basketball game, or a music reinforcement system. Many gymnasiums hang a center cluster of speakers, aimed to optimally cover the entire room, and then they use a separate sound reinforcement system for musical performances and presentations.

Youth Group

Most successful churches have thriving youth groups. Young people in every generation like to push the boundaries of acceptability. Their music usually needs to be louder, more aggressive, and more controversial than the adult music they hear in regular church. In fact, the less the adults relate to it, the better it is in many cases.

Youth group members are also very technically and artistically savvy—they've grown up with the very best imagery and multimedia excellence in history. They know what's good, and they know what's…uhh…not good. Excellent video imagery and sound quality are very attractive to young people. When these factors are delicately interspersed with getting them turned on to God, great things can happen that radically change their lives.

One of the primary concerns in youth ministry is that the youth group members realize the proper balance between their relationship with God and their infatuation with the pretty lights and sound. An excellent youth pastor can skillfully guide the group through this maze, producing kids who love God first, but who still get excited about excellence in music and great performances.

Watch for the kids who love sounds and lights in the youth group. Their aptitude for technical tasks is amazing. Some of my very best all-time staff members have been youth group kids who grew up in church and loved to help.

Permanent System

The permanent system in the youth room should be impressive-sounding; it should look like a concert system that kids see their friends' bands using—except a lot cleaner and less thrashed—and it should be well-organized and documented. It should include a subwoofer and provide a clean and clear version of the music that's run through it—even though some of the sound might not be all that clean and clear.

Portable System

Active youth group members like to get out of the house, and that usually means taking their sound system with them. A portable, such as a Mackie ONYX, Yamaha, or Soundcraft analog mixer, with larger powered cabinets and subwoofers, should function very well for this application. Be sure to purchase road cases for everything. An excellent system for storing cables and mics, as well as well-organized racks, is a must.

SKB and Anvil provide excellent cases and racks for most needs. In addition, there are a couple of online companies (www.portablechurch.com and www.churchonwheels.com) that provide excellent solutions for the portable sound systems, trailers, cases, and so on.

Evaluate Your Existing System

If you've walked into the middle of a preexisting system, you'll need to spend some time evaluating what you have, what works, what's broken, and what's been borrowed or loaned out. If you find the system in disarray, you'll need to proceed in an orderly manner. Make a list of what you have and what you need—this is an excellent task for a computer spreadsheet.

Any time you redesign a system, start at the core and work outward toward less-important components. Fortunately, most systems will contain a certain percentage of high-quality and functional ingredients, but you might need to replace some key ingredients so that you have a solid foundation upon which to build. It is usually best to build a system in a calculated order. For example, if your speakers are blown and malfunctioning, they should be replaced before you buy new microphones. It wouldn't matter if you had the best microphone ever made if the speakers are inadequate.

It is fundamentally important that you have an acoustical analysis performed on your sanctuary. If the room has inherent problems with reflections, standing waves, and other acoustical anomalies, those problems will still be inherent no matter what you do to the sound system. You might be able to get a little better coverage out of a new system, but the acoustical problems will still dominate any sound in the room until the problems have been resolved.

Here is a list of core system ingredients in a subjective order of importance:

*   Acoustic treatment
*   Amplifiers
*   Speakers
*   Crossovers and system processors
*   Speaker wire
*   Mixer

- Cables connecting the mixer to the amplifiers
- Snake from the mixer to the stage
- Monitor system
- Microphones and cables
- Direct boxes
- Effects processors and so on

Sometimes it's not practical to replace system components according to their value to the core system. Certain needs arise, and flexibility should be considered in implementing a plan. However, fundamental system integrity depends on the basic ingredients being capable of producing an excellent sound.

Planning for Growth

Always develop a plan for your system. Determine your current and future needs and desires, and write them down. Things written on paper are almost always easier to understand than things just rattling around in your brain. When you can see a list of everything together, it's easier to distinguish between components you need and ones that you just want.

Sound-system growth should be based on congregational needs. If the entire congregation isn't hearing the message clearly, if there are intermittent components that constantly distract from the intent of the message, or if the system is working so hard to keep up with the demands placed on it that it distorts and unsatisfactorily reproduces the sound of the music team, something needs to be done. Either an upgrade or a replacement should be procured immediately. The message preached in church is of the utmost importance; it needs to be heard clearly and projected faithfully.

Make a list of reasons for sound system development. Just because something is newer and gets a lot of media hype doesn't mean it will do a much better job in your setting. If you base new equipment purchases solely on need, you'll probably find that you buy fewer new items. I usually buy upgrades to all my software as soon as they're out because I like to make sure I'm being as efficient as possible at my musical and writing tasks. I'm always disappointed when I install the latest and the greatest version of a very cool piece of software and it is nearly identical in its performance for the task I perform. Often that's the way new equipment is—it might have a couple hundred new features, but if you don't use those new functions at all in your daily work, purchasing the new equipment is essentially a waste of money. Be sure you need what you buy, and that you only buy what you need; however, if it really increases your efficiency and effectiveness, buy it as soon as you can afford it.

Growing the Plan

If you determine that there are many legitimate needs that will help your sound system become the ministry communication tool that it should be, make another list. List all the equipment needs in order of importance. It's a good idea to ask the advice of your pro audio retailer—they'll probably have some insights that you hadn't though about, and their advice could save you time and money. If they're interested in a long-term relationship, they'll be honest and helpful.

Once you've settled on a plan for developing your system, divide everything into phases. Depending on the number of changes you need to make, it usually is a good idea to develop a three-, six-, nine-, or twelve-month plan. Include your church financial officer in the plan. Accountants love it when there is a plan that can be budgeted for.

Don't be fearful about presenting your needs, even if they seem large and out of control. As long as there are excellent ministry reasons for your suggestions, they'll be considered carefully. Present a plan that includes everything it will take to do the job the right way. Be financially prudent—don't overspend. However, don't under-spend—it's a waste of money to do the job shabbily. You'll only need to go back and redo it the right way all over again.

Analysis Tools

Acoustical analysis is a complex process. It's important that you understand the fundamental processes involved in acoustical analysis; however, detailed analysis is performed by expensive devices, and the data they produce is complex to decipher.

RTAs

The real-time analyzer, which we covered in Chapter 15, provides a rough idea of the overall frequency characteristics of the room at the location of the calibrated microphone. Equalization adjustments are made to the system equalizers in an attempt to flatten out the curve. Other analysis tools quantify the reverberation character of the acoustical space on a three-dimensional graph, displaying the initial sound and its decrease along the z-axis timeline.

The goal of acoustical analysis is to discover echoes, dead spots, and hollow spaces, and then to solve them through the addition of absorbers or diffusers.

TEF Analysis

A commonly used analysis tool is the TEF analyzer. TEF stands for *time, energy, frequency*—the three components of a sound wave. To really be able to tell what goes on with sound in any acoustical environment, you must quantify the frequency spectrum of the source and how it changes in energy level over time.

TEF analysis lets you see the initial sound, whether it's a clap or a guitar, plus the decays as they happen over the course of time, until they disappear.

The TEF analyzer also takes measurements of the polar response of speakers; acoustical measurements, such as 3-D arrival time; and musical acoustics factors. Intelligent analysis of the TEF data along with an understanding of acoustical principles, computer modeling, and implementation of a solid plan should result in acoustics that support excellent speech clarity and musical expression.

More Analysis Tools

There are a few excellent software programs that provide excellent acoustical analysis. SMART Live, Fuzz Measure Pro, and TEF analysis provide valuable information that helps the sound operator and acoustician determine the best course of action regarding control of reflections, echoes, standing waves, and so on.

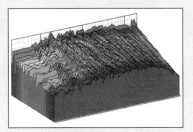

The waterfall chart displays the acoustic decay on a 3-D graph.

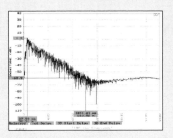

This chart illustrates the acoustic decay on a 2-D graph.

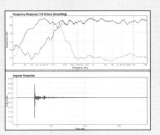

This chart compares the reaction of the left and right channels to a computer-generated chirp.

chapter 17

MONITOR SYSTEMS

The monitor system, sometimes referred to as a *foldback system*, is designed to provide an adequate musical mix for the musicians on stage. They need to hear what the rest of the singers and instrumentalists are doing so that they can accurately perform their portion of the whole performance.

A good monitor mix for all musicians is not optional—it's a must. Musicians are inspired by the sound of their instrument, whether it is a voice or a bassoon. If they hear what they need, their performances will be more musical, better in tune, and much more together with the rest of the musicians.

Monitors typically come in four different configurations:

- Floor wedges
- Side fills
- Small format
- In-ear

Feedback

Feedback happens when the sound system is turned up so loud that frequencies that are acoustically dominant loop between the speakers and the microphone until a loud squeal or howl builds to a climax. The likelihood of feedback increases with each active microphone. The experienced and adept sound operator will mute any inactive mics to decrease the chance of feedback. Some mixers even include an automatic muting feature, similar to an expander, that automatically keeps inactive channels turned down or off.

Feedback is a primary concern in any monitor system, except when using in-ear monitors. The closer the microphones are to the speakers, the greater the chance of feedback. When the system is tuned and voiced for optimum functionality and sound quality, the sound operator is trying to provide a mix that both sounds good and provides the most gain possible before feedback occurs—this is typically referred to as "gain before feedback."

It is best if the monitor system is capable of providing more gain before feedback than the performers really need. If all stage monitors are at their maximum possible levels, there are more potential accumulations that can quickly blossom into a full-blown feedback outburst. In addition, monitor systems that are too loud dramatically degrade the sound of the main FOH mix.

Directional Characteristic

Floor monitors should be very directional. If they provide a wide coverage area, such as that produced by many enclosures designed for FOH applications, there might be adverse effects on the feedback characteristic of the monitor system. It is always best if the monitors can aim at the musicians' ears.

Mic Polar Response

Consider the polar graph (directional characteristic) of the musician's microphone. When the microphone exhibits a true cardioid directional characteristic, take advantage of the 180-degree off-axis discrimination and position the floor monitor either directly in front of the musician or slightly right or left around 170 to 190 degrees off-axis. The specified point of off-axis discrimination is the position at which the least feedback will occur.

If the musician is using a microphone with a hypercardioid directional characteristic, move the monitors into the point or points of off-axis discrimination. Positioning the monitor directly in front of a musician using a hypercardioid microphone will cause more feedback problems because of the increased sensitivity directly off-axis.

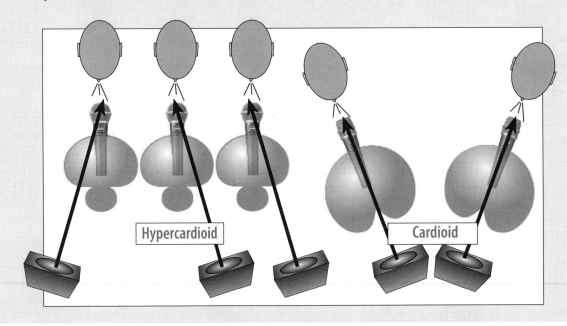

Feedback and Polar Response Characteristics

Some musicians prefer to keep the floor monitor directly in front of them—others prefer the monitor off to the side slightly. Select the microphone with the polar response characteristic matched to the monitor placement for minimal feedback. A true cardioid characteristic exhibits minimum sensitivity directly behind the mic. A hypercardioid characteristic is sensitive directly behind the mic but is least sensitive around 170 and 190 degrees off-axis.

Hypercardioid

Cardioid

Monitor Connections

Monitor systems are typically fed from auxiliary buses on the main FOH console, except in very large staging configurations in which a completely separate monitor mixer is used. Simply connect the line-level aux output from the mixer to the amplifier that is connected to the monitor speakers. If you're utilizing powered monitor cabinets, simply connect the line-level aux send directly into the line input of the powered monitor.

Ringing Out the System

Once the monitor system is completely operational and all mics are turned up roughly to their performance settings, the monitors need to be rung out. Ringing out the monitor system involves the use of a sweepable equalizer with control over four or more narrow frequency bands, typically a quarter of an octave or narrower. This equalizer should be inserted between the aux send from the mixer and the power amplifier connected to the monitors being adjusted.

If your setup uses multiple separate monitor mixes, there should be an equalizer for each aux send, and each set of monitors needs to be rung out separately. Equalizers designed specifically for this type of monitor control typically provide a level cut for each frequency, but there is no boost control. In addition, the bandwidth is fixed and very narrow (less than a quarter of an octave).

Monitor Connection Diagram

Snake boxes provide additional connection jacks, typically labeled A, B, C, D. These jacks are usually used to send the main mix to the amp rack and the auxes to the monitor amplifiers.

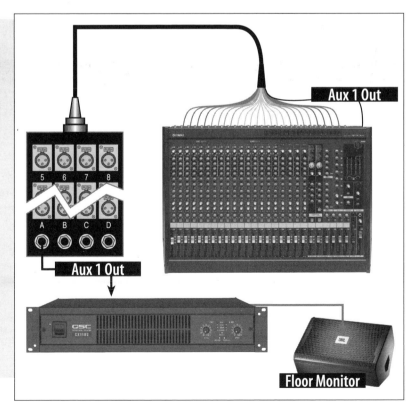

The Procedure

Ringing out any system is a fairly simple procedure.

- Be sure that all mics are on and up to their approximate performance levels.
- With the equalizer in front of you, turn the monitor system up slowly until it starts to feed back. The pitch that you hear first is the most likely frequency to feed back.
- On the first EQ band, set the boost/cut control for an extreme cut.
- Sweep the frequency selector through the suspected frequency range until the feedback disappears.
- When the feedback disappears, turn up the cut level until the feedback returns, and then cut about 3 dB below the level that the ring is revealed.
- Once the feedback disappears, turn the system up further until another ring (feedback tone) appears. If the original ring returns, increase its cut by an additional 3 dB and proceed in boosting the overall level until a new ring appears.
- On the second band, set up another cut and sweep the frequency selector until the new ring disappears.
- Continue until you reveal and remove another one or two ring frequencies.

Once you've completed the ring out procedure, you should be able to reduce the level to a normal level and trust that your monitors will be relatively free from feedback. It is ideal if the selected frequencies are only cut by the necessary amount, which isn't necessarily the maximum amount.

This ring out procedure is very efficient, but often the frequencies that ring first are the ones that provide clarity. If they've been too dramatically reduced, the sound of the monitors might seem lifeless. If you ring out the monitor system and the sound quality seems lifeless, try reducing the cut levels on each selected frequency. Add 1 dB to each band and reassess the sound quality. Continue this process until you reach the proper balance between feedback susceptibility and sound quality.

Video Example 17-1

Ringing Out the System

The Feedback Eliminator

A feedback eliminator automatically adjusts the monitor system equalization to compensate for feedback that occurs at any time. It acts as an automatic electronic equalizer, sensing ringing frequencies and reducing the corresponding frequency until the ringing tendency disappears.

A popular feedback eliminator is the dbx AFS 224. (AFS stands for *automatic feedback suppressor*.) It is a stereo device that contains 24 very narrow-bandwidth, automatic filters per channel. When activated, the device senses feedback, knows the frequency, and turns that frequency down to eliminate the feedback. The bandwidth of the selected frequency can be as narrow as 1/80 of an octave. This tool has very few user-selectable controls and does almost everything automatically.

The selection of the narrowest bandwidth (1/80 of an octave) produces the best results in a high-quality music application. The other bandwidth settings (1/20, 1/10, and 1/5 octave) work best for less complex music setups and voice applications. Wider bandwidths are less processor-intensive, so they respond quicker to developing feedback, though they color the sound more than narrower bandwidths.

This device also provides a user setting that determines whether the filters release the cut once the potential for feedback is gone. It will turn the previously cut frequencies back up to full level within 10 seconds, 10 minutes, or 60 minutes of the disappearance of the feedback threat.

These devices are very powerful and extremely effective, but because they are so completely in control of the system equalization, they can remove life from the sound of the system, whether monitors or FOH.

Singers Can Help

Musical instrument mics are sometimes the cause of feedback problems, but most of the time feedback issues are a result of vocal mics. Because the vocal mics are much more open and mobile, they provide more opportunity for predominant frequencies to loop into feedback. The way singers handle the mics is very important in the elimination of feedback. Here are a few microphone dos and don'ts for vocalists.

Sing Close to the Mic

Timid and inexperienced singers tend to shy away from the microphone, which makes the sound operator turn up the mic so that the vocalist can be heard, which increases the chance of feedback. Singers should stay within an inch or so of the microphone unless they're singing really loudly, in which case they can back off to within a few inches. Experienced singers learn how to work the microphone so that they can sound intimate and close in the quiet sections and blended and balanced in the loud sections. A good singer with great mic technique will do most of the mixing for the sound operator.

Hand Over the Mic

I don't know what it is about the human species, but when we hear a squeal, we tend to put a hand over a microphone. This is precisely the wrong thing to do because putting a hand over the mic increases the amount of feedback. In fact, if you are holding a microphone that

Avoiding Feedback

Getting the open microphone close to the monitor increases the likelihood of feedback, no matter what. However, learning to point the back of the mic at the speaker greatly decreases the chance of feedback.

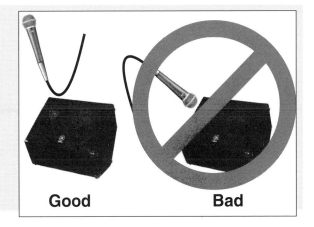

Good　　　　**Bad**

hasn't fed back all day long, simply put your hand over the mic and you'll probably hear a whole bunch of feedback—that is, of course, unless your automatic feedback suppressor is working overtime and getting rid of all the feedback.

Experienced vocalists know that you shouldn't put your hand over the mic when feedback happens, but they only know that because they probably put their hand over the mic the first 100 times it happened to them. If feedback happens, step away from the mic—as long as it's in the stand. It's never good to just drop the mic and run.

Hold the Body

Hold the body of the microphone. If your hand gets too close to the mic capsule or if it surrounds the capsule, leaving just the top exposed, you have turned the nice cardioid mic into an omnidirectional mic—omnidirectional mics feed back a lot. Some singers think this looks cool and they like how it feels to look like a gangster rapper, but it's a really bad idea.

Don't Point

Inexperienced singers like to hold the microphone, and they get very casual about where the front of the mic is pointing. All of a sudden they're talking and joking around with the other singers, and their hand falls to their side, which might be comfortable, but it points the microphone straight into the floor monitor. That's a really bad idea. It provides an unimpeded path from the mic to the speaker, which should cause a huge amount of feedback—again, unless your automatic feedback suppressor is working overtime and getting rid of all the feedback.

Leaning Down to the Monitor

Instruct the singers and vocalists in the proper ways to lean down near a monitor. Ask them to point the back of the mic at the monitor if they must lean down close—this is will create the least likelihood of feedback. Using the off-axis discrimination that is built into the mic design is intelligent, fairly easy to remember, and it helps rid the musicians, sound operator, and audience of the pain and suffering caused by feedback.

More Mics Cause More Feedback Problems

With one open mic on the stage, there is a quantifiable amount of gain before feedback. Placing more microphones in the same proximity as the first mic increases the feedback potential, reducing the gain before feedback by 3 dB with each doubling of the number of mics on stage.

Floor Wedges

Floor wedges come in all shapes and sizes. Some have a 10-inch woofer and a horn for the highs; others have a 15-inch woofer with a horn for the highs. These monitors are quite capable of loudness. In fact, depending on the acoustic space, floor monitors are completely capable of producing enough volume to fill most rooms beyond the desired FOH level. As a sound operator, it is frustrating to fight for a clean sound in the room, and then to realize that turning the master level to the FOH speakers off makes no difference in the sound of the mix—everything has been coming from the monitors.

Look behind the Group

When placing the floor monitors, always consider the directional characteristic of the microphones near them. In addition, pay close attention to the surroundings onstage. If there is a hard wall behind the musicians, the sound from the monitors will bounce directly back to the audience, directly back into the microphones, or up at the ceiling and then back down at the musicians or the audience. Any of these scenarios is bad news for the sound quality of your mix.

Thankfully, the most appropriate rooms for floor monitors are very large, and they typically have an actual stage with lots of empty space backstage and heavy curtains directly behind the musicians. If the group is on a small stage with low ceilings, it will be very difficult to get a sufficient monitor level for the singers and instrumentalists and still achieve a great-sounding mix in the room.

The Angle of Ascent

Floor wedges usually have two or three ways they can sit on the ground—each way angles the speakers differently. Sometimes it is necessary to angle the speakers nearly straight up; other times, a less extreme upward angle is more appropriate.

Get Back, Jo Jo

To most musicians it would seem that the closer the monitor is, the better it will be heard. This isn't always the case—in fact, it is rarely the case. If you move the monitors back about four to six feet from the musicians, set the proper angle, and aim the speakers at their ears, they

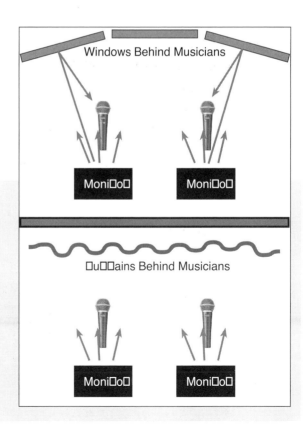

Looking Behind the Group

Soft surfaces behind the musicians help to minimize feedback. The sound operator should be as concerned about the surfaces behind the musicians as he or she is about the other surrounding surfaces. The illustrations below provide markedly different sound. Windows and other glass surfaces behind the musicians cause strong reflection from the monitors back into the microphones. Soft surfaces behind the team deaden the sound and minimize reflections from the monitors back into the mics.

will typically hear much better than if the monitor is close to them and pointing directly up at their ears.

With the monitors back away from the musicians, it is a little easier to aim them at their ears, plus the moderate distance gives the components a chance to develop their full energy together.

The Escalation of Desires

All musicians want to hear themselves best, but they really want to feel like they're surrounded by the band. Musicians like a full sound, and they typically like it loud. In addition, they're almost all convinced that everything is about them, so they want their monitor loud. When you get a whole stage full of these folks, the monitor level just creeps up and up. When it's almost right, one person turns his or her monitor level up, and that causes everyone else on stage to respond by turning their monitors up. It gets out of control very quickly.

When this happens, the sound operator needs to take control and turn everyone's monitor level down to a more reasonable level. There's a good chance that everyone will actually hear better with the monitors turned down a little bit. If the sound of the monitors and the basic mixes is pretty well set, try turning the monitors off and asking the group to play a song without monitors. They'll probably hear a little better than they thought they would. Once everyone is slightly comfortable playing without monitors, ease the level of the monitors back in while the group plays. Have the musicians let you know when the monitor level is just right. There's an excellent chance that the band will be completely satisfied with a fraction of the previous monitor level. In addition, the sound that they hear will be largely from the house, which is a very large and impressive sound. They'll probably get into the sound more and enjoy their performance experience just a little bit more.

To Wedge or Not to Wedge

A very large room with a large and traditionally configured stage area is the best place to use floor wedge monitors. Any time the stage area absorbs rather than reflects the monitor sound, its influence on the FOH mix is reduced. Additionally, a large room with a large sound system composed of many speaker enclosures can more easily dominate the audience area with direct sound from the FOH system, mostly overriding the monitor system leakage.

Side Fills

A large stage, in a large venue, often requires the use of side fill monitors, especially when there is a lot of movement around the stage.

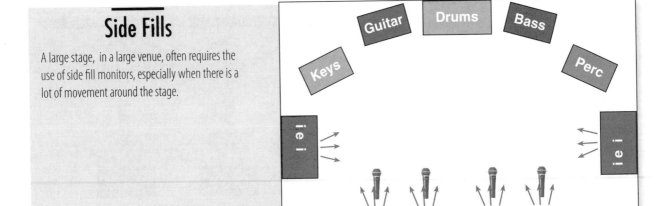

Small rooms and small stages are not conducive to the use of floor monitors. Because of the reflections from the stage area—along with the sheer amount of sound generated by the floor monitors—the stage volume easily dominates the entire room with monitor level, leaving little space for the main FOH speakers to add intelligibility and definition to the mix.

Dealing with Floor Monitor Leakage

The problem with stage monitor leakage into the house is that it really has no definition or intelligibility. It is mostly muddy sounding, providing an accumulation of mid and low-mid frequencies. This component, when in competition with the house mix, makes the FOH mix sound thick and often out of control. This is especially frustrating in situations in which there are volume-control issues.

In many situations, the monitors provide a higher sound pressure level in the room than is desired from the entire system. All you can do in this situation is convince the musicians to use less monitor level and to turn their onstage instrument levels down.

Often, the primary offender is the drummer. Acoustic drums can easily dominate the onstage volume, causing the rest of the musicians to require loud monitors just to hear over the drums. A Plexiglas shield around the drums can help reduce the acoustic drum stage level, especially when used in conjunction with absorption and diffusion behind the drums.

Side Fills

Particularly large stages can bury a large group, amazingly enough. When a performing group consists of a lot of singers, dancers, and instrumentalists, it's very difficult to create a monitor system that covers every portion of the stage in a way that they can all hear well.

Because floor wedges are typically very directional, we often use *side fill* speakers to fill in the gaps and create even monitor coverage across the entire stage. Side fill enclosures are typically rectangular enclosures, just like FOH speakers, that set about head height at the sides of the stage and aim across the stage. Be sure that they aren't aimed in a way that they spread into the front rows of the audience.

Small-Format Monitors

Generally, the smaller the stage size, the smaller the monitors should be. The advantage to small-format stage monitors is that they are positioned within a few feet of the musician and at an elevated level. They're usually placed atop a microphone stand or on a table or other flat surface near the musician. Because of their close proximity to the musicians, they typically have less of an adverse effect on the FOH mix. They simply can't—and really don't need to—generate the massive air movement that a typical floor monitor can.

Often, these small-format monitors are powered. Most of them come with multiple inputs and tone controls. Many include an XLR microphone input so that the monitor can double as a PA system for small and intimate settings.

The leakage that comes from these small speakers is much thinner and less intrusive than the leakage that comes from large floor wedges—it is much easier for the FOH system to overcome. In addition, because they produce less dramatic air movement, these monitors cause fewer onstage acoustical reflection problems.

The Galaxy Audio Hot Spot series has been the workhorse of the small-format stage monitor market for many years. Galaxy Audio offers monitors with many different configurations, which perform very well in small to medium-sized performance venues. The Fender 1270 series and the Yamaha MS101II and MS202II also function very well as small-format stage monitors.

In-Ear Monitors

In-ear monitor systems have become the most popular monitoring system. They essentially feed a mix—often a user-controlled mix—to the musicians through headphones or ear buds. This is by far the best way for musicians to hear exactly what's going on in the group. Many musicians who are stuck in the old school hold a certain disdain for in-ear systems—they simply prefer floor monitors. Some things just are what they are. As the sound operator, you need to provide the tools necessary for the musicians to do their best work. Even though you or I might think in-ear systems are far superior to wedges, the singers and instrumentalists need to be convinced, too. If they need floor wedges to do their best and most musical performance, then that's what they should have. Hopefully, the rest of the musicians prefer in-ear monitors, so at least the stage level will be limited to the one or two monitors that need to be there.

A pure in-ear system, in which everyone onstage uses headphones to monitor themselves and the rest of the group, gives the FOH sound operator complete control and flexibility to create a great-sounding mix—monitor leakage is no longer an issue.

In many groups, the only remaining stage volume is the drums and guitar and bass amplifiers. As the in-ear systems have gotten better and better and as amp modeling has become more realistic, many bassists and guitarists are eliminating their amplifiers altogether Because keyboardists typically run direct anyway, it is quite common that the acoustic drums are the least remaining stage volume issue. Some drummers have been willing to switch to electronic drums so that they, too, are completely controlled by the FOH mixer. Granted, there is an advantage to electronic drums, but there is a musical advantage to acoustic drums. Depending on the style of music and the performance attitude of the musicians, acoustic drums might be the only choice, or electronic drums might be completely acceptable.

Ear Buds with Optional Ear Pieces

The most comfortable in-ear headphones are known as ear buds. They are comfortable, they work well, and they don't mess up your hair. High-quality ear buds, like the Shure E-4s (pictured), provide a full low end and an excellent high end. Notice that there are several different interchangeable ear pieces. They all sound different, feel different, and some even accentuate different frequency ranges. Listen to them all and decide which ones are best for your unique needs.

One of the sets in this picture was custom-molded for my ears. This is the optimal way to use ear buds—they effectively seal off outside sounds and they are very comfortable. Check out www.sensaphonics.com for more information about custom-molded ear pieces.

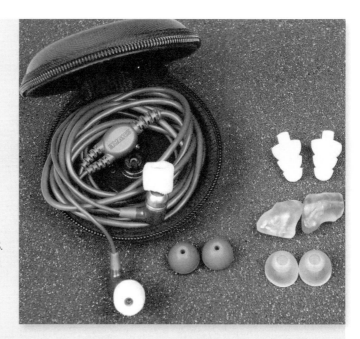

In churches in which volume control is the main concern, electronic drums are often necessary. You need to decide in your church, along with your pastor and leaders, whether acoustic or electronic drums are appropriate.

Headphones

Some musicians prefer to use traditional headphones. Be sure that the headphones cover the musician's ears completely and that there is a seal around them. Even though the headphones are close to your ears, the sound from the room can easily interfere with the in-ear mix if the ear pads are foam or small and non-obstructive.

Part of the reason for sealed earpieces is to eliminate leakage, but the more important part involves volume control for the musicians. Because the speaker elements are so close to the musician's ears, there is a potential for hearing damage if the headphones are constantly extremely loud. If the maximum amount of outside room sound is blocked by the headphones, the musician has the luxury of setting the volume at a modest level and still hearing the mix full and strong.

Studio drummers usually use headphones that are constructed from the same earpieces that the ground control people wear when they're working around jet engines. These earpieces seal off almost all of the outside noises. Because drums are very loud, these headphones let drummers receive almost all of their drum sounds through the phones, even at low headphone volumes; however, in a live setting, these phones are definitely visually obtrusive.

I prefer to use the Sony 7506 series phones. They seal off the outside sound pretty well, they have a strong high end, and they have a reasonably tight low end. In addition, they are very durable, they're a standard in recording studios, and Sony sells all of the parts separately so they're easy to repair.

Ear Buds

Most musicians prefer to use ear buds for their in-ear phones. They are small, visually unobtrusive, and capable of adequately blocking out the outside sounds. Most of the time, the audience is unaware of the ear buds, especially when the user has custom-molded earpieces.

If you're serious about your in-ear monitors, get some custom-molded earpieces. Simply visit an audiologist and have him or her make molds of your ear canals. These molds are then sent to a company that uses those molds to make synthetic earpieces that fit exactly in your ears. These molded pieces are very efficient at blocking outside sounds, and they're comfortable in a performance setting, where they must be worn for an hour or so at a time.

The earpieces simply connect to the ear-bud drivers. Each ear piece must be manufactured to fit your make and model of ear-bud driver. The most popular drivers for in-ear systems are the Shure E series ear buds. The E-3, E-4, and E-5 have become the most common drivers because they sound great and they're durable. The ear buds also come with several standard earpieces, which work well, but not as well as custom-molded earpieces. If you're using the standard earpieces, listen to them all—they all sound different and work better for certain types of source material. Shure also provides a few different sizes of each different material so that users can get the best seal against outside sounds.

Aux-Based In-Ear Systems

Sometimes it is most convenient to feed the in-ear monitoring system from the FOH mixer aux sends. This leaves some of the control with the sound operator and provides a simple way to combine instruments to one mono send. For example, when using a small in-ear system, it might not be practical to send each guitar to a separate in-ear system channel. In this case, simply create a satisfactory mix of all electric and acoustic guitars at the FOH mixer aux send and connect that send to one of the in-ear system inputs.

In the illustration, notice that auxes 1 – 4 are connected to the in-ear hub; however, the keyboard is connected directly to the in-ear system from the DI outputs rather than from a mixer send—any single instrument that is connected through a DI can be connected to the in-ear system in this manner. The send is still going to the FOH system through the DI, but connecting from the DI's quarter-inch output to the in-ear hub gives full monitor control to the musicians at their personal monitor mixers.

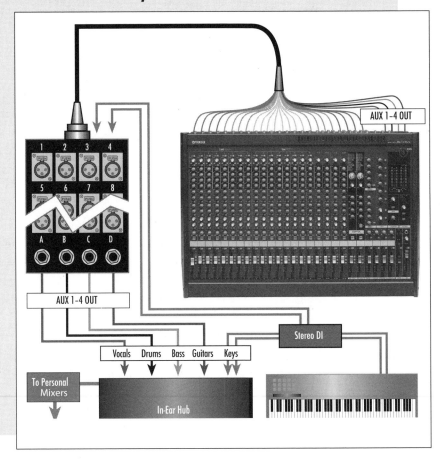

ButtKicker

When using any in-ear system, especially in the absence of a real bass amplifier, guitar amplifier, and acoustic drums, it is common for the musicians to feel as if they've lost much of the power that they feel in truly acoustic live performances. If the house volume is low, this feeling of powerlessness is exaggerated.

The Guitammer Company has come to the rescue with a device called the ButtKicker. This device is a specially designed low-frequency driver that attaches to the ButtKicker platform for those who stand while performing. They also make the drum throne mounting bracket so that the drive can easily be mounted to any drum throne. The ButtKicker really does make the musicians feel as if they're playing in a large venue at a high volume level. Because music is such an emotional expression, a simple tool such as the ButtKicker can help the musicians play better and more expressively together.

Recommended In-Ear Systems

There are a few different in-ear monitor theories, and the systems that serve each school of thought are similar in functionality, although different in capacity.

Aux-Based Systems

Some systems are built with between four and ten inputs. They assume that the FOH mixer will still send separate mixers through the onboard auxes to the headphone system. This leaves the control of the mix in the hands of the sound operator while still eliminating the need for floor monitors.

There are different ways to use these systems. They all make sense, but some might make more sense for your application.

1. Use a small system, such as the Furman HDS-6. This device accepts four mono sends and one stereo send at the hub. Each musician has his or her own personal listening station with level controls for all four auxes and the stereo input. The simplest approach involves the sound operator—he or she should build a custom mix to send to each of the four inputs. These mixes can be used by four individual musicians or shared by multiple musicians.

2. Again, use a small system, such as the Furman HDS-6. With this method the sound operator should develop a mix of vocals only on aux bus 1, drums only on aux bus 2, bass guitar on aux bus 3, and guitars on aux bus 4. He or she should then send them all to the four mono inputs of the system hub. Connect a split from the keyboard DIs to the stereo input of the hub. With this approach, each musician builds a custom mix from all these ingredients. Any monitor system channel that contains one instrument alone can easily be fed from the stage by splitting the DI output into the hub input—this frees up more auxes for the sound operator to use for other tasks.

Channel-Based System

Aviom Technologies has been the dominant force in the development of this technology. Their systems are based on 16-channel hub systems connected to personal monitoring stations that each have full level and pan control of all 16 channels. This is an amazing performance tool for instrumentalists and vocalists.

Direct Outs and the In-Ear System

In-ear systems with increased capacity, such as those manufactured by Aviom, provide greater flexibility for the musician while requiring less interaction from the sound operator. Many of the inputs in these systems are fed by the direct outputs from the FOH mixer channels—they can also be looped through the in-ear system hub on the way to the FOH mixer. These systems connect using standard CAT5 cables, so it is common to position the system hub at the FOH mixer position, running inexpensive CAT5 cable to the stage, and then distributing the audio signals to the personal monitor stations.

The sound operator should make creative use of available auxes and channel direct outputs. In the illustrated connections (right), the first channels are connected to the hub through the direct outputs of the first 12 channels. Auxes 5 and 6 are providing a sub-mix of the drum overheads, toms, and hi-hat. Aux 2 is providing a sub-mix of the backing vocals, and Aux 1 consists of a sub-mix of the lead vocalist and other announcement mics.

Everything that is connected to the in-ear system hub is available at the personal mixing stations.

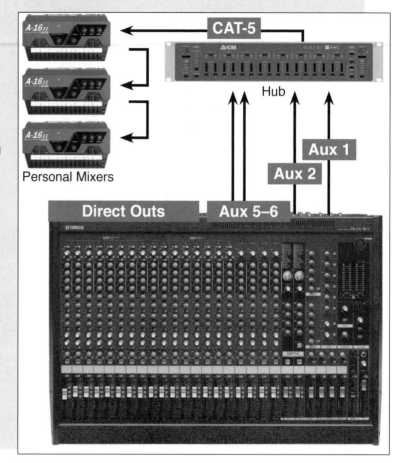

With this system, musicians can adjust the mix so that they hear exactly what they want to hear. Aviom devices carry high-quality analog signals along with high-definition digital audio up to 192 kHz, 24-bit. Their Pro64 series also provides for the transmission of up to 14 channels of non-audio control data, such as MIDI, RS-232, and GPIO.

These devices deliver all 16 audio channels along with the other communication data on standard CAT5 computer networking cable, so connections are efficient and the cable is readily available and relatively inexpensive.

Sixteen channels are obviously not enough to accept every channel on a large-format console with 48 or more channels; however, in most applications 16 channels is sufficient to build an excellent monitor mix. Simply patch the channel direct output for as many channels as is practical into the hub. In addition, set up sub-mixes using the FOH mixer aux buses and feed them into the hub. With a little adjustment over time, most groups find it very satisfactory to use this type of blending of the direct channel outputs and some of the aux sends.

Interface with FOH

There are a few different ways to interface with the house system—all make functional sense, although certain theoretical signal integrity considerations shouldn't be overlooked. One way or another, the microphone and channel line input signal must be split so that they route simultaneously to the mixer and the monitor system core.

Direct Outputs

Many mixers provide direct outputs for every channel. These direct outs are a simple passive split off the mixer's incoming signals—they are perfect for use as the input source for the in-ear monitor system, such as the Aviom. Simply patch the channel direct outputs to the line inputs on the hub.

The channel direct outputs often provide a PRE/POST switch that selects the split point.

- In PRE position, the signal is split just after the input preamp. This signal is affected by the trim adjustment alone—equalization and fader levels don't make a difference on the direct signal.

- In POST position, the signal is split after the EQ section and is typically affected by fader movement. When POST is selected, the direct out to the monitors is therefore affected by the gain trim, fader settings, and equalizer settings.

In the majority of circumstances the direct outputs must be selected as pre fader so that the fader and equalizer settings aren't constantly changing the monitor levels. It is best if the FOH mix and the monitor sends are completely independent of one another.

There are specific scenarios in which the use of the POST setting on the direct outputs is worth considering. Any instrument that requires substantial EQ for the FOH mix probably requires the same EQ to sound good in the headphones. If the equalized signal helps the musician feel better about his or her instrument or voice, it will probably inspire a better performance.

The mixer subgroups can help facilitate a practical method for providing a post-fader send to the musicians. For example, once the drums have been equalized for the perfect sound and the levels have been adjusted, assign the entire kit to a stereo subgroup. Be sure to take the drums out of the main mix at each channel, and be sure the subgroup master is assigned to the main mix. With this setup, the sound operator should leave the individual channel levels set, adjusting the overall drum-kit level with the subgroup master. As long as the sound operator and drummer understand that the channel fader and EQ movements will affect the drum mix, but that the level should stay constant unless it really needs to change for the good of the FOH mix, this system works very well because the drums sound so much better with EQ than without.

System Hub Splits

Many in-ear systems provide a splitter in the hub device. Using this tool, the microphone and line inputs connect to the in-ear hub—the signal split is performed in the hub before it can be connected to the FOH mixer. This system works well and is very convenient in situations in which the FOH mixer doesn't provide direct outputs. Because these signals enter and exit the in-ear system hub before they even get to the mixer, all signals are pre-equalizer and -fader; therefore, some sounds might not be as polished in the in-ear system as they are in the house.

Be sure that you're very convinced that the signal split in the in-ear hub maintains the original signal integrity that it receives at its input. It's always best to connect the stage feeds directly to the FOH mixer so that there are the fewest possible opportunities for intermittent and degraded signals. However, splitting through the in-ear hub is often the solution that makes the most sense when all things are considered.

Splitter Snake

The splitter snake is used frequently to split the stage feeds for separate sends to recordings or a separate monitor mix system. When using a splitter snake, the connection box on one end receives the stage microphone, line-level, and DI signals, which are sent to the monitor-control device. Also on the box are XLR outputs for every input—these outputs are used to connect to the FOH mixer. The splitter snake is very effective, and it provides a passive split off of each input it receives.

The Splitter Snake and the Dedicated Monitor Mixer

In a large and complicated system for a complex stage presentation, a dedicated monitor mixing console is often used. A sound operator, separate from the FOH operator, does nothing but provide custom mixes for every instrumentalist and singer.

Simply connect all stage inputs to the snake box and then run one snake to the FOH mixer and the other snake to the monitor mixer—all of the splits are wired inside the snake box.

In a system where the signal is split several times, it is best to use an active splitter, which electronically regains the signal lost during the splits and verifies optimal impedance matching for all connections.

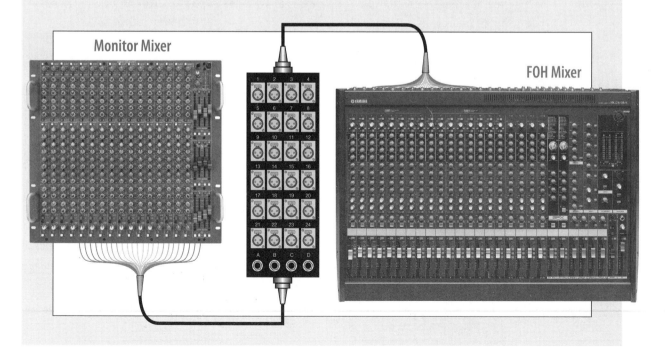

Active Versus Passive Split

Each signal-splitting system so far has utilized a passive split, which is essentially the same as using a simple Y cable for each connection. The nature of input and output impedances is such that splitting a signal for distribution to a few different inputs doesn't cause significant signal degradation, so passive splitters provide acceptable results. If, however, the signal must be split more than a few times, an impedance mismatch is created between the inputs and outputs, and there is potential for signal degradation.

The active splitting device uses electronic circuitry to maintain the signal impedance and integrity so that the split is inconsequential. If your situation demands that you split the stage feeds for sends to the FOH mixer, monitor system, recording setup, overflow mix, and so on, you might need to include an active splitting device to regain optimum signal integrity.

A device such as the Whirlwind AS8X4 provides one passive and four active splits for each input channel. Although the active splitting devices are much more expensive than passive splits, they are necessary in many setups to guarantee the best possible signal at each destination.

Aux Bus

In addition to simple signal splits, most in-ear systems require the incorporation of aux sends from the FOH mixer. Because the systems, such as the Aviom, provide 16 channels of control and because most FOH mixers provide 24 to 48 channels, there must be some obvious compromises and combinations. Auxiliary bus sends can easily be used to combine channels that function together musically. For most purposes, the following channel groups can be combined through the aux buses to provide an adequate in-ear mix:

- Choir mics
- Pastors, announcers, and worship leaders
- Drum overheads and toms
- All mics used for dramatic presentations and skits

Monitor Mixer

A separate monitor mixer, sometimes called the *monitor mix matrix*, is often placed onstage with the musicians. Sometimes it's operated by one of the musicians; however, in a large stage production there is usually a dedicated sound operator whose primary task is setting up and operating the musician's in-ear or floor monitor mixes.

This system is unlike the Aviom system because the musicians don't have control over their personal mixes—they need to ask the onstage monitor operator for changes. This is actually preferable in many performance situations in which staging activity and logistics make the physical positioning of the personal mixer cumbersome. Also, a professional-quality monitor mixer has an identical channel configuration to the FOH mixer, so each channel is mixed at the musician's preference without the limitations of summing through aux sends or eliminating ingredients due to a lack of available channels.

A good monitor mix matrix provides pan controls for each channel, as well as adequate equalization, trim adjustments, and master send levels. Some monitor mixers are rack-mounted and function together to provide adequate track configurations; however, most professional monitor mixers look a lot like the FOH mixer. They often utilize rotary pots for channel levels, rather than faders. This helps free up control surface space for more aux sends. In addition, channel level controls aren't any more important than the aux send levels on this type of mixer. Monitor mixers often use 12 or more aux sends that are connected to the amplification systems for onstage or in-ear monitors.

Getting Wireless

Musicians who are in fixed locations, such as drummers, keyboardists, and often guitarists, bassists, horn sections, and backing vocalists, can typically function uninhibited when wearing wired headphones. However, the leaders and active musicians should consider wireless receivers for their in-ear link to the monitor mix. With this type of system, the mix output is connected to a wireless transmitter, which sends the mix to the body pack receiver that connects to the musician's belt or a clothing item.

Systems such as the Shure PSM and Sennheiser 300 series wireless in-ear monitors provide exceptional frequency agility along with excellent fidelity and flexibility.

Characteristics of a Good Monitor Mix

The musician using the monitor system determines whether his or her mix is good or bad. Unlike a mix in the house or on a CD—where most people will agree when something sounds great or blatantly bad—the monitor mix just needs to provide the sound that inspires the musician to perform world-class music. However, as a seasoned sound operator, you can help the musicians through the process of building their own mix. Sometimes they are so inexperienced at mixing music that they don't recognize the ingredients that are getting in the way of the sound they want to hear—they just know they can't hear themselves, and they don't like it.

Listen Up!

When setting up the monitor system, be sure that you get in position to hear what the musicians hear. If they're using wedges or small-format monitors, make your way up to the stage to listen to their monitors. Listen for the band and the voices; be sure that singers are getting enough of the primary pitched instrument so that they'll have an accurate frame of reference for pitch—how they hear determines whether they'll sing in tune or struggle to find the correct pitches.

If the musicians are using in-ear systems and they're struggling to hear well, grab some headphones and plug into their mix. Keep a stereo headphone Y cable around for this purpose so you can plug into their system and they can still hear as you help them with the mix. Sometimes the problem they're having has to do with the sub-mixes you're sending from the FOH mixer auxes. These auxes can typically be soloed at the mixer, so check them occasionally just to see whether the mix they're getting is satisfactory.

SYSTEM CONFIGURATION AND LAYOUT

Whether you're the sound operator in a church, a local club, or a major arena, be mindful that all venues are different and they all require special attention to details concerning sound system placement and tuning. This chapter provides several important considerations regarding various setup and design scenarios.

Mixer Placement

If the sound operator isn't in a prime listening position in the room with the congregation or audience, the decisions he or she makes about the sound quality and balances will be based on bad information. Too many churches put sound operators at a huge disadvantage by sticking them in the balcony, trapping them in the very back of the room, or even worse, putting them in a separate booth—the list goes on and on. Churches are too often designed with the maximum possible number of seats for the congregation, leaving the sound operator position at or below the bottom of the list of priorities.

If you want the best possible results in any live setting, the sound operator must be in a listening sweet spot. You will dramatically improve the sound quality by simply moving the sound operator into the room with everyone else, about three-quarters of the way toward the back of the room, in most cases, and lined up with the correct monitoring position for the specific panoramic configuration.

Mono

If your system is mono, meaning exactly the same mix is sent simultaneously to the left and right speakers, the mixer should be placed directly in front of the left or right speaker. Rather than occupying the center position, in which the sound operator hears the same source from left and right anyway, a more accurate—or at least as accurate—mix for the room can usually be achieved when directly inline with one side.

This position also helps create less of an obstruction for those behind the sound operator.

Stereo

Most stereo mixes in a live sound application are fundamentally mono mixes with stereo effects. Keyboard and guitar effects sound wide and impressive in stereo, yet the same amount of energy is really coming from the left and right sides—the delay and reverberation effects simply vary parameters between left and right to help achieve a spacious sound.

Depending on the source, there might be some occasions when mix ingredients are panned to one side or the other as a special effect, but because the entire room full of listeners must receive a correct balance, it's impractical to pan various instruments left or right. For example, panning the guitar left and the keys right would give one side a mix with lots of guitar and no keys, and the other side a mix with lots of keys and no guitar.

If there are reasons to position mix ingredients across the panoramic spectrum, the mixer should definitely be centered in the room. Even if the stereo aspect of the mix consists of simple choruses, delays, and reverberation effects across the stereo spectrum, it's a good idea to position the sound operator and mixer in the center of the room, near the back. This provides the best perspective from which to make valid mixing decisions.

If space is an issue or if the sound engineer prefers, it's acceptable to position the mixer either off-center or inline with one side, as long as the mix is utilizing stereo effects, yet isn't truly stereo.

LCR

It is very common in large installations to take advantage of the LCR (left-center-right) capabilities of modern mixers. In this configuration a center cluster is used to fill the center image with primarily voices (pastor, announcements, worship leader, and so on) with the left and right channels providing the musical instrument component of the mix.

Like the stereo mix, the LCR mix commonly uses the left and right sides to create width and depth in sounds that are otherwise fundamentally mono; however, it is still preferable for the mixer to be positioned in the center, directly in front of the center cluster and between the left and right sides. Even in the case of great primarily mono sound with stereo effects, it's best to stay in the middle of the LCR image. Deviating more than roughly 15 percent from the center in either direction will create a problem for the sound operator in creating a mix that is satisfactory for the entire room in its relationship between voices and instruments.

Walk the Room

It's very important that no matter what configuration is being used, the sound operator walks around the room to verify the sound quality. A sound operator who sits at the mixer and doesn't wander the room can't have an accurate perception of what the rest of the room is hearing. In addition, walking the room can help alert the sound operator to developing malfunctions within the system. Large and complex acoustical environments could have several subsystems within the overall acoustic space. Each subsystem should be assessed regularly.

Rehearsals and sound checks provide excellent opportunities to walk the room, but there is nothing quite as enlightening as moving around the room during a performance with an audience in place. Find an assistant or a trusted team member to operate the board while you see what it really sounds like around the room. Don't leave the musicians or presenter with an unattended mixer; however, it is worthwhile to trust a less experienced person at the mixer while you get an accurate and realistic perspective on the sound quality.

Speaker Placement

Speaker placement is critically important to the effectiveness of any sound system. There are some important concerns when you are positioning speakers in a live sound environment:
- Cover the audience.
- Don't cover the stage.
- Don't obstruct the view of the stage.
- Blend into the décor.
- Ensure minimal double coverage of dispersion.

Front Line Boundary

Any scenario that involves a microphone pointing at a speaker is risky. To best avoid feedback, avoid positioning any speaker behind the musicians, facing toward the audience and into the microphone.

In a small-stage setting, speakers should be set on stands, about head-height, and they should be at or in front of the mic line. The mic line is simply the imaginary plane that runs across the front of the stage parallel to the back wall and even with the mic closest to the audience.

In a large sound system configuration, even though the speakers might be behind the front line boundary, their coverage area should begin in front of the mic line.

There is one possible exception to this rule. When using a very small system at a low volume, it is sometimes acceptable to position one speaker behind the musicians as a monitor and the other speaker in front of the musicians as the primary reinforcing speaker. If there is one speaker behind the group and one in front, the rear speaker should be set quiet enough that feedback is not a problem. Also, the fact that the speakers aren't on the

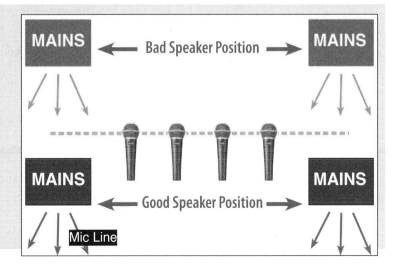

Keep Mics Behind the Front Line Boundary

Keep the microphones behind the coverage area of the main FOH speakers to decrease the likelihood of feedback.

same plane could cause phasing problems, so the actual position of the speakers might need to be adjusted by a few feet until the sound is solid and free from comb-filtering effects.

Critical Distance

The *critical distance* describes the distance from the speakers where the direct and reverberant sounds are equal in volume. Many permanent installations utilize a cluster, or multiple clusters, of speakers hung high enough that they cover the entire room. Deploying from above the audience and aiming the speakers down is an effective method of optimizing the dispersion of the speaker components; however, the cluster must always be inside the boundary of the critical distance measurement.

The precise critical distance is dependent on the room characteristics as well as the coverage pattern of the speakers. Speakers with narrow coverage patterns have a longer critical distance than speakers with wide coverage patterns. In addition, rooms with lots of hard, reflective surfaces have a much shorter critical distance than rooms with lots of absorptive surfaces.

Calculating the critical distance of simple acoustic spaces involves knowing some of the measurements in the acoustic space and running the calculations. An online critical distance calculator is available at http://www.mcsquared.com/java.htm.

Coverage Pattern

Each enclosure specifies its designated coverage area in degrees, radiating horizontally and vertically. This coverage pattern is critical to speaker choice and placement. This specification is not random or ambiguous in any way, which means that cabinet selection is also not a random or ambiguous choice.

It's important to determine the type of deployment that suits your acoustic space, and then to build the speaker complement that fits the room. For example, the JBL VP7212 series speaker cabinets are available in two different coverage patterns: 40 x 50 degrees and 50 x 90 degrees. Another enclosure, the JBL VRX932A, exhibits a coverage pattern of 100

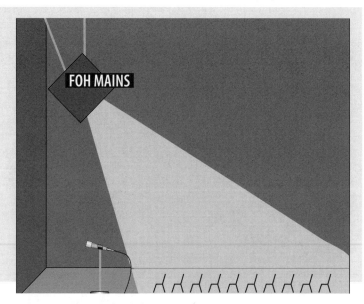

Speaker Coverage Pattern

Be sure to calculate the coverage area for your FOH speakers when calculating their placement. It's important that the coverage area starts in front of the mic line. It's also important that the coverage area is focused on the intended region without spilling onto the sidewalls and other surfaces that will reflect into the room, causing confusion of the mix image.

FOH MAINS

x 15 degrees. These two different enclosures fill completely different needs and should be deployed in dramatically different ways.

Create a computer-generated drawing of your room with accurate dimensions to scale, determine whether you want the speakers head-height or hung from above, and draw the angles that define the potential speaker's coverage patterns, horizontally and vertically. Once you have a few drawing ingredients together, it's fairly simple to get an idea of how well the speakers will cover the room. Be sure to pay attention to surrounding hard surfaces, avoiding dramatic reflection back into the main coverage area. These reflections can cause tone coloration and cancellations.

Connecting Multiple Speakers

Most small acoustic spaces are covered very well by either a single pair of speakers (left and right) or by a pair of speaker stacks, consisting of any complement of horns, cone speakers, and subwoofers. Large acoustical environments typically require multiple speakers arranged in clusters, or groups of speakers covering different zones of the same acoustic space.

The most efficient way to provide power to high-quality speaker cabinets is with a single bridged amplifier for each enclosure. Though this method is more expensive than simply using the left and right channels of a stereo amplifier to power different enclosures, it is the most efficient procedure.

Modern professional power amplifiers accept balanced low-impedance signals at the input and then pass the same signal to the output jack. The crossovers provide the frequency bands for each component of a multi-way system. In systems that utilize multiple cabinets for each frequency range, simply daisy-chain a group of amplifiers together. Each amplifier receives the same signal and feeds the speakers with the same frequency band.

Every time two identical speakers are aimed at the same listening location, the listeners hear an increase in dB SPL of between 3 and 6 dB, depending on the coverage angle of the speakers and the intensity of their audio focus.

Time-Aligning Components

In a large install, especially concerning a multi-way system, the components are often spaced out over a considerable distance. There are frequently zones to cover, including side rooms, balconies, and acoustically hidden areas, and often the back of the performance space is so far from the stage that speakers must be positioned in the back of the area for coverage. Outdoor concerts are a perfect example of this because the audience could be between five and a thousand or so feet away from the front of the stage.

In any system, the speaker components must be placed on the same time plane through the use of delay lines. Imagine a glass wall in front of the stage and inline with the FOH speakers. This glass wall (vertical plane) represents the starting point of the mixed audio—all other speakers must be considered in reference to this plane.

Sound travels at a rate of about 1,130 feet per second, which easily computes to about 1.13 feet per millisecond. Because the speed of sound is somewhat dependent on humidity, temperature, and elevation, it is typically accepted that sound travels about one foot for every millisecond. With this in mind, a speaker that's about 25 feet in front of the main

FOH stack would need to be delayed by about 25 milliseconds to be heard at the same time as the FOH stack.

It is important that all speakers in a system are time-aligned so that the audience hears them working together rather than against each other—this is true whether the speakers are one foot apart or 500 feet apart. When there are large separations between speakers, such as 500 feet, we consider that each foot between components is equal to about 1.1 milliseconds.

Center Cluster

Many systems designed for voice reinforcement utilize a single center cluster, which consists of enough speakers to adequately cover the room, all strategically aimed and carefully placed. The cluster typically hangs several feet directly above the stage, front and center. The actual height of the cluster depends on the enclosure design, the absorption character of the room, and the resulting critical distance. High speaker placement lets the designer cover a room more evenly and still work within the critical distance.

Arrays

An *array* is essentially a combination of speaker enclosures working together to create a uniform sound throughout an acoustical space. The grouping that is commonly associated with the term "array" contains multiple speakers with relatively wide horizontal coverage and relatively narrow vertical coverage, hung very close together vertically, with splayed enclosures designed to allow the proper positioning for the maximum uniformity of coverage. The individual speaker enclosure in this arrayed configuration provides focused coverage on a specific zone. Each enclosure covers a narrow vertical area—when working together, multiple arrayed enclosures provide a more intimate, closer sound than enclosures designed to cover a larger and more general area.

In actuality, multiple configurations of arrays are designed. Some utilize speakers that are split apart; others utilize speakers that are hung closely together. In all cases the intent is to provide a uniform sound in which all enclosures join to create a sound that is controlled, efficient, and typically more powerful than any of the enclosures used alone.

Arrays provide an opportunity for uniform coverage of a large and complex acoustical environment; however, the fact that multiple enclosures work together, slightly overlapping coverage areas in an attempt to blend their coverage patterns, also brings some potential for degradation. Arrayed speakers typically hang as close together as possible, but there is still a significant amount of space between enclosures. By the nature of the distance between the speakers mounted in each enclosure, any overlapped coverage areas are likely to contain frequencies that are in phase and out of phase. The relative frequencies that sum and cancel between the two contrasting points of origin combine to create a comb filter. In the extreme, comb filters can create a very hollow, inaccurate sound. Well-designed arrays are expertly positioned to minimize the adverse impact of severe comb filtering.

The integrity of the audio energy created by any system is dependent on the excellent design and intelligent combination of components, as well as their physical deployment and electrical connections. An array can provide uniform frequency coverage and consistent levels throughout the room when the components are perfectly matched and supplied with adequate and appropriate amounts of power. However, achieving this goal is not a matter of random choices—it's a matter of excellent design and careful planning.

There are several kinds of arrays that are commonly in use in professional sound reinforcement applications.

Point Source Arrays

Point source arrays utilize speakers positioned in the closest of proximities to each other, positioned to create a large coverage area with uniformity of sound and consistency in level. Typically, the backs of the enclosures are touching, with the speakers splayed apart to create the desired coverage; however, there are multiple arrangements of enclosures commonly used to create single-point source arrays. Many enclosures designed to be used in array combinations utilize enclosures with splayed top and bottom angles that allow perfect positioning for multiple enclosures and angles that match the coverage patterns of the speaker components.

Common point source arrays include single-point source, parallel, and crossfire arrays.

- The point source array positions the speakers so that their drivers are as close together as possible and so the sound projects from a single point behind the cluster. Sometimes this involves the use of rectangular boxes, angles apart so the backs of the drivers aim at the same spot, but the fronts are splayed apart. In this arrangement, the enclosure's coverage angle must equal the splay angle. Point source arrays are used in mono center clusters in which intelligibility is of primary importance. They are also commonly used in large stereo and split mono deployments because they provide a very uniform and balanced coverage of specified area.
- The parallel array utilizes square or rectangular enclosures stacked together in a line. Often parallel arrays are stacked in large concert settings. Although it looks very impressive, this type of array is not effective in the creation of uniform frequency response characteristics or consistent levels throughout the acoustical environment. Because the enclosures are touching and facing forward, their coverage patterns overlap substantially, creating undesirable phase interactions plus dramatic hot spots in the combined coverage areas.
- The crossfire array utilizes enclosures touching in the front and separated in the back, with the speakers aimed inward. This is a vintage design concept that provides too many opportunities for comb filtering and sonic inconsistencies across the coverage area.

Split-Point Source Arrays

Any split configuration simply involves positioning speakers apart from each other, typically facing forward, at identical elevations and positioned along the plane that extends across the front of the stage.

There are a few common positioning options for split-point source arrays. The speakers, such as those on either side of a stage, can be aimed apart from each other, facing directly forward from the stage, or aimed toward the center of the room. The choice of split configurations should be determined by the desired coverage area and the coverage potential of the enclosures.

The Speaker Array

Speaker arrays provide excellent coverage and increased intimacy. The array pictured below utilizes three cabinets and a subwoofer hung on each side, running in stereo. Each cabinet produces a very narrow coverage vertically (15 degrees) and a relatively wide horizontal coverage (90 degrees). All cabinets hang together to form a tight and clean overall coverage pattern.

- When separate enclosures are aimed apart from each other, their coverage patterns should slightly overlap at the center of the coverage area. This configuration is fundamentally an extension of a point source array—imagine the angle of the speaker position coming to a point behind the stage area. This split configuration could be used to augment a point source array, as long as the split-configuration enclosures are positioned in the line projecting from the point source directly toward the back of the coverage area.

- When the speakers are aimed directly forward from the stage and parallel to each other, the coverage pattern must be considered to achieve uniform coverage throughout the space. The distance between the speakers is not a random variable. This configuration, when poorly implemented, can create a very uneven coverage. The potential for extreme anomalies exists in overlapping coverage areas, as well as from the influence of reflections from surrounding surfaces. Careful consideration must be taken regarding the distances between the enclosures, as well as the distances from the side walls and other surrounding surfaces.

- A point destination array uses speakers aimed at the same point. This is not ideal for live sound reinforcement unless it is carefully designed and deployed because it multiplies the potential for phase anomalies and massive coverage hot spots. However, this is the typical configuration for studio reference monitors and some home hi-fi applications, in which a single listener or a small group of listeners receives direct sound from the speakers in a stereo setup. A setup using a point destination configuration assumes that the listener's ears are along the path to the destination, but that the actual destination is well behind the listener, at the point of intersection of the line of sight from the speakers through the listener's ears.

Zone Coverage

Zone coverage is very important in a complex acoustical environment. Many rooms contain nooks and crannies that are hidden from the main FOH speakers. It is important that these areas receive the best possible sound coverage. Balconies, under balconies, overflow rooms, and foyers are often covered with the FOH program. All of these zones should seem like one unified system to the participant or congregation member.

In a well-designed system, the FOH mains will cover all of the seats in a specific area without random spillover into adjacent areas or onto surrounding sidewalls. This being the case, zone coverage should be devoid of direct coverage from the FOH speakers. Each zone should be considered a separate area that receives controlled coverage over the seats contained therein. Even though reflections into the zones are inevitable, there should be minimal interaction between the direct sounds coming from the discrete zone systems.

Time Align

It is as important that the component coverage areas are carefully considered in all zones. It is desirable that the sound heard in each zone seems to come from the FOH mains. This is only possible through the proper and precise time alignment of the FOH and zone systems.

Time alignment involves delaying the send to zone systems according to the distance from the FOH speakers—the farther the zone from the FOH speakers, the longer the delay. As stated previously, general delays for large time alignment settings are easily calculated as about 1.1 milliseconds of delay for each foot between the two systems. For example, if the balcony is 80 feet from the FOH enclosures, the time alignment delay for the balcony should be about 88 milliseconds.

Electronic and acoustic analysis tools can easily determine the precise delay from the source to the zone. It is important that these delays are as accurate as possible because small deviations provide opportunities for destructive comb filtering and other interactions; however, because the zones have width and depth, each listener receives a varied opportunity for adverse effects from the audio combination. Therefore, delays based on commonsense rules of thumb are often acceptable for the accuracy of zone coverage.

Keep It Neat and Simple

Every installation should be clean and simple. In a permanent installation, cables should be custom-made to connect to each device with only the desired amount of slack. It's sometimes desirable to leave enough extra cable for a rack to swivel or for easy access to specific components, but most of the time cables should be constructed so that there is no extra cable coiled up around the gear.

Often, equipment is added to a system and stock off-the-shelf cables are used to make the connections, leaving several feet of excess wire laying behind the racks or mixer. This should be rectified as soon as possible by constructing custom cables that eliminate excess in order to avoid any accumulation of clutter—simply cutting the stock cable to the correct length and adding another connector is acceptable.

Bundles

As multiple cables are used for various connections, use cable ties to bundle related wire, or utilize custom-designed snakes to keep the wires neatly bundled. Multi-pair snake cable is readily available and should be used wherever appropriate. Keep line-level runs bundled together so that they can be separated out if necessary at a later date. When routing AC cables, be sure to cross line-level bundles at a 90-degree angle to minimize noise and interactions with audio signals. Bundling AC cables with the line-level audio cables can result in unacceptable noise, especially when bundled with unbalanced cables.

Digital Snakes

Several manufacturers offer digital snakes. These snakes utilize interface boxes at both ends of the snake, connected together with cable designed to carry data rather than audio. Digital snakes can carry impressive amounts of audio, in excess of 64 channels, across cable that is typically inexpensive and small enough to be unobtrusive in almost any setup.

- The Aviom system carries 64 full-bandwidth audio channels across standard CAT5 network cables.

- The Roland RSS digital snake provides 40 channels with mic preamps, input level adjustments, phantom power, and scene recall of parameter adjustments. It also provides two cable connections with redundancy that automatically switch to the backup cable in the case of loss of the primary cable.

- The FiberPlex LightViper digital snake provides 32 audio channels and multiple data format capabilities.

Digital snake connections allow for complete network integration across the entire digital system while providing ample connectivity with small, inexpensive cables.

Fine-Tuning the System

When the system is set up and fully functional, it's time to find out whether your well-planned complement of gear actually sounds good. Most of the time the best of plans produce the best of results—time-tested theories and choices based on wisdom and experience tend to produce stellar results. In the event that something hasn't quite realized its theoretical potential, we have some commonly used analysis tools to help us quantify system strengths and weaknesses.

This chapter is written to help you understand the tools and techniques used to quantify the integrity of a live sound system; however, the study of analysis methods and devices is vast and demands specialized study regarding the specific analysis tools.

Ear or Gear

Your most valuable tools in the assessment of any sound system are on either side of your head. It doesn't matter if everything looks perfect on every analyzer at your disposal—if the system doesn't sound great, it's not a great system. Conversely, when used correctly and when considered in the proper perspective, analysis tools are very valuable. They can be essential in the process of setting up a system, especially when you are adjusting large and complex combinations of speakers and amplifiers deployed throughout multiple zones.

So, really you need your ears and your gear to get a great sound out of a system. When the system looks right on your analysis tools, it is probably capable of producing high-quality audio—the sound operator can simply adjust the channels to suit his or her taste.

Voicing the System

Many sound operators use analysis tools to flatten the system EQ and to adjust the zone levels, and then they'll play a favorite musical CD through the system to see what it really sounds like. Sometimes an experienced sound operator makes small changes to the system EQ to establish a comfortable and familiar frame of reference. Adjusting the system by ear is referred to as "voicing the system." Many sound operators—especially those with a technical mindset—prefer to leave the system in its analytically perfected state. On the other hand, many sound operators—especially those with a more musical mindset—prefer to use their ears to fine-tune the sound of the system.

Speaker manufacturers also use the term "voicing" in reference to setting the crossover points in a multi-way system. They are trying to get the best sound from the components they're using together.

Voicing is very subjective, in contrast to analysis, which is very objective—there is a fine balance between the technical and musical approaches to sound. Both have great value and, as you develop a refined ear and a definite opinion about sound quality and musical integrity, you'll probably find yourself relying on your ears and your gear. In the big picture, system design is all about creating a great sound reinforcement system from a subjective viewpoint, while maintaining excellent technical integrity and intelligent implementation.

Analysis Methods and Devices

Analysis tools help the sound operator quantify the audio system's performance. It is advantageous to adjust the system so that it subjectively sounds good, but verification of signal integrity and acoustical interactions helps the sound operator avoid unnecessary problems with feedback, phasing issues, acoustical anomalies, and so on. Analysis tools also help provide a consistent mixing environment from month to month or from venue to venue.

Many sound operators function very well with a rack-mounted RTA and a calibrated microphone. Others use software-based analysis packages that are capable of displaying EQ characteristics as well as 3D decay characteristics, and various phase components in the electronic and acoustic domains.

The Sound Level Meter

The sound level meter (SLM), also called the *sound pressure level meter* (SPL meter) or the *decibel meter*, provides a simple reading of acoustical energy. The decibel meter can be set to respond to average levels or peak levels. It can be set to various weighting scales, each favoring a different band of frequencies. C-weighting displays full-bandwidth readings, whereas A-weighting focuses on the speech and intelligibility band between about 1 and 4 kHz.

For our simple evaluations of overall level, we typically set the sound pressure level meter to read average levels at a C-weighting. This provides the best overall assessment of sound energy for our purposes.

Rack-mounted SPL meters utilize a calibrated microphone. The microphone is not wireless, but it can be moved throughout the room to verify acoustic energy in multiple areas within one zone, or within multiple discrete coverage zones. The fact that the cable is connected to the mic makes this system impractical for making evaluations during a performance; however, rack-mounted SPL meters provide a very convenient way to evaluate the sound level at the mixer location in a permanent install or during an event.

Handheld sound level meters contain the calibrated microphone at the end of the device, typically positioned so that the sound operator can easily read the meter when the mic is aimed correctly. These meters are very portable and provide a convenient way to verify the sound levels at various locations throughout the room.

There is a correct way and an incorrect way to hold the SLM when evaluating levels—the wrong way is the most natural.

- **Incorrect**: A new sound operator will usually stand facing the sound source, pointing the SLM microphone at the source, holding the meter just above waist height, and looking down to read the level. This tends to provide inaccurate reading—typically too high—because the direct sound and the reflections off the sound operator combine at the microphone in an unpredictable and unreliable phase relationship.

- **Correct**: Because the calibrated microphone is omnidirectional, it doesn't need to be pointed at the sound source—it simply needs to be out in the open so there are minimal sound reflections from surrounding surfaces. The sound operator should stand facing 90 degrees away from the source, holding the SLM at arm's length, at about head height. The area around the meter should be free of reflective surfaces (including the sound operator), and the line of sight to the speakers or sound source should be clear. This procedure will produce readings that are reliable and repeatable.

The Rack-Mounted Real-Time Analyzer

The rack-mounted RTA is simple, fast, and easy to use. When the system is set up, simply connect the pink noise generator to a mixer channel, turn up the level to about 100 dB, and adjust the system EQ for the desired curve.

The typical hardware-based RTA displays the energy content of 31 frequency bands. These bands correlate to the sliders on a 31-band graphic equalizer—the most common format of equalizer placed inline between the mixer's main output and the crossover or power amplifier inputs.

A typical real-time analysis follows a fairly consistent routine similar to the following:

The Inline 31-Band Equalizer

The 31-band graphic equalizer works in conjunction with the RTA to quantify the frequency character of the sound system. The line outputs of the mixer connect to the line input of the 31-band EQ. The line outputs from the EQ connect to the line inputs of the power amplifiers. Finally, the powered output from the power amplifier connects to the speaker cabinets.

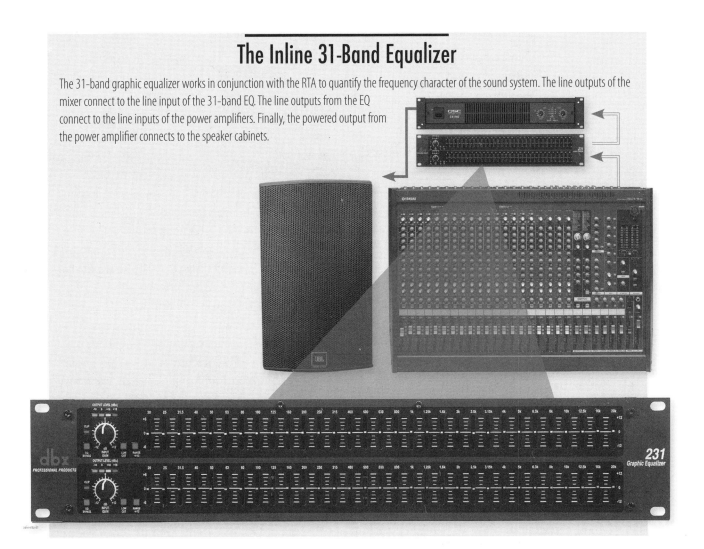

- Ring out the monitors.
- Feed pink noise to the system.
- Verify the functionality of all FOH speakers.
- Adjust levels of individual enclosures so that they are correct for each zone. (The entire room should typically be within 2 or 3 dB of the same SPL.) Start with the enclosures covering the front of the room. When evaluating a system with several enclosures covering multiple zones, leave each speaker on as you turn up the additional speakers. This helps provide an accurate assessment of the system level—there will inherently be an overlapping of coverage areas, and simply turning the speakers up one at a time won't provide an accurate picture of the entire system level throughout the room.
- Connect the calibrated microphone to the RTA input and set it at the desired location (typically in the mix position to start).
- Route the pink noise send from the RTA to the FOH system and set the pink noise level so that it reads between 95 and 100 dB on the SLM (when C-weighting is selected and it's set to read average levels).
- Adjust the RTA input level so that the 31 meters are mostly at the zero line (center of the vertical meter).

- Adjust the system's 31-band equalizer to achieve the desired reading on the RTA—depending on the room and the system the RTA should typically read flat throughout the frequency spectrum with a low-pass roll-off above 16 kHz or so and a high-pass roll-off below 60 Hz or so.

- With the pink noise still on and up in the FOH system, route the RTA pink noise to the on-stage monitor system and increase the send until the monitor level is around 95 dB. (Obviously, this doesn't apply to in-ear systems.)

- Notice the change in the readings of the RTA. This will reveal the inherent EQ problems inflicted by the floor monitors. There's a good chance that the monitor system will create a boost in the EQ characteristic of the system between about 200 and 1,000 Hz, depending on the room, the monitors, and the size and shape of the stage area. If the monitors are constantly active—and if there is a full bandwidth monitor signal most of the time—it's a good idea to simply adjust the FOH EQ for a flat reading when the monitors are up. However, if the monitors are primarily used for vocal reinforcement, it might be best to leave the FOH EQ alone. Each situation is different, and the sound operator must decide what to do to provide the best possible sound in the room. The value in this lesson is in the awareness of the inherent influence the monitors have on the FOH sound—the educated sound operator must adjust according to each scenario. If you find that the EQ on most channels is cut in the low midrange frequencies in order to achieve a clean sound, it is probably due to the adverse effect of the stage monitors on the FOH sound. In such a case, the pink noise should be turned up again, and the FOH EQ should be set for a flat RTA reading while the monitors are up.

- Set the calibrated microphone at various audience locations to verify system accuracy throughout the listening area. Discrepancies between zones should be adjusted through adjustments to the crossovers feeding the zone systems, amp levels, or variations in the relative levels of a multi-way system. In extreme cases, separate equalizers should be used to adjust the frequency balance of discrete zones; however, a well-designed system probably won't require it.

Following the previous procedure will result in the establishment of a technical baseline. The system will be verifiably adjusted, and the sound operator will be able to mix with the confidence that comes from trusting that the system is set up properly.

The Software Real-Time Analyzer

Many software-based analyzers provide all of the normal RTA and SLM functions, along with several other acoustical and electronic signal analyzers. In addition, they come complete with elegant and functional graphical user interfaces and information storage capabilities.

Packages such as FuzzMeasure Pro 2 by SuperMegaUltraGroovy (available for Mac) include a comprehensive set of analytical tools, including:

- **Advanced audio hardware support**: You capture audio on multiple channels simultaneously and perform audio device correction automatically.

- **Delay finding**: The sound operator can precisely measure the delay between the mixer output and the time when the sound hits the audience's ears, compensating for any built-in delays in the audio hardware.

- **Graphing**: A definite strength of software-based analyzers is their ability to provide print-quality graphs of all data, comparisons, and stored information, along with amazing on-screen graphics that provide real-time representation of audio measurements.

- **Pink noise and sweeping sine wave generators**: Besides the pink noise generator required for RTA adjustments, a sweeping sine wave generator is provided for other tests and measurements.

- **Waterfall plots**: When trying to locate resonant frequencies in an acoustic space, a simple 2D plot won't cut it. A waterfall plot shows a 3D graph of the frequency content and its decay over time. A full-bandwidth burst is generated, and then the decay time is plotted on the 3D graph, providing a visual image of the EQ characteristic of the reflection component of the room sound.

- **Comparisons**: It is simple to compare measurements that are made over a period of time or as a result of component changes or other acoustical and electrical adjustments.

- **Frequency domain analysis**: The software offers suggestions as to what the data and measurements indicate.

- **Impedance measurements**: In addition to acoustical measurements, software packages provide electrical impedance measurement tools, such as the Thiele-Small parameter estimation program, to help the sound operator quantify and verify the integrity of speaker and amplifier connections.

- **Reverberation time measurements**: Reverberation time measurements are crucial to understanding the performance of an acoustic space.

- **Time domain analysis**: It's important to see a detailed close-up of your impulse response in the time domain. Software-based analyzers typically provide this type of data in various forms.

The SmaartLive analyzer by SIA Software, available for Mac and PC, includes a similar list of features, especially when working together with its companion package, Acoustic Tools, which focuses on calculating quantitative acoustical values and objective speech intelligibility indices. SmaartLive by itself acts as a powerful RTA, including such features as:

- **Spectrum analysis**: Real-time signal analysis is available with multiple simultaneously viewable measurement modes. A two-channel RTA is available with selectable bandwidth settings down to 1/24-octave. Also, a single-channel spectrograph provides a 3D display of the signal's spectral content over time.

- **Sound pressure level**: SPL measurements read out with Flat, A-, and C-weighting options and standard fast and slow integration times, as well as the option to display an SPL histogram with maximum, minimum, and average values over the measurement period.

- **Spectrum data logging**: Data can easily be logged and stored for later retrieval.

- **Transfer function**: The transfer function provides precise dual-channel frequency response measurement using any test signal, including the program material from the mixer during a performance or an event.

- **Impulse response**: Impulse response measurements provide precise delay time and system acoustic response quantification.

- **Note ID**: A cursor is available in spectrum and transfer function modes that indicates the nearest musical note corresponding to a frequency from the standard 12-tone scale. This is a useful feature for helping bridge the communication gap between sound engineers and musicians.
- **Weighting curves**: Standard A-, C-, and user-definable weighting curves are available for application in spectrum and transfer function modes.
- **Internal signal generator**: The internal wave generator outputs standard measurement signals, including pink noise, sine wave, dual sine, and looped wave (*.wav) files.

The SIA Acoustic Tools add a powerful set of functions. The Acoustic Tools offer the capability to:

- Capture the impulse response measurements
- Determine reverberation and early decay times
- Calculate objective speech intelligibility metrics
- Analyze the frequency characteristics of reflections and other trouble spots in octave, 1/3-octave, or narrow-band resolution
- Study decay characteristics for specific frequencies based on time- or frequency-domain data
- Display wave file data as a spectrograph
- Compare time- and frequency-domain representations of two wave files
- Record random-length mono or stereo wave files at sampling rates up to 48,000 kHz with 8- to 24-bit sample resolution
- Print out plots and data tables in report-style format
- Export data in ASCII format for use in other applications

These powerful analysis tools provide a means to quantify the sonic integrity of any live sound reinforcement system.

Video Example 18-1

Software Analysis Tools

The Inadequacy of the RTA

Analytical tools are very useful and provide some excellent information for the sound operator and system designer; however, most of the measurements must be performed in an empty venue, simply by virtue of the way they're carried out. It is unprofessional and unkind to subject any audience to loud pink noise and audio bursts for even a short duration. Therefore, all measurements are subject to change as the presence of people in the venue changes the time and frequency characteristics of the acoustic space.

In the big picture, with all things in proper perspective, it is still very useful to set up the system properly, even in the absence of an audience. With each new setup and live sound experience, the operator will gain new insights as to how a system tuned without an audience translates to the same system with the audience present.

Documentation

Every system should have a comprehensive documentation package. The package should include all equipment manuals and detailed diagrams of how the system is connected. This is often difficult to accomplish, especially when several volunteers are serving as sound operators. Eventually, someone will add a device to the system—there's a good chance that the addition will not be documented—and then someone else will add a device, and so on. Before you know it, an otherwise stable and fully functional system has had several components changed, substituted, or added. When a problem arises and the documentation is required, several things can occur: no one knows where the documentation is, it doesn't exist, or it is in one of several boxes of leftover parts and pieces in a back closet.

One person should be in charge of documentation. In a church setting, the lead sound operator should faithfully take on the task of maintaining or developing the system documentation. If all volunteers understand the plan for a documentation package, they can help by placing the materials where they're supposed to go after they use them, but one person should be responsible for verifying that the documentation is up to date and accurate.

Be sure to date all system changes and additions. Also, provide information about who made the change, where the equipment was purchased, and any other pertinent information. Some of these points might not seem important in the heat of the moment, but once you need the information it will seem very important, and you'll be thankful that you were conscientious enough to document it.

Three-Ring Binders

Get all manuals three-hole punched so they will store easily in a three-ring binder. Be sure that all diagrams and pertinent documents are neatly labeled and organized in the binder. Because complex systems might require multiple binders full of documentation, develop a detailed table of contents. Use dividers to separate the binder into logically grouped sections, such as:

- Microphones
- Mixer
- Effects and Signal Processors
- Drive Rack
- Amplifiers
- Speakers
- Analysis Tools
- System Schematics and Diagrams
- Training Aids
- Articles
- Notes and Comments
- Receipts
- Warranty Information
- Technicians, Repair, and Manufacturer Contacts

A tool such as this is very valuable to the system users, but it provides an invaluable service to newcomers. This documentation should be readily available for review, but it should not be given out for off-site perusal—too many bad things can happen to documents in the hands of the well-meaning helper.

Computer Documentation Files

It is impractical to scan all equipment manuals and store them on a computer; however, almost all manufacturers post manuals in PDF format on their websites. Develop a file that includes all PDF documents that pertain to the sound system and organize it into a set of files with the same labels as the three-ring binder.

This virtual documentation package serves as an excellent backup in case of a catastrophe with the hard-copy version. It also provides an exceptional means of providing new sound operators access to system documentation. The file is likely to be a little large for e-mail transfer, but it will easily fit on a CD or DVD. Keep extra copies on hand and up to date.

Build a Documentation System with a Plan

When a system is being built in stages, it is very important to document the completed system along with the status during each phase of development. Not everyone can afford to build their dream system all at once—it usually needs to grow over time. Documenting the complete plan on paper helps everyone involved to know what the next phase consists of. It is much easier to intelligently plan a system with the final result in mind than it is to simply buy a new gadget every time there's an extra few hundred dollars lying around.

Document All Connections and Settings

All connections and settings should be documented. Be especially careful to document the settings of all major components once the system is set up and ready to go. Settings for all crossovers, amplifiers, equalizers, dynamics processors, and so on should be carefully recorded. In the likely case that an overzealous volunteer jumps in and makes a bunch of changes, a lot of time will be saved if all of these fundamental settings can be quickly restored to their original status. Following some carefully taken notes is much quicker and easier than setting all of the analyzers back up and readjusting the system.

Also, keep track of all repairs and idiosyncrasies. If there is a running log of problems and concerns, it provides an excellent means of pinpointing real issues that require service.

Keeping Track of Gear

In a complex system, it is important that all components are logged in a database. Serial numbers are very useful. It's ideal if every piece of equipment is marked with a permanent metallic stick with a unique serial number and the name of the organization that owns it.

Record the serial number in a database, along with the location of the device, where and when it was purchased, the cost of the device, and the name of the service technician or installer. Leave a column for notes about implementation, repairs, and so on.

If the entire system is logged into a database, and if someone takes on the task of keeping it up to date, components will be easily accessible, and there will be less frustration over lost gear and fewer confusing accounts of the system's status.

ACOUSTIC CONSIDERATION

In any live sound situation, the acoustical characteristics of the surrounding environment play a critical role in the overall sound quality. So many times, especially in a church setting, the sound operator is doomed to failure before he or she even gets started. Sanctuaries are typically very large, they're designed to hold the maximum number of people in a contained space, and the ceilings are typically very high—after all, it is a house of God.

The side walls are usually parallel and long, the front and back walls are usually parallel, and the ceiling usually goes up to a point (it's a concave angle). All of these traits together are a recipe for disastrous sound.

Parallel side walls provide the perfect reflective surfaces for frequencies that fit evenly between the walls to set up a pattern of reflection, called a *standing wave*, *mode*, or *axial mode*. When acoustic energy creates a pattern, reflecting back and forth between two surfaces along the same path, the energy typically increases by nearly double. Although there is a slight decrease in energy as the sound wave reflects off the wall, it still retains much of its original energy. The accumulation of in-phase reflection results in a near-doubling of the amplitude. If the waveform reflects out of phase, the result could be marked phase cancellation at the reflecting frequency.

Dealing with Acoustics

A room that is acoustically live contains a lot of hard, flat surfaces that reflect sound waves efficiently, causing substantial ambient reverberation. A room that is acoustically dead contains a lot of soft surfaces that absorb most of the reflections. Live acoustics are good when they've been designed to enhance the acoustic properties of a voice or an instrument. However, uncontrolled acoustical reflections are potentially destructive and must be managed.

There are many acoustic principles and treatment products that you can apply to your situation. Let's take a look at some factors that affect the way sound interacts with the acoustic environment.

The Room

It is possible to build a room that sounds great without much acoustic hocus-pocus at all. The way sound waves interact as they reflect between major surfaces and as they're diffused around the room is very much influenced by the dimensions of the room as well as by absorption and material density.

If you construct the room from scratch, you can choose dimensional relationships that work together to create an even and consistent frequency response and coverage in each space. Creating an acoustic space that inherently sounds good is fundamental to the presentation of excellent sound. Most of the time, we're forced to deal with and compensate for existing acoustic flaws. This is always difficult, but it's not insurmountable.

In any room designed for music reproduction, it is best to avoid parallel surfaces. A rectangular room, although common and cost effective to build, provides us with three sets of parallel opposing surfaces: two sets of side walls and the ceiling and floor.

Typically, more randomness is better in an acoustical environment. Designing rooms with non-parallel opposing walls often accomplishes this in the construction phase. Study any excellent acoustical environment. You'll notice interesting angles and unique shapes. These are designed to randomize the reflections of sound waves in the room.

Diffusion and Absorption

As well as manipulating room dimensions and/or angles, we must consider the importance of diffusion and absorption. Diffusing a waveform causes it to careen off its otherwise sonically destructive path. Absorption diminishes the waveform's ability to reflect. It decreases the acoustic energy, potentially to the extent of completely eliminating reflection,

Acoustic Design Principles in Action

The church illustrated (right) was designed by the Walters-Storyk Design Group for Le Noirmont Catholic Church. It includes several excellent design principles that help control reflections. Notice the slightly non-parallel side walls and the convex angle protruding into the front and rear of the room. Any acoustic space with such an intelligent physical design is likely to exhibit very few problematic acoustical anomalies.

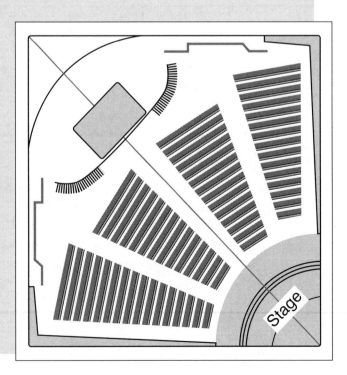

especially in the upper-mid and high frequencies. The amount of absorption accomplished depends on the frequency of the waveform, the amplitude, and the absorption properties of the reflecting surface.

It's really not all that great for our purposes to operate in a highly absorptive environment. Although it calms the reflections down, too much absorption also robs the sound of life and sparkle. We're typically much better off if the ambient sound is controlled through a combination of minimal absorption and well-designed diffusion.

Even a large room with an unnaturally short decay time feels lifeless and small. In an overly zealous attempt to provide controlled acoustics, the inexperienced acoustician might include too much absorption and not enough reflection. It takes a certain degree of reverberation to help the sound retain its personality and character.

A complex diffuser breaks up the standing tones that occur when a large room has parallel surfaces, turning them into a more controlled-sounding reverberation.

Mode/Standing Wave/Resonance

Modes are simply reflections that set up a pattern between surfaces in an acoustic space. The problems that modes create are destructive to the quality of the sound heard in the room. Modes are also called *standing waves*. The term "standing wave" provides an accurate mental image of this acoustical problem. If a sound wave reflects back along the same path from which it came, it will reflect again once it reaches its originating surface, then back again, and so on. In essence, the sound wave forms a pattern that, if visible to the naked eye, would seem to stand still in the room as it reflects back and forth. Such a standing wave poses multiple potential problems.

Reflections

This diagram shows just a few of the more predominant reflections in a simple acoustic space. Even with these basic reflections, the picture quickly becomes complex. Imagine the reality of omnidirectional sound generation and the thousands of active echoes in a live acoustical environment. Though it seems very overwhelming to consider the incredible number of possible reflections, it's better for our purposes to have many random, yet controlled, reflections. It's when a pattern forms and repeats itself that we start to realize problematic acoustics. These problematic patterns typically occur between the side walls, the end walls, and the ceiling and floor.

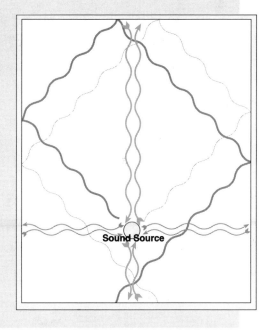

Sound Source

- If the sound wave follows the identical path on its reflection, increasing the size of the crest and trough of the affected frequency, a resonance (an increase in energy from that frequency) occurs.
- If the sound wave reflects out of phase with the original sound wave, cancellation occurs.
- If the same modes occur between multiple opposing surfaces, they work together to increase the overall anomaly within the space.

There are three types of standing waves that we must consider: axial, tangential, and oblique.

Calculating the Frequency of Standing Waves in a Room

A study of sound waves reveals that they exhibit physical, as well as aural, characteristics. One of those characteristics is wavelength, abbreviated by the symbol Lambda (λ). Because we easily calculate the wavelength of a specific frequency using the equation Wavelength (λ) = Speed (1120 feet/sec.) ÷ Frequency (Hertz), it's a simple matter of cross-multiplication to deduce the equation to calculate the frequency of a specific wavelength: Frequency (Hz) = Speed (1120 feet/sec.) ÷ Wavelength.

Consider any set of room dimensions. We can easily calculate the possible standing waves between any two opposing surfaces because waves create modes only when they fit evenly between two surfaces. Whereas the equation Hz = 1120 ÷ λ provides the frequency of the complete sound wave that fits between the two surfaces, our actual first possible problem frequency is twice as large as the distance between the opposing surfaces—180 degrees into the wave cycle also fits evenly between the surfaces, offering the potential to create a standing pattern. The equation for the first standing frequency is, therefore, Hz = 1120 ÷ 2 λ.

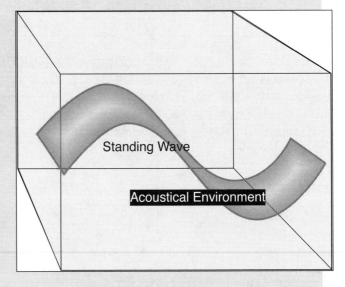

Standing Waves/Modes

The waveform that fits evenly between two opposing surfaces will create a pattern as it reflects back and forth between the walls. This illustration demonstrates a complete waveform between two surfaces; however, half of this waveform also fits between these surfaces (one crest or one trough). In addition, whole-number multiples of that crest or trough also fit evenly between those same surfaces and, likewise, create a standing pattern.

These standing waves, also called modes, are what determine the "sound" of a room. Any room which lets the same frequency stand between multiple surfaces (side walls, front and back walls, and ceiling and floor) will typically need acoustical treatment to enable reliable and musical recordings.

Standing Wave

Acoustical Environment

After we've calculated the first standing frequency, we simply multiply that frequency by consecutive whole numbers to develop a list of the other frequencies that fit evenly between the surfaces.

Axial Standing Waves

Axial standing waves (axial modes) stand between two surfaces: opposing walls as well as the floor and ceiling. These modes cause the most acoustical damage. When a sound wave reflects back and forth between two surfaces, there is ample opportunity for energies to accumulate and cancel. This simple reflection must be diffused or trapped in order to be eliminated as a problem.

The basic calculations for determining potential acoustical anomalies consider axial modes below 300 Hz. Frequencies above 300 Hz are easier to deal with, using absorption and diffusion, than those below 300 Hz. Frequencies below 300 Hz contain enough energy that they're minimally affected by absorption and even some minor nonparallel construction techniques. In acoustic design, the goal is to minimize the number of problematic standing waves; however, complete elimination is unrealistic in most existing constructions.

Because we know standing waves are an issue in most acoustical spaces, we can use knowledge and experience to help predict potential problems. We then must choose to select treatments to minimize those problems or, given the knowledge of their existence, electronically compensate.

Tangential Standing Waves

Whereas axial modes stand between two surfaces in a consistent pattern, tangential modes reflect off four surfaces. For instance, a sound wave that reflects off all of the walls travels around the room as a tangential standing wave. Once an axial mode is turned into a tangential mode, it is no longer considered as much of a problem because there is less likelihood that the pattern will stand in the room. Substantially more energy is required to enable a waveform to repeat its pattern when reflecting off four surfaces.

Oblique Standing Waves

An oblique standing wave must touch six surfaces—for example, all four side walls as well as the ceiling and the floor. Oblique standing waves offer few problems in acoustic design because of the increased amount of energy required to repeat the pattern throughout all six reflections and the unlikely chance that the pattern will actually stand.

In acoustic design, it is beneficial to transform axial modes into tangential or oblique modes through construction and diffusion techniques.

Calculating Problematic Modes

Our study of acoustics is relatively simple and is designed for your understanding of basic concepts. In the real world, the science of acoustics is very complex. Accurate prediction and compensation techniques typically involve many complex calculations, along with the use of acoustical analysis tools, which help the acoustician determine the proper course of action for a given acoustical space.

Axial, Tangential, and Oblique Modes

There are three types of standing waves.

- Axial standing waves reflect between two surfaces.
- Tangential standing waves reflect around the room, touching four walls.
- Oblique standing waves touch all six of the primary room surfaces (side walls, front and rear, and ceiling and floor).

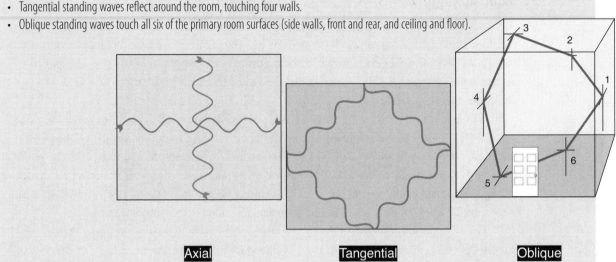

Axial   Tangential   Oblique

We can, however, develop a solid understanding of the basics of this complex science through the analysis of simple acoustical environments.

Consider All Opposing Surfaces

We use the equation previously mentioned (Hz = $1120 \div 2\lambda$) to determine the first mode in any dimension. First, consider the first mode in all three dimensions (length, width, height). Multiply the first mode by consecutive whole numbers (1, 2, 3, 4, and so on) up to 300 Hz. Make a list of the results of these calculations for each dimension.

For example, consider one room with a length of 14 feet. Calculate the first mode by plugging 14 into the equation (Hz = $1120 \div 2\lambda$). Hz = $1120 \div 28 = 40$ Hz. Once you have calculated the first mode, create a list of whole-number multiples up to 300 Hz (40, 80, 120, 160, 200, 240, 280).

Now perform this calculation for width and height. Next, combine all three lists into one sequential list that contains all results from each dimension in ascending numerical order.

Red Flags

Once the list is created, we are looking for coincident standing frequencies (frequencies that occur in more than one dimension). When the same frequency occurs in more than one dimension, there's an increased likelihood that the modes will accumulate, resulting in a resonant frequency.

When we study basics of equalization, we'll discovered that cutting a specific frequency range exposes the adjacent frequency bands. If the mids are cut, it often sounds like the highs and lows are boosted. That same principle applies to our study of acoustics. Once we've listed all possible modes, in addition to noting coincident frequencies, we must also

Calculating Axial Standing Waves

Using an equation to calculate modes is efficient, but it's more efficient to use a spreadsheet with all calculations built in. This way you can easily see the impact of a slight dimensional change.

Notice the coincident mode between all three surfaces at 140 and 280 Hz, both isolated by 35 Hz on the high and low sides. This room has problems and needs some well-thought-out treatment.

Room Dimensions		Ratio
Height	8	1.00
Width	12	1.50
Length	16	2.00

Plug in values and then the run macro "Create Sorted List"

Height Modes	Width Modes	Length Modes		Height Modes	Spacings
70.00	46.67	35.00		35.00	-11.67
140.00	93.33	70.00		46.67	-23.33
210.00	140.00	105.00		70.00	0.00
280.00	186.67	140.00		70.00	-23.33
350.00	233.33	175.00		93.33	-11.67
	280.00	210.00		105.00	-35.00
	326.67	245.00		140.00	0.00
	373.33	280.00		140.00	0.00
	420.00	315.00		140.00	-35.00
		350.00		175.00	-11.67
		385.00		186.67	-23.33
				210.00	0.00
				210.00	-23.33
				233.33	-11.67
				245.00	-35.00
				280.00	0.00
				280.00	0.00
				280.00	-35.00
				315.00	-11.67
				326.67	-23.33
				350.00	-23.33
				373.33	
				385.00	
				420.00	
				420.00	

Equation for modes Hz=1120÷2λ

note frequencies that are isolated on either side by more than 25 Hz. In other words, on our sequential numeric list of modes, we must take note of increments of 25 Hz or greater.

Obviously, any room with two or three dimensions the same provides multiple opportunities for coincident modes and isolations.

Flutter Echo

Flutter echoes can be heard after the sound source ceases. They sound like a ringing or hissing sound. As mid frequencies stand, especially between sidewalls, they resonate after the source is through. Try clapping your hands in an empty room with a lot of hard surfaces. The ringing after you clap is probably flutter echo. The flutter is a repeating echo back and forth between opposing surfaces, so in smaller rooms it is faster, and in larger rooms it is slower. Rooms with dimensions larger than about 50 ms exhibit flutter echoes that sound like distinct delays fading away over time.

Flutter echoes are easily controlled using absorption or diffusion. Take note of opposing hard, flat surfaces. Something needs to break up the natural reflection back and forth between parallel walls in order to eliminate flutter echoes.

Your Location in the Room Matters

As you notice problem frequencies, you must be cognizant of potential problems in those frequencies. Because these problem frequencies are standing between surfaces, the degree of the effect is dependent on where the listener is in relation to the cumulative energy of the standing wave.

A node is the point at which the crest turns into the trough. At this node, there is decreased acoustic energy for the specific frequency. If you are located at a node point, you'll hear little of that frequency. However, if you move a few feet closer to or farther from the source, you might realize a huge increase in the amplitude of that same frequency. Your physical position in the acoustical environment determines the frequency content of the sound you hear.

Everything Else in the Room Matters

Sound waves are reflected and absorbed, to some degree, by everything in the room. The number of people and their dispersion throughout the room affects the sonic characteristics. For reasons such as this, acoustics and the prediction of sound interaction with the acoustic space become very complicated.

It is true, however, that a sound will only be effectively isolated by an object larger than the wavelength. In our calculations of potential problem frequencies, we consider that frequencies above 300 Hz can be controlled through absorption and diffusion—we concentrate on frequencies below 300 Hz. These lower frequencies contain more energy potential because of their size and the amount of air they move. In addition, as we consider the length of a 300-Hz sound wave (3.73 feet), and as we look around most rooms, there aren't many objects between opposing surfaces that large or larger—there just aren't many common physical objects that will control those frequencies.

The Rectangle

Unfortunately, in its basic state, a rectangular room is ineffective for creating reliable audio recordings. Flutter echo and standing waves combine with the direct sound in a way that decreases intimacy, reliability, and sonic integrity. So, we must use knowledge and creativity to create an adequate working environment.

The Goal

What are we really trying to accomplish in controlling the sonic character of an acoustical space?

- In their simplest form, we want to randomize patterns of energy within the environment. Sound waves that reflect, forming a pattern in the room, cause problems—we want to avoid them.

- We must also realize that standing waves happen, especially in any rectangular room. Our goal is to manipulate the environment in such a way that there are as few coincident axial modes as possible when we consider all dimensions.

- We must do what we can to avoid isolation of specific frequencies via larger than acceptable spacings between axial modes (greater than 25 Hz).

Solutions to Acoustical Problems

If you're fortunate enough to construct your own venue, there are a few suggested proportional equations when considering basic room dimensions. These ratios act as guidelines for determining dimensions resulting in acoustic environments with minimal peaks and nodes.

Dimensional Proportions

The Golden Section or Golden Ratio (1:1.6:2.33) is often referenced as a ratio between length, width, and height that produces a room with minimum anomalies. When you determine the ceiling height, this ratio determines the other dimensions. If the ceiling height is 8 feet, the width and length are 12.8 feet (8 x 1.6) and 18.64 feet (8 x 2.33).

There are several other suggested dimensional ratios that result in the minimum number of resonances and nodes:

- 1:1.9:1.4
- 1:1.9:1.3
- 1:1.5:2.1
- 1:1.5:2.2
- 1:1.2:1.5
- 1:1.4:2.1

Adjusting Angles

As we noted before, axial modes hold the most potential for problematic acoustical phenomena. One of the reasons acoustic designers avoid parallel surfaces is to minimize axial modes. This technique alone is not enough to control sound reflections in an acoustical environment.

It takes a fairly extreme angle to deflect an axial mode into a tangential or oblique mode. The typical room can't afford the decrease in square footage required to skew walls enough to make a huge difference.

Concave Angles

Concave angles focus sound waves at a specific point. In the recording studio, contrary to the effect of the concave angle, we traditionally make an effort to randomize and confuse audio sound wave focus for our acoustic purposes.

Whereas we must avoid focusing and providing the opportunity for sonic patterns to accumulate, the large concave design is the basis of the amphitheater. In a concert setting, musicians positioned in the "bowl" enjoy the fact that their performance is focused on the audience.

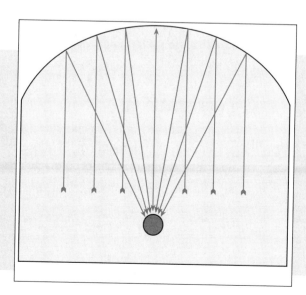

Convex Angles

Convex angles help redirect potential axial standing waves into much less destructive tangential waves. The proper implementation of convex angles can help transform an otherwise unusable acoustical environment into a reliable creative studio.

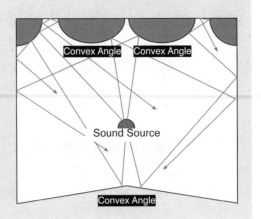

We must keep in mind that for every flat surface, there are two corners—one at each end—plus, there are angles at the floor and ceiling. As walls, ceilings, and floors meet, they form one of two different types of angles: concave and convex. Concave angles come to a point, or bend, away from you. Convex angles come to a point, or bend, toward you.

Concave Angles

A concave angle, such as a typical corner in a room, focuses reflections in a particular direction. Because most of the flat surfaces in any acoustical environment form concave angles at the connection points with other surfaces, there's no way we can avoid concave angles. However, we must realize the effect they have on room acoustics and adjust our setup accordingly.

Control of the focusing power of concave surfaces can be achieved by constructing convex angles and diffusers on opposing surfaces. For example, the focus across the room, caused by a concave angle, will be diffused, redirected, or confused by a convex angle at its point of focus.

Convex Angles

Convex angles diffuse reflections. A convex angle comes to a point or bends toward you, leaving only one point where the sound wave can bounce directly back in the direction from which it came. All other points on the convex angle deflect the sound wave. In construction, convex angles offer an excellent means to transform axial modes into tangential or oblique modes.

Treating Surfaces

Most acoustic designs incorporate a blend of hard and soft surfaces, controlling reflections rather than eliminating them. This provides a very comfortable and creative environment. A recording studio that has been overly deadened, using lots of very absorptive surfaces, lacks life—it is a very unnatural listening environment and typically feels uncomfortable to most listeners.

Absorption Coefficients

This chart indicates the reflective properties of various types of materials. Remember, an absorption coefficient of 1.00 is ultimately capable of absorbing sound waves.

Material	125 Hz	250 Hz	500 Hz	1 kHz	2 kHz	4 kHz
Window	.35	.25	.18	.12	.07	.04
Concrete Block	.10	.05	.06	.07	.09	.08
Gypsum Board on 16" o.c. 2 x 4s	.29	.10	.05	.04	.07	.09
Heavy Carpet on Concrete	.02	.06	.14	.37	.60	.65
¾" Acoustical Tile	.09	.28	.78	.84	.73	.64
Concrete Floor	.01	.01	.015	.02	.02	.02
Wood Floor	.15	.11	.10	.07	.06	.07
Linoleum on Concrete	.02	.03	.03	.03	.03	.02

Absorption Coefficient (Absorbers)

As we implement absorptive material, we must be aware that each material has a unique ability to absorb specific frequency bands with varying efficiency. The quantification of this absorptive trait is called the *absorption coefficient*. This reaction to sound waves is typically rated at a specified frequency and is noted in terms of absorption effectiveness.

An open window into outer space is the typical image representing 100% absorption because all sound enters and none returns. We think in terms of percentage and speak in terms of two decimals. A material that absorbs half of the energy at the specified frequency is said to have an absorption coefficient of .50. An open window is said to have an absorption coefficient of 1.00.

All materials exhibit an absorption coefficient: wood, fabric, glass, marble, and so on. As we consider, for example, our list of standing waves, if we notice coincident modes at 240 Hz, isolated by 35 Hz from adjacent modes, it might do us a lot of good to include materials in the design that exhibit a high absorption coefficient at about 240 Hz. In this way, we can use the existing space effectively, controlling the problematic tendencies rather than simply dealing with the problems. This very simple example is the basis for much of what designers consider when they select dimensions and materials.

Diffusion Panels

Diffusion panels randomize reflections. Their physical design, in a seemingly random pattern, effectively disperses otherwise focused waveforms.

The absorption coefficient is specified at six frequencies: 125 Hz, 250 Hz, 500 Hz, 1 kHz, 2 kHz, and 4 kHz. Remember, the closer the absorption coefficient is to 1.00, the greater the absorption. Notice, on the previous chart, the difference in absorption between lower and higher frequencies. Many soft surfaces, which one might guess would be very absorptive (carpet, acoustical tile, and so on) exhibit little to no absorption at 125 Hz.

Diffusers

It's often most desirable to control interfering reflections using diffusion instead of absorption. A diffuser confuses and randomizes reflections rather than absorbing them. This tool provides control without robbing the room of its ambient life.

There are several different designs used to construct diffusers. Some are constructed of wood, and some use synthetic materials; however, the purpose is the same, regardless of the physical characteristics. Diffusion panels are placed at any position in the acoustic space that requires reflection control.

The Ceiling

Ceiling design is a critical factor in acoustic design. Standing waves between the ceiling and floor provide ample opportunity for tonal coloration and destructive sonic interaction. Most large rooms, such as the church sanctuary, include a high ceiling that slopes up to a point. Although this helps focus the sound back down to the congregation, it might be 20 to 50 feet high, which delays the sound coming back to the congregation by between 40 and 100 milliseconds. Delays of this length can easily form flutter echoes and delays that confuse sonic integrity and intelligibility.

Typically, angled panels suspended from the ceiling above the audience are used to diffuse and deflect the focused and standing waves. These panels, called *clouds*, effectively help shape the feel of the acoustic space. They can transform a large and cumbersome acoustical space into a room with a much tighter and more intimate feel.

Corners

Obviously, each corner provides the same focusing power as the concave ceiling. The corners in a rectangular room, which meet at 90-degree angles, are often closed off by simply

Clouds

Often, the ceiling is acoustically problematic. Ceilings that rise to a center point focus reflections right back into the middle of the room. If the ceiling is flat, it contributes dramatically to any standing wave problems. A common ceiling treatment involves flying a series of clouds across the length of the audience. Clouds are typically nothing more than a series of angled panels that deflect standing waves and reflect sound evenly down to the audience. Depending on the height and angle of the clouds, it is possible to tame the wildest of venues. Clouds also let the audience hear their own interactions with, and reactions to, the performance.

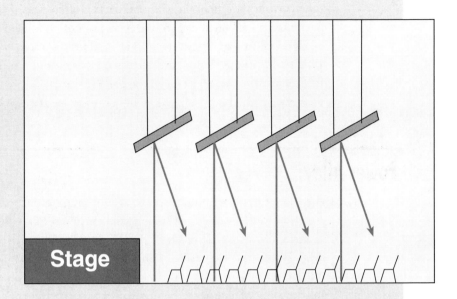

placing a wall across the angle. Notice that most well-designed large acoustical environments don't have fully-realized corners that come to 90-degree concave angles.

The Bass Trap

By now, it's probably pretty clear to you that low frequencies pose the most looming problem in our quest for perfect sound. In large acoustical environments, the low frequencies are typically the most destructive. The fact that they contain impressive proportional energy and that they are dominant in the acoustical environment makes them very difficult to overcome.

If the resonances and nodes are controlled in any acoustic environment, the sound is much more intimate, and the mixes created are more pleasing and less imposing. The bass trap is designed to address these issues.

There are basically three types of bass traps:
- The Helmholtz resonator
- The panel trap/membrane resonator
- The broadband bass absorber

The Helmholtz resonator is simply a large, tuned box with a hole in it that resonates at a specific frequency. Much like a flute, pan flute, or pop bottle, this resonator responds to one frequency. The fact that it resonates at a predetermined frequency, which ideally corresponds to a problem acoustical frequency, helps minimize that frequency in the ambient room sound. Because it takes energy to resonate the box, that energy is diminished in the room. The Helmholtz resonator can be designed for one specific frequency or for a broad band of frequencies.

The panel trap is essentially a wood frame with a thin plywood (or similar material) front panel. Fiberglass insulation is often placed in the panel to help increase absorption. The frequencies affected by the panel trap are dependent on size and material density. It's typical to incorporate panels that absorb frequency bands centered at 100 and 200 Hz. This is an efficient way to control modes in the trouble frequency range below 300 Hz.

The broadband bass absorber utilizes a rigid fiberglass material that is about four times denser than fiberglass insulation. It is often placed behind a wall or other structure. I have used an effective design that incorporates an angled back wall made from varying sizes of wood boards spaced at somewhat random distances from each other (between 0.5 and 1.5 inches). In this design, the boards act as random resonators, the heavy fiberglass (often Rockwool) absorbs very efficiently, and the slatted design allows low frequencies to enter the void behind the boards. The low frequencies are effectively trapped and absorbed.

Power Alley

Any time the same audio is reproduced from different speakers, positioned at different locations in the room, the listener will hear the resulting audio combination. At each location in the room, the sonic result of these combinations of speaker outputs is slightly different. In other words, the mix might sound very different at one location compared to another within the same venue. Because of the differences in relative phase between the speaker outputs, the frequency content of the mix varies with the listener's location.

This phenomenon is most extreme in the low frequencies. Whereas mid and high frequencies exhibit the same principle, it is much less extreme because the sound waves are small, relatively directional, and more easily diffused and absorbed than low-frequency sound waves. When subwoofers, positioned on the left and right, are part of a system, the low-frequency sound waves they reproduce often combine dramatically at the mid-point between the cabinets—this mid-point is typically down the center of the listening area. This area of increased low-frequency amplitude, created by the combined outputs of the subwoofers, is called the *power alley*.

The outputs of the subwoofers, typically below 150 Hz, are omnidirectional. Therefore, the low-frequency audio information they each produce is active throughout the listening area. Every listening position receives some interaction between the left and right subwoofers, especially when reflections are taken into consideration. The listening position down the center of the room (the power alley) is where the least amount of destructive interaction occurs between the left and right subwoofers. It also where the direct, sonic outputs from both subs is unhindered, in phase, and most powerful in relation to the rest of the mix. In addition, because low frequencies often form standing patterns, a walk from the stage to the back of the room will probably reveal dramatic peaks and dips in response throughout the low-frequency range.

Software analysis tools can easily build a model of the frequency content throughout the listening space to help pinpoint frequency anomalies such as the power alley. These models can be built from acoustical assessments or from enclosure specifications and room dimensions. Often, in the field, a sound operator must listen and assess the sound throughout the room with his or her ears, a simple decibel meter, or an RTA. Awareness of the potential existence of the power alley effect is important. Varying the distance between the subwoofers changes the severity of the phenomenon. Also, positioning the subs in the center (or at least at the same location) along the plane defined by the left and right speaker stacks helps eliminate these low-frequency problems. If the subs are flown, it might be easy to position them in the center, but when they're on the floor, they often need to be positioned to the left or right.

Power Alley

In its simplest form, it is easy to see that each location in the venue provides a different tonal balance for the listener because of the changing phase relationship between the left and right speakers (left). A more complex analysis of the speaker outputs (right) typically reveals an abundance of low frequency information down the center of the room (power alley) as a result of the combination of left and right subwoofer outputs.

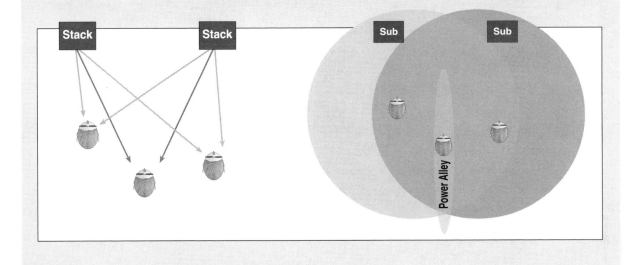

Using Your Knowledge of Acoustics

It is important to understand some of the basic principles of acoustics. It helps the sound operator understand why building a mix is sometimes very difficult and other times pretty easy. However, the study of acoustics is very complex—design and implementation are best left to the professionals who have excellent track records. Working with a professional who has designed and built venues that look and sound great provides the relative assurance of ending up with a functional and inspiring room.

The knowledge you have can help you understand the potential problems in your church. Even though experienced designers should handle costly construction projects, it is instructive to assess the environment and to develop some theories about how to solve the acoustic problems.

Draw a diagram of your facility. Consider the corners, ceilings, and walls—imagine the potential interactions between them and devise possible solutions. If you think you have some great ideas to help improve the acoustics in your venue, invest some money and have them reviewed by a qualified and experienced acoustician.

New Construction

If you're part of a building project that includes a sanctuary or auditorium in which music and audio are important, include acousticians at the beginning of the design phase. Too many people design facilities based on seats rather than on the quality of the experience for the people in the seats. Simple dimensional adjustments at the beginning of the design phase can make the difference between building a room that enhances the communication experience and building a room that distracts from the communication experience.

chapter 20

MIKING THE GROUP

I n order to create a great-sounding house mix, two things must be verified: (1) The instruments must be in excellent working order—they have to sound good to start with. (2) The connection to the mixer, whether a microphone or direct box, must be intelligently implemented so that the excellence presented at the source translates well to the amplification system.

In other words, *good source + good mic technique + good system = good sound.*

Electric Guitar

Guitar is fundamental in the establishment and authenticity of a good-sounding mix. Anyone who has been around music for a while will testify to the cyclical nature of trends in guitar sounds. As the era of electronics in music grows, we can expect that the sounds we love today will become dated or irrelevant tomorrow. With equal certainty, as we continue to move forward, history will repeat itself. Someone will rediscover the beauty of an era gone by, revive it, and reclaim it. The bottom line is: Save your toys. They will be cool again!

The guitar is a great instrument and offers definitive sounds that crystallize the stylistic feel of the song. In order for your rock band, worship team, jazz band, or bluegrass group to sound right, the guitar sounds must be right. If your guitar player is entrenched in a bygone era, you'll have a tough time getting your mix to sound current and culturally pertinent. If this is the case, be honest with the guitarist, supply some current-sounding CDs, and ask that he or she try to emulate the guitar sounds on the recordings.

With any instrument, especially guitar, much of the sound that comes from the instrument is a result of the musician's touch, aside from any effects or mic technique. Therefore, much of the quest for the best guitar sounds relies on the guitarist liking the type of music the group is playing and understanding the mindset behind it.

Guitar Fundamentals

A guitar isn't really a full-range instrument. Although it offers a full sound, which ranges from ear-piercing to soul wrenching, its musical and sonic strength lies in the efficiency of its range rather than its breadth.

The frequency range of the fundamental pitches in the guitar is fairly narrow. The fundamental frequency of the lowest guitar note is E2 (82 Hz). The fundamental frequency of the highest note on a standard 22-fret six-string guitar is D6 (1.174 kHz).

Standard tuning for the six-string guitar is E2 (82 Hz), A2 (110 Hz), D3 (147 Hz), G3 (196 Hz), B3 (247 Hz), and E4 (330 Hz). It's interesting that most of the frequencies we accentuate in a guitar recording are well above the majority of its fundamental frequency range. It's typical procedure to roll off the frequencies below about 100 Hz; fullness is usually around 250 Hz, and boosting a frequency range between 3 and 5 kHz adds clarity. We should deduce from these facts that the harmonic content of the guitar is what we really find most appealing about its sound.

Tone and Timbre

Tone and *timbre* are terms that are often used interchangeably; however, closer inspection of the guitar and its tonal characteristics leads us into a study of the subtle differences between these two terms. Whereas the tone controls on the electric guitar decrease high frequencies or low frequencies, in actuality timbre has more to do with the instrument's harmonic content.

Individual guitars sound different for various reasons. The type of wood used is a factor, especially in the instrument's sustain. More dense materials tend to produce instruments with a greater capacity for natural sustain. Softer woods tend to provide a warmer tone, often at the expense of sustain. The physical composition of each instrument influences its harmonic content or, in other words, its timbre.

Tone and timbre are also influenced by the musical expression and nuance that comes from the guitarist. A ten-dollar guitar in the hands of a master guitarist can sound like a

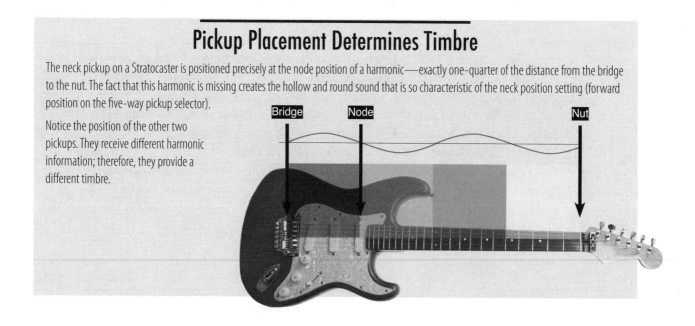

Pickup Placement Determines Timbre

The neck pickup on a Stratocaster is positioned precisely at the node position of a harmonic—exactly one-quarter of the distance from the bridge to the nut. The fact that this harmonic is missing creates the hollow and round sound that is so characteristic of the neck position setting (forward position on the five-way pickup selector).

Notice the position of the other two pickups. They receive different harmonic information; therefore, they provide a different timbre.

Bridge   Node   Nut

million bucks—a million-dollar guitar in the hands of a beginner can sound like ten bucks. An excellent instrument in the hands of an excellent musician sets the stage for musical excellence and a great-sounding house mix.

Pickup Placement

The position of each pickup on an electric guitar is very important to the sound of the instrument. The reason why the bridge pickup sounds thinner than the neck pickup is explained by the harmonic content at each location.

For a simple illustration of this concept, let's consider the neck pickup on a standard Stratocaster. Sonically, this pickup provides a deep, warm, and somewhat hollow sound— much warmer than the other pickups. If we look at the physical position of the pickup, we'll see that it is exactly one-quarter of the distance from the bridge to the nut.

Most guitarists know about harmonics on the guitar neck, especially at the twelfth fret, where lightly touching the string while plucking divides the string into two halves, producing a harmonic one octave above the fundamental. If we were to position a pickup at the twelfth fret, we would not hear much of the harmonic because the string is essentially still at that location—we've stopped the fundamental and allowed the first harmonic to ring freely in the two halves of the string. This point at which there is no vibration is called a *node*.

There are also natural harmonics at other locations on the neck. One of these is at the fifth fret, which is the point where the string is divided into four segments. This harmonic produces three nodes between the bridge and nut—one at each quarter of the length of the string. A standard Stratocaster neck position pickup is positioned directly below this node, explaining its hollow and deep tone. Whereas this pickup hears next to nothing from this harmonic, it receives a strong fundamental and first harmonic. In fact, the first harmonic (stopped at the twelfth fret) is strongest at this point. Like equalization, in which reducing one frequency band accentuates the bands on either side, the same concept applies to harmonic content. The virtual elimination of this midrange harmonic results in a sound that is full in lows and rich in highs.

This simplified illustration of harmonic content provides a clear delineation between tone and timbre. Timbre consists of the sound created by the relative levels of harmonics. In this Strat neck position, no amount of tone control—frequency boost or cut—can add back in what is harmonically missing because it simply isn't an available timbral ingredient.

In actuality, there is a constantly varying balance of timbre as the guitarist moves up and down the neck. As the length of the string changes during the fretting process, the balance of harmonics changes at each pickup position. This becomes part of the inherent sound of the instrument, which provides musical inspiration for the virtuoso and, at times, frustration for the novice. Each song provides a new opportunity for sonic inspiration as the key changes and the relative typical string length controls the available textures at each pickup position.

Tone Controls

Most guitars offer passive filters for tonal control. When the bass or treble controls are all the way up, they are providing 100% of what the pickup has to offer for that range. Backing off the treble control simply trims off the frequencies above a certain point—functioning as a variable low-pass filter. Backing off the bass control trims off frequencies below a certain point—functioning as a variable high-pass filter.

Guitars with active circuitry typically contain amplification circuitry that provides boosted treble or bass frequencies.

Frequency Content

Most guitars and guitar amplifiers exhibit a fairly modest high-end rolling off above 5 or 6 kHz.

An electric guitar can be adjusted to include massive amounts of low-frequency content. This might sound great when you're alone in your bedroom, but in the context of a band, it will probably sound muddy. In order to fit the guitar perfectly into a mix, you must decide which frequency range supports the song best, then adjust the tone accordingly.

Running Direct into the Mixer

Modern guitar effects processors provide excellent digital amp models, speaker simulators, microphone modelers, and every imaginable dynamic and delay effect. In addition, they provide subtle sonic characteristics that imitate the sound of a well-miked speaker cabinet.

Electric guitar can easily sound very good running directly into the mixer. This helps eliminate stage volume from the guitar amp—and we all know how much guitarists like to turn up their amps—and it lets the sound operator shape the guitar sound perfectly for the house.

If your guitarist is willing to run direct, spend some time to find the effects unit that provides the best sounds for your musical preference. The Line 6 company helped pioneer this digitally modeled guitar sound—they provide some excellent devices. Also, other companies offer viable options. It's up to the guitarist and the sound operator to make sure that the guitar and effects sound great in the mix.

This is a perfect opportunity to test your relationship with the local music retailer. Try a few different devices through the house system with the music team. Make your choice based on the sound, not on the name brand or what someone else uses. Be discriminating about the sound—use your ears!

Patching the Electric Guitar Direct

The procedure used to patch the electric guitar directly into a mixer depends on the input type. There are two basic scenarios:

1. If you have access to the mixer line inputs, it's usually good to plug the guitar straight into the mixer channel line input, without a direct box (DI). This works well most of the time, but success and signal quality are dependent on pickup signal strength and impedance of the mixer line input. Most modern guitars and mixers get along quite well.

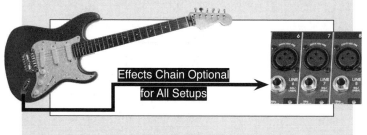

Effects Chain Optional for All Setups

2. If you're plugging into a mic input, you must use a DI to match impedance and to optimize levels.

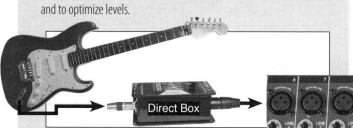

Direct Box

Line Output and Speaker Output

Some amplifiers have a line output that can be plugged into the mixer line input. This technique helps maintain some of the character of the amplifier electronics.

Caution! Do not plug the guitar amplifier's speaker output directly into the mixer line input.

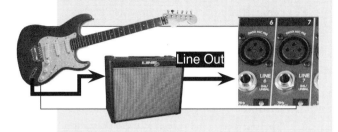

Be careful! Never plug the powered speaker output from your amplifier directly into any device that you're not completely certain is designed to accept it.

Some DIs provide a switched input, letting you select INSTRUMENT or AMPLIFIER as the signal source. This technique can work very well. It adds the characteristic sound of the amplifier EQ and effects along with the tone of the amplifying circuitry.

Always be sure the DI is selected to accept an input from the speaker output when using this technique.

Miking the Speaker

Although running direct into the mixer provides unsurpassed control, some guitarists will go straight into a coma when presented with the option. On one hand they'll probably be surprised at how great the guitar can sound in the mix when run direct; on the other hand, their coma is partially justified.

Much of the tone and musical character comes from the amplifier. Each amp provides a tonal ingredient that is difficult to completely emulate through a digital model. Even though industry hype might say that the digital model is just like the real thing, when compared side by side the real thing wins. However, the benefit of running direct and the ability to shape the tone through myriad available settings make the digital modeling effects device a very attractive option.

If the guitarist really wants to use his or her amplification system, there are a couple of good compromises.

• Place the amp in a side room off the stage and mike it.

• Place the amp in a box constructed specifically to house it, or use a specially constructed isolation speaker cabinet.

Both of these options provide excellent results while taking advantage of the superior tone quality of an excellent guitar amplifier. An isolation cabinet, such as the Randall Isolation 12 speaker cabinet, provides authentic speaker tone while controlling the stage volume—it lets the sound operator mike the speaker inside the isolation cabinet with minimal stage volume.

The isolation cabinet contains a microphone mounted inside the cabinet, with $\frac{1}{4}$-inch jacks to access the speaker and an XLR connector to connect the microphone to the mixer.

Mic Techniques

• Turn the amp up to a fairly strong level. This doesn't have to be screaming loud, but most amps sound fuller if they're turned up a bit.

• Next, place a moving-coil mic about one foot away from the speaker. Most guitar amps will have one or two full-range speakers. These speakers are typically 8 to 12 inches in diameter. Moving-coil mics are the preferred choice for close-miking amplifiers because they can handle plenty of volume before they distort the sound. Also, the tone coloration

Aiming the Microphone at the Speaker

Pointing the microphone at the center of the speaker produces a tone that is edgy and contains more high-frequency content than other mic locations.

Pointing the microphone away from the center and toward the outer edge of the speaker cone produces a warmer, smoother tone with less high-frequency content than other mic locations.

Before using equalization and other processing, use mic choice and placement to define the desired sound. There is a broad range of sound available at this initial phase of recording.

of a moving-coil mic in the higher frequencies can add bite and clarity to the guitar sound.

• If the amp you are miking has more than one identical speaker, point the mic at one of the speakers. Point the mic at the center of the speaker to get a sound with more bite and edge. Point the mic more toward the outer rim of the speaker to capture a warmer, smoother sound.

• Audio Example 20-1 demonstrates the sound of an amp with the mic placed six inches from the speaker and pointed directly at the center of the speaker.

Audio Example 20-1
Mic at the Center of the Speaker

• Audio Example 20-2 demonstrates the sound of the same amp, same guitar, and same musical part as Audio Example 20-1. Now the mic is aimed about one inch in from the outside rim of the speaker, while maintaining the distance of six inches from the speaker.

Audio Example 20-2
Mic at the Outer Edge of the Speaker

Video Example 20-1
Changing Mic Position on the Speaker Cabinet

Combining the Direct and Miked Sounds

This technique lets you blend the direct and miked sounds, including the intimacy of the direct sound with the warmth of the miked sound. Blend both of these sounds onto one mono track or record them to two separate tracks—saving the blend for mixdown.

1. Plug the guitar into the effects (optional), then into the 1/4-inch DI input.
2. Plug the THRU (Out to Amp) jack on the DI into the amplifier input.
3. Patch the DI XLR output into the mixer XLR input.
4. Mike the speaker, connecting the mic to a separate mixer input.

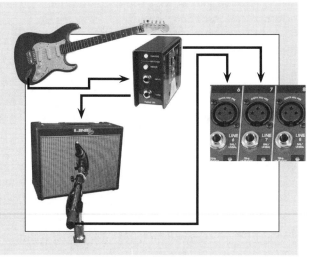

Electric Guitar Levels

The electric guitar should be set for peaks around 0 VU or between –12 and –18 FSM—there is no real advantage to pushing the levels hotter. In addition, it's not necessary to set levels conservatively—transients aren't usually an issue because typically the guitarist's effects processor heavily compresses sounds.

Pickups

Guitar pickups all work on the same essential principle. Their purpose is to convert the string vibrations into an electrical signal. This is performed by a combination of a magnet and copper wire coiled around the magnet. The string motion produces a variation in the magnetic field, resulting in an electrical image of the string vibration rate and intensity. The variations between pickups have become greater through experimentation with the amount of wire wrapped around the magnet and combinations of close-proximity pickup assemblies, but the fundamental structure is very consistent between pickups.

Increasing the amount of copper wire coiled around the magnet produces a stronger signal. In fact, some pickups create such a strong signal that they naturally overdrive the amplifier input. With super-high output pickups like this, the only way to get a clean sound is to reduce the instrument volume control. Some players prefer these stronger pickups because they provide an aggressive sound and feel. Some players prefer the natural clarity of a weaker pickup.

The main downside to extra-strength pickups is that they can reduce sustain. If the pickup configuration is too strong, the strings actually stop vibrating sooner than normal because of the magnet pull induced by the pickup.

Most players who experiment with pickups end up using a stronger pickup at the bridge position—because the signal is weaker at the bridge—along with a more natural-sounding, yet full, pickup in the neck position.

Pickup Position

Many people credit the pickup for the sound of the instrument. In actuality, that is partially true; however, the position of the pickup is often more influential on the sound of the instrument than the actual pickup design.

As I explained earlier in this chapter, the position of the pickup determines the harmonic content of the sound. If the pickup is placed at a node of one or two harmonics, it will sound vastly different than if it's placed at the precise point where all harmonics are working together optimally.

There is definitely an inherently different sound between single- and double-coil pickups, but their placement determines the essential sonic timbre.

Pickup Types

There are two basic types of guitars: single-coil and double-coil. Additionally, some modern pickups contain active circuitry.

Single-Coil Pickup

This is the approximate size of most single-coil pickups. Sometimes they're hidden by a plastic or metal cover.

These pickups are common on guitars such as the Fender Stratocaster, and they provide a clean, transparent ringing tone.

Single-coil pickups are the most susceptible to noise. If you have a problem with noise when recording a guitar with single-coil pickups, try moving the guitarist to a different location in the room. If the noise persists, try having the guitarist face different directions. There's usually somewhere in a 360° radius where the noise and interference are minimal. Keep the guitar away from computers, drum machines, or other microprocessor-controlled equipment for minimal noise.

Single-Coil

Single-coil pickups have a thin, clean, and transparent sound, but they can be noisy, picking up occasional radio interference. These pickups, typically found on a stock Fender Stratocaster, are usually about 3/4-inch wide and 2-1/2 inches long.

Audio Example 20-3

Single-Coil Pickup

Double-Coil Humbucking

Double-coil pickups have a thick, meaty sound and are the most noise-free of the pickup types. They get their name from the fact that they have two single coils working together as one pickup. These are wired together in a way that cancels the majority of noise that is picked up. These can also be called *humbucking pickups*. Double-coil pickups are common on most Gibson guitars, such as the Les Paul.

Audio Example 20-4

Double-Coil Pickup

Many guitars have a combination of single- and double-coil pickups. It's common for a double-coil pickup to have a switch that will turn one of the coils off. This gives the player a choice between single- and double-coil.

Double-Coil Pickup

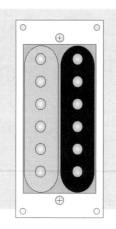

This pickup configuration uses two single-coil pickups working together as one. They're wired together in a way that minimizes noise and radio interference.

Sometimes both pickups are visible, and sometimes they're hidden by a gold, chrome, or plastic cover.

The double-coil sound is fuller and less shrill than the single-coil sound. This design is used on the standard Gibson Les Paul.

Active pickups contain amplifying circuitry to boost the pickup output level. Therefore, the active pickup typically uses fewer windings, resulting in decreased impedance and magnetic draw in the string vibration. Decreased impedance results in a signal that is less affected in the high frequencies by cable influences.

An active guitar pickup is typically capable of very clean sounds while also providing a strong signal for aggressive and characteristic distorted sounds.

Equalizing the Guitar

There are certain EQ ranges that add specific qualities to guitar sounds. Depending on the type of guitar and style of music, EQ changes can have varying results. Here are some good starting points for equalizing a guitar.

Boosting 100 Hz can add a good, solid low end to most guitar sounds. Boost this frequency sparingly. It can be appropriate to turn this frequency up, but most of the time a boost here will conflict with the bass guitar. I end up cutting this frequency quite often on guitar. Listen as I turn 100 Hz up and down on the guitar sound in Audio Example 20-5.

Audio Example 20-5

Boost and Cut 100 Hz

Two-hundred Hz tends to be the muddy zone on many guitar sounds. A boost here can make the overall sound of the guitar dull. A cut at 200 Hz can expose the lows and the highs so that the entire sound has more clarity and low-end punch. Cutting this frequency can help a double-coil pickup sound like a single-coil pickup. Audio Example 20-6 shows the effect of cutting and boosting 200 Hz.

Audio Example 20-6

Boost and Cut 200 Hz

The frequency range from 250 Hz to 350 Hz can add punch and help the blend of a distorted rock sound. Notice the change in texture of Audio Example 20-7 as I boost and cut 300 Hz.

Audio Example 20-7

Boost and Cut 300 Hz

The frequency range from 500 to 600 Hz often contains most of the body and punchy character. Try to hear the body of the sound change as I cut and boost 550 Hz on the guitar in Audio Example 20-8.

Audio Example 20-8

Boost and Cut 550 Hz

The frequency range from 2.5 kHz to about 5 kHz adds edge and definition to most guitar sounds. I'll boost and cut 4 kHz on the guitar sound in Audio Example 20-9.

Boosting 8 kHz to around 12 kHz makes many guitar sounds shimmer or sparkle. These frequencies can also contain much of the noise from the signal processors, so cutting these frequencies slightly can minimize many noise problems from the guitarist's equipment. Listen as I boost and cut 10 kHz on the guitar sound in Audio Example 20-9.

Audio Example 20-10

Boost and Cut 10 kHz

Video Example 20-2

Equalizing the Guitar Sound

Delay

The use of delay on a guitar sound has the effect of placing the sound in a simple acoustical space. Delays of between 250 and 350 ms can give a full sound for vocal and instrumental solos (especially on ballads). This is a very popular sound. It's usually most desirable if the delay is in time with the music in some way. Audio Example 20-11 was recorded at a tempo of 120 bpm. I've added a 250-ms delay, which is in time with the eighth note at this tempo.

Audio Example 20-11

The 250-ms Delay

A slapback delay of 62.5 ms is in time with the 32nd note, at 120 bpm, and gives an entirely different feel to Audio Example 20-12.

Audio Example 20-12

The 62.5-ms Delay

Regenerating a longer delay of about 200 to 350 ms can really smooth out a part. All of these effects usually make the guitarist sound like a better player than he or she really is. Guitarists love that! This enhancement can be advantageous to all concerned, but don't overdo the effects, or the part will get lost in the mix. It might lose definition and sound like it's far away.

Electronic Doubling

Doubling a guitar part is a very common technique whenever the sound system is in stereo—in mono this technique tends to make the guitar sound muddier. Doubling can smooth out some of the glitches in the performance and can give the guitar a very wide, bigger-than-life sound. Pan the double apart from the original instrument, and you'll usually get a multidimensional wall of guitar that can sonically carry much of the arrangement. Doubling works well in rock tunes in which the guitar must sound huge and impressive.

Electronic doubling involves patching the instrument through a short delay, then combining that delay with the original instrument. To set up an electronic double, use a delay time between about 7 and 35 ms. Short doubles, below about 7 ms, don't give a very broad-sounding double, but they can produce interesting and full sounds and are definitely worth trying. Pan the original guitar to one side and the delay to the other.

Audio Example 20-13 demonstrates a guitar part doubled electronically using a 23-ms delay and no regeneration.

Audio Example 20-13
The 23-ms Double

Chorus/Flanger/Phase Shifter

Chorus, flanger, and phase-shifting effects are very common and important to most styles of electric guitar. A smooth chorus or flange can give a clean guitar sound a ringing tone. It can add richness that's as inspiring to the rest of the musicians as it is to the guitarist. Listen to the chorus on the clean guitar part in Audio Example 20-14.

Audio Example 20-14
Chorus

A smooth phase shifter can add color to a ballad or interest to a funky rhythm guitar comp. Notice the interest that's added to Audio Example 20-15 by the phase shifter.

Audio Example 20-15
Phase Shifter

The chorus effects are often part of a solo guitar sound used together with distortion, compression, and delay. The guitar in Audio Example 20-16 is plugged into the compressor first, then the distortion, next the delay, and finally the chorus.

Audio Example 20-16
Multiple Effects

Reverberation

Reverb is used primarily to smooth out the guitar sound when it must blend into the mix. Too much reverb can spell disaster for the clarity and definition of a good guitar part. On the other hand, most electric parts sound good with a bright hall reverb sound, a decay time of about 1.5 seconds, a predelay of about 80 ms, high diffusion, and high density. This kind of setting offers a good place to start shaping most guitar reverbs. Audio Example 20-17 demonstrates a guitar with this set of effects.

Audio Example 20-17
Hall Reverb

There are several other types of reverb that can sound great on many different musical parts. Experiment. Often, the sound of the guitar is so interesting with the delay, distortion, and chorus that there's really no need for much (if any) reverb. Always base your decisions regarding reverb on the natural acoustic reverberation time. Many acoustical spaces are so active that very little reverberation or delay effects are needed—in fact, they often create a sound that is excessively muddy and messy-sounding.

Try adjusting the predelay to add a different feel to the reverb sound. Longer predelays that match the tempo of the eighth note or quarter note can make the part sound closer to the listener and as if it was played in a large room. Listen as I adjust the predelay during Audio Example 20-18.

All-in-One Effects

These multi-effects processors are unbeatable for their convenience and power. Companies such as Line 6 and Roland have really led the way with great-sounding and powerful effects. Internal routing capabilities often let the user determine which order the signal path takes through the various effects. The ability to save a custom sound to memory for recall later is irresistible to most players.

As these devices have gotten more powerful and as the sounds have become more accurate, all-in-one effect pedals have become useful at any level. There is, however, an occasional trade-off, in which the simple convenience of the system tempts us to compromise the sound we really want for the music.

Whenever you're recording guitar through a multi-effects processor, try to save the application of reverberation and delay for mixdown. Effects are so easily applied and sounds are so quickly built, and they can sound so good with the track, that you might want to just print the sound with delay and reverb. Once you print the sound, there's no way to minimize the effects during mixdown. If you truly love the sound of the reverb and delay in your guitar rig, you can use it as an outboard effect during mixdown. Simply build an effect with reverberation and delay, feed it with an aux bus, then return the device outputs to the mixer channels.

Almost all of these effects pedals provide stereo outputs. On certain effects these outputs act as one big mono output; on others there is a distinct stereo image. Sounds that include chorus, phase shifting, and flanging effects are almost always stereo. In addition, most reverberation effects are stereo. Many amp simulation algorithms offer multiple speaker outputs. If you're using basic compression and overdrive sounds, there is little benefit to recording the stereo outputs—mono will work well.

If you have the tracks available, go ahead and record the stereo outputs. It won't hurt anything, and you might create a nice psycho-acoustic effect created by panning the channels during mixdown.

Stomp Boxes

There has been a movement toward the all-in-one guitar effects boxes and pedals because of their ease of use as powerful sound-shaping tools. However, the all-in-one devices don't always sound the best for certain applications. Hence, the single guitar effects pedal—the stomp box—still lives.

These small and inexpensive effects pedals let the musician shape a more characteristic and unique sound. Although the most popular multi-effects pedal is often inspiring for a player, so are vintage stomp boxes. Many of these small, single-purpose devices have become classics because their tone has inspired many great guitarists. They've frequently defined an era or genre. The MXR Dyna-Comp still has a sound that I can't find on other devices. The Ibanez Tube Screamer is the same way. Even in a world-class studio, where the room is full of ultimate gear, engineers and producers are turning to some of these inexpensive vintage tools in the quest for a unique sound.

Digital Modeling

One of the exciting developments of the digital era is digital modeling. When I plug my Les Paul into my old Fender Deluxe amp, it produces a very characteristic and recognizable sound. That combination not only makes a characteristic sound that I could recognize a mile away, it also creates a unique and recognizable waveform that can be duplicated and repeated. It's a simple matter of mathematics to calculate the difference between the waveforms of the sound coming directly from the instrument and the sound after it has gone through the amplifier and out of the speakers. Once we calculate the difference between the direct and amplified signal, that formula can be applied and added to any direct instrument sound. In this way, the sonic character of nearly any amplification system can be cloned with incredible precision and accuracy.

Among others, the folks at Line 6 and Avid have created amplification systems that use this modeling principle in a very effective way. They have modeled the sounds of many different guitar amplifiers—the original Fenders, Marshalls, Rolands, Hiwatts, and so on. They've also modeled the classic guitar effects, including specific types of compression pedals, delays, choruses, and reverbs. They've even modeled the sound difference between various speaker cabinet configurations, not to mention giving the user the ability to select the microphone used to capture the speaker sound. I've played through most guitar

Modeling Devices

Avid and Line 6 tend to lead the way in the amp-modeling world. They each offer a way to build very realistic and musical guitar sounds. With either of them, the guitarist can build sounds that are extremely similar to the sounds produced by high-quality vintage and modern amps. A guitarist who has a good ear for guitar tones and who is willing to spend some time building sounds can really get a lot of good out of either of these devices.

The Eleven Rack is particularly useful for those guitarists who spend a lot of time recording to Pro Tools; plus, it comes with a copy of Pro Tools! It assists with processing during recording and can be disconnected from the recording system and taken to the gig; therefore, the exact same sounds are available in the studio and onstage.

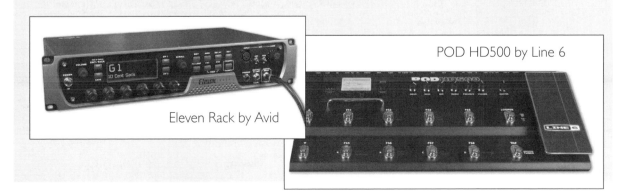

POD HD500 by Line 6

Eleven Rack by Avid

Onscreen Control of Guitar Processors

The simplest way to build modeled guitar sounds is through the use of a software editor for Mac or PC. These editors allow for virtually all possible configurations of amps, mics, speakers, and effects.

Line 6 and Avid each provide a visual user interface that makes a guitarist feel at home. The dynamics, amps, mics, and effects processors look like the classic tools that we're all used to. The knobs look the same, they're in the same location, and they provide the same kind of tone variations as the original classic devices after which they are modeled.

POD HD500 by Line 6

Eleven Rack by Avid

amplification setups and I'm amazed at how accurate these models are. Most guitar effects manufacturers currently offer their own rendition of the modeling system.

Listen to the guitar sounds in Audio Example 20-19. I have a Gibson Les Paul plugged straight into the device, and then running directly into the console. Those who have played guitar through these amps should recognize the sounds as very accurate and authentic. Because the modeling is so accurate, the amplification systems on these units must be clean and sonically transparent enough to faithfully reproduce the modeled sounds.

Audio Example 20-19

The Fender, Marshall, Roland JC-120, and Vox Sounds

As a purist, one might or might not appreciate the accuracy of these models. There is something special about the sound of the real thing that's difficult to quantify mathematically. However, there's no denying that these modeling systems provide vast flexibility to the home recordist who probably doesn't have all the classic amps and instruments readily available.

Integration of Hardware and Software

When hardware, such as the Line 6 POD500 or Avid Eleven Rack, is combined with its companion software editor, creating sounds becomes a much easier task. However, it is still always up to the artist or engineer to be able to recognize the components in the sound and be comfortable enough with the equipment to create authentic and musical tones.

Software Control of Modeling Systems

Building sounds and accessing controls are much easier when using the proprietary patch editing software for Mac or PC. The front panel maze on the device faceplate isn't impossible to navigate, but it's a lot more cumbersome than its onscreen counterpart.

Editing software typically accompanies the device in the documentation packaging but the newest versions are almost always available online. Connecting via USB or FireWire, all controls and routing options are communicated between the internal processor, computer processor, and online. Sounds, and banks of sounds, can be easily and quickly shared online. Simply upload your favorite bank of sounds to a friend or download one that he or she sends you.

Video Example 20-3

Modeled Guitar Sounds

Bass Guitar

Bass is an interesting instrument. Its function in most musical settings is to give a solid rhythmic and harmonic foundation for the rest of the arrangement. As the sound operator, it's your job to get a great bass sound that blends with the rest of the song. It's the bass player's job to play a solid rhythmic part with the correct musical touch. Without both player and sound operator working together in any recording situation, things can become difficult.

Live players often develop a brutal performance approach to bass. They get used to playing as hard as they can to help themselves feel the beat and to just get into the music. This heavy touch results in a lot of string buzzes and rattles that are usually covered up or forgiven in a live performance. However, an experienced bassist controls his or her touch and provides a full tone that defines the low-frequency tone and rhythmic definition. Any player with a touch that's too light or too heavy provides a sound that is difficult to mix. The sound operator can try and try, but the sound won't be right until the bassist learns to play with a consistent and full tone.

A player with a solid but controlled touch can get a great rhythmic feel and a clean sound. Listen to Audio Example 20-19. The bassist is playing very hard and causing buzzes and clacks that will distract from the song.

Bass Guitar through the DI

1. Plug the bass guitar into the direct box input.
2. Plug the XLR output of the direct box into the mic input of the mixer.
3. Send the bass guitar signal to the recorder using the bus assignments.
4. Set the levels and record.

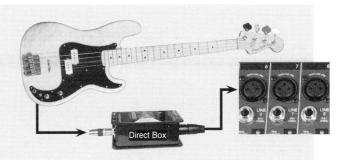

Audio Example 20-21 demonstrates the same bass part performed with a controlled touch.

Audio Example 20-21

The Controlled Performance

A good player, playing with a solid touch that's completely controlled, makes the sound operator's job easy. A player who consistently overplays and is out of control makes the sound operator's job difficult.

Direct Box/Direct In (DI)

Bass guitar typically sounds best when run directly into the mixer, either through a direct box or by plugging directly into the line input.

If you're using a direct box, plug into the direct box, then plug the XLR out of the direct box into the mic input of the mixer. This approach usually produces the best sound and offers the advantage of long cable runs from the direct box to the mixer.

A passive direct box simply transforms the high-impedance output into a low-impedance signal suitable for the mic input of the mixer. The bass in Audio Example 20-22 is running through a passive direct box directly into the mic input with no EQ or dynamic processing.

Audio Example 20-22

Bass through the Passive DI

An active direct box contains circuitry that, besides matching impedance, enhances the signal. Active direct boxes typically have more high-frequency clarity and more low-end punch. Audio Example 20-223 demonstrates the same bass as the previous example run through an active DI.

Audio Example 20-23

Bass through the Active DI

Miking the Bass Speaker

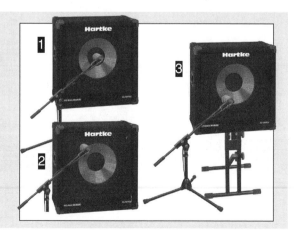

1. Pointing the mic at the center of the speaker produces a sound with more high-frequency edge.
2. Pointing the mic at the outer edge of the speaker cone, away from the center, produces a warm, smooth sound with less edge.
3. When miking the bass cabinet you'll almost always get a tighter, more controlled sound if you get the cabinet up off the floor. Try placing the cabinet on a chair or other type of stand. This approach will help control the low frequencies.

Bass into DI and Amplifier

- Plug the bass into the direct box input.
- Patch the THRU (out to amp) jack into the amplifier input jack.
- Patch the low-impedance XLR output of the direct box into the mic input of the mixer.
- Place a mic in front of the speaker and connect the mic cable to a separate mic input on the mixer.

This setup lets you use the mixer to blend the direct and miked sounds. Record both of these sounds onto one tape track in the desired proportions or record each signal to a separate tape track. It's best to record the miked and direct signals to separate tracks if you can spare them. That way, you give yourself freedom in the mixdown to blend the sounds in the way that supports the music best.

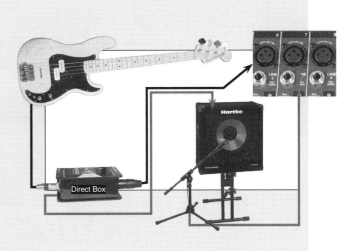

Pickup Output Level

It's almost always best to set the instrument's volume control at maximum. This sends the hottest signal to the board, resulting in a better signal-to-noise ratio.

Many of the newer pickups, especially pickups using active electronics, have very strong output levels. These strong levels can overdrive the circuits of some direct boxes and mixers. If the active bass electronics are overdriving the direct box or mixer inputs, try turning the volume down at the bass until the sound is clean and distortion-free.

Plucking Styles

Most bass parts are played with the first two fingers of the player's right hand. This technique produces a solid low end with good definition in the attack of each note.

Audio Example 20-24

Finger Plucking

Some players pluck the bass with their thumb only. This usually gives a sound that is fuller in the lows with a twang in the highs.

Audio Example 20-25

Thumb Plucking

Some parts are played with a pick. Using a felt pick produces a sound that's similar to the sound produced by using the fingers, but without quite as much low end.

Audio Example 20-26

The Felt Pick

Using a regular plastic pick produces a sound that has a lot of attack and a clear sound.

When playing with a pick, the player might mute the strings slightly with the heel of the right hand, producing a good attack with a tight, solid low note. This is a common sound in country music.

It's important for the sound operator to understand the difference created by simple technique changes on the musician's part. If the bassist isn't playing in a way that supports the music and fits perfectly with the groove, the sound operator is at a complete loss to make the mix sound great. However, if the sound operator asks for a simple technique change, such as using fingers instead of a pick, the entire mix could come together perfectly.

Asking the bassist to adjust his or her right hand position is a great way to get the raw sound that you need for a song. If you need a smooth, sustained, low-end bass sound, you could EQ all day and never get the effect of simply asking the bassist to pick the strings a little closer to the neck. Or, if you need the bass to cut through the mix a little better, the best solution could be to simply ask the bassist to pluck a little closer to the bridge.

Audio Example 20-29 demonstrates the sound of a bassist plucking back by the bridge. This sound generally works best for punchy rock, fast country, or some R&B songs.

Audio Example 20-30 demonstrates the sound of a bassist plucking up by the neck. This sound is generally best for slow ballads that need a full, sustained bass sound to support the rest of the arrangement.

Most bass parts sound best if plucked somewhere between the two extremes demonstrated in Audio Examples 20-29 and 20-30. Being aware of this sound-shaping technique can save you a lot of time and energy. Audio Example 20-31 demonstrates the sound of the same bass as the previous two examples, plucked about halfway between the point where the neck joins the body and the bridge.

Levels for Bass

The bass level should peaking at about 0 VU. If the sound is strong in the low end, it's usually okay to push the bass level to +1 or +2 VU. If the bass sound is particularly thin with lots of transient snaps and pops, set the level at −1 to −5 VU to compensate for the transient attacks.

Compressing the Electric Bass

Bass is usually compressed. There's a big difference in level between notes on many basses. Some notes read very hot on the VU meter, and some read very cold. Because the compressor automatically turns down the signal above the user-set threshold, it helps keep the stronger notes under control.

If the bass notes are evened out in volume by the compressor, the bass track stays more constant in the mix and supplies a solid foundation for the song. If the bass is left uncompressed, the bass part can tend to sound especially loud and boomy on certain notes and disappear altogether on others. More consistent levels from note to note typically provide the best foundation for most recordings.

A compressor becomes especially useful if a player snaps a high note or thumps a low note because the level changes can be extreme. Not only does the compressor help control the louder bass sounds, but it also helps the subtleties come through more clearly. If the loud sounds are turned down, the entire channel can be boosted to achieve a proper VU reading. As the channel is turned up, the softer sounds are turned up, which makes them more audible in the mix.

The bass part in Audio Example 20-32 is not compressed. Notice the difference between the loudest and softest sounds.

Audio Example 20-32

Non-Compressed Bass

Bass Compression - Insert and Line Input

You can insert the compressor into the signal path by patching the insert send from the mixer into the input of the compressor, then the output of the compressor to the insert return of the mixer.

It's best to insert the compressor as close to the beginning of the signal path as possible. The best patch point for a compressor is just after the mic preamp, as the signal heads to the equalizer.

You can also plug the bass directly into the compressor, then plug the compressor into the line input of the mixer. The success of this technique depends on the bass pickups, the sensitivity and impedance of the compressor input, and the level compatibility between the compressor output and the mixer's line input. When you use all compatible ingredients, this approach offers good control and a clean-sounding bass track. Be sure you've matched operating levels (+4, -10) and impedances.

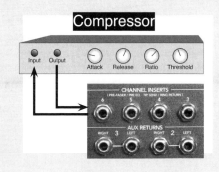
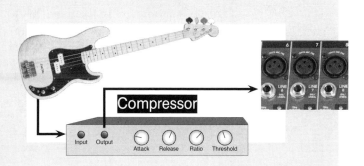

Procedure for Compressing Bass

1. Set the ratio (typically between 3:1 and 7:1).
2. Set the attack time. The attack time needs to be fast enough to compress the note but not so fast that the attack of the bass note is removed. If the attack time is too fast, the bass will sound dull and lifeless.
3. Set the release time. Start at about 0.5 second. If the release time is too slow, the VCA will never have time to turn the signal back up after compressing. If the release time is too fast, compression might be too obvious because the VCA reacts to each short sound by turning down, and then back up.
4. Adjust the threshold control for the desired amount of gain reduction (typically about 6 dB at the strongest part of the track).

Audio Example 20-33 uses the same bass used in Audio Example 20-32 This time the bass is compressed with a ratio of 4:1 and gain reduction of up to 6 dB. This example peaks at the same level as the previous example, but notice how much more even the notes sound.

Audio Example 20-33

Compressed Bass

If the bass part is very consistent in level and the player has a good, solid, predictable touch, you might not even need compression. I've been able to get some great bass sounds without compression. This only happens when you have a great player with predictable and disciplined technique, a great instrument, and the appropriate bass part. Aside from these factors, most bass parts need compression.

If the bass part includes snaps and thumps, consider limiting. With a limiter, most of the notes are left unaffected, but the snaps and thumps are limited. Limiting is the same as compression, but with a ratio above 10:1.

If the limiter is set correctly, the bass part can be totally unprocessed on everything but a strong thump or snap. The thump might exceed the threshold by 10 dB, but if the ratio is 10:1 or higher, the output of the compressor won't show more than a 1-dB increase.

Follow this procedure to correctly adjust the limiter:

- Set the ratio control to about 10:1.
- Set the attack time to fast.
- Adjust the threshold so that gain reduction only registers on the snaps and thumps.

The bass part in Audio Example 20-34 isn't limited. Notice how much louder the snaps are than the rest of the notes. Also, note that the normal level is low to keep the snaps from oversaturating the tape.

Audio Example 20-34

Snaps Not Limited

The bass part in Audio Example 20-35 is limited. Notice that now the snaps aren't much louder than the rest of the notes, and the entire part sounds louder because the limiter has squashed the peaks.

Equalization of the Bass Guitar

It's often appropriate to roll off the frequencies below about 40 Hz. This can get rid of frequencies that might never be heard but are adding to the overall level of the mix. To add a good low-end foundation to a bass sound, try boosting between 80 and 150 Hz. This frequency range will produce a very solid feel, and these frequencies can be heard on almost all systems. On the bass in Audio Example 20-36, I boost at 80 Hz, then sweep the boost from 80 Hz up to 150 Hz.

If the bass sounds muddy and thick in the lower mids or upper bass, try cutting at a frequency between 250 and 500 Hz. Cutting these frequencies can help a stock P-Bass sound like a bass with active electronics. This is one of the most common requests from bassists. Cutting in this range and running the bass through an active direct box can usually produce the desired effect. In Audio Example 20-37, I'll cut at 250 Hz, then sweep the cut from 250 to 500 Hz.

In Audio Example 20-38, I'll boost at 250 Hz, then sweep the boost from 250 to 500 Hz.

The frequencies between about 700 and 1,200 Hz contain the sound of the bass string being plucked, plus the harmonics that can help the listener recognize the pitch of the bass notes. Listen to Audio Example 20-39 as I boost, then cut at 1,000 Hz.

The upper clarity and string noise on a bass usually resides in the frequencies between 2 and 3 kHz. Listen to Audio Example 20-40 as I boost at 2 kHz, then sweep from 2 kHz to 3 kHz.

On most bass sounds, the frequencies above 3 or 4 kHz don't add much that's usable. Even though the upper frequencies contain important harmonics, they aren't usually boosted because they also contain most of the string and fret noises. The key to getting a great sound is in determining what sound best complements the mix. Always compare the bass sound to the kick drum and shape the bass to work with, not against, the kick. If the kick is heavy in one particular low frequency, avoid that frequency on the bass.

Panning the Bass

Bass is panned straight down the center of the mix. Bass frequencies are omnidirectional, so panning isn't effective from a listening perspective. The upper frequencies of the bass are directional and can indicate placement, but panning the bass track is not good for the stereo level of the mix. The bass needs to be centered to distribute the low-frequency energy equally to the left and right sides of the mix and to provide a solid foundation for the rest of the group.

Drums and Percussion

A good drum sound must be appropriate for the musical style. It's very important that the sound operator listen to a lot of recordings of popular music in the genre of the live performance. Without an accurate sonic imprint on which to base mix decisions, it is unlikely that the sound will be creatively satisfying.

Good drum sounds will almost always have:

- Clean highs that blend with the mix
- Solid lows that blend with the mix
- Enough mids to feel punch
- Not so many mids that the sound is muddy
- A natural, warm tone
- Dimension, often sounding larger than life
- Believably appropriate reverberation
- Excellent balance and blend in the mix

Drum Conditioning

To get good drum sounds, it's necessary to be familiar with drum tuning and dampening techniques. A bad-sounding drum is nearly impossible to get a good recorded sound from. A good-sounding drum can make your live sound experience much more enjoyable.

If the drum heads are dented and stretched out, it will be very difficult to build a great drum sound. New heads can make a world of difference in the sound of each drum.

If the drums aren't high-quality instruments, there's a good chance that the shells aren't smooth and level, and there's a possibility that the drums aren't even perfectly round.

If this is the case, the heads won't seat evenly on the drum shell, and there'll be a loss of tone detracting from the drum sound.

Tuning

Often, the difference between a good-sounding drum and a bad-sounding drum lies simply in tuning. The standard approach to tuning involves:

- Tuning the top head to the tone you want
- Making sure the pitch is the same all the way around the head by tapping at each lug and adjusting the lugs until they all match
- Duplicating the sound of the top head with the bottom head

If the head isn't tuned evenly all the way around, the head won't resonate well. You'll probably hear more extraneous overtones than smooth tones.

Drumsticks

The drummer's choice of sticks and their condition can make a big difference in the sound of the drums. Nylon-tipped sticks have a brighter-sounding attack than wood-tipped sticks, especially on cymbals. Hickory sticks have a different sound than oak sticks, and they both sound different than graphite or metal sticks. Heavy sticks have a completely different sound than light sticks.

Most experienced drummers carry several different types of drumsticks with them, even though they each probably have an overall favorite.

Muffling Drums

There are several techniques for muffling and dampening drum tone. Trends shift with time and genre. Whereas drums of one era and style are highly controlled and dampened, drums of the next are open and free. It's your job to stay in touch with current trends, adjusting your techniques accordingly.

Don't use muffling as a substitute for a well-tuned drum. It's hard to beat the sound of a great drum with great natural tone. I try to use a dampening technique to lightly control unwanted overtones that I know will be difficult to deal with while I'm building the mix.

Try each of the illustrated dampening techniques to hear the difference in the sound of each approach. I've gotten great sounds by using self-adhering weather stripping and a product called Moon Gel, applying the amount of material in the positions that create the sound I want. Moon Gel is a Jello-like solid rectangle, approximately 1 x 1.5 x .125 inches. This is a very flexible approach. Both Moon Gel and the foam weather stripping are easy to move around for the desired sound, and they provide an even, natural sound and appealing dampening. Moon Gel sticks to the head much like weather stripping, without the sticky residue.

Hardware

Hardware matters. It provides a stable and solid foundation for drum tone. The drum hardware is a good indicator of overall product quality and attention to detail. Drums with excellent hardware will almost always sound better than drums with substandard hardware.

Free-Hanging Tom Mount Systems

Toms are typically the most difficult of the drums to find the perfect sound for. The tone is often full of dominant overtones, or even just a lack of tone. This is especially true of drum sets that use a mounting bracket that attaches to the tom shell and is held in position by a tube that penetrates the bracket and clamps into place. Attaching any solid bracket to any drum (kick drums or toms) chokes the sound and provides an unimpressive tone.

Mounts like the one pictured, which let the drum hang freely, allow the drum to vibrate naturally, providing an excellent tone with much less tinkering by the drummer and engineer.

Shells that are true and hoops that are meticulously crafted make your job a lot easier. If the head does not make even contact around the shell, it's very difficult to get good tone from the drum. If the drum is well-made, it is much easier to get a great drum sound and to find a pure drum tone.

Mounting hardware is also crucial, especially tom mounts. Any mount system that lets the drum float with no screwed-in hardware provides the drum the opportunity to sing. Imagine screwing a large bracket to the face of an acoustic guitar—it's easy to imagine there might be a change in the sound. The same is true for drums. A company called RIMS makes a great mount system that holds the drum by a set of lugs, and the difference in sound is notable. Most major drum manufacturers currently have free-floating tom mount systems.

Avoid mounting anything on the kick drum. Toms and cymbals should be mounted on separate stands, not on brackets screwed into the kick drum. The more freedom any drum has from contact with any other solid object, the better it will sound.

Room Acoustics

This size and shape of the acoustical environment is very influential on the sound of the drums. If drum kits, identical in every way, are placed in two different rooms, they might very well sound great in one room and terrible in the other. Larger rooms help the drum sounds realize their potential. In a small room there are often so many immediate reflections and negative phase interactions that the sound quickly becomes thin and choked-sounding.

Rooms with lots of hard, reflective surfaces provide a mixing challenge for the sound operator because the drums excite the abundant acoustical reverberation tendencies. In other words, the drums are usually too loud and they tend to leak into every other mic in the room, creating a mix with little intimacy and intelligibility.

Theories of Drum Miking

Most of the drum sounds you hear on albums are achieved through the use of several microphones recorded separately to several tracks that are blended and balanced during the mixdown. This is ideal. Practically speaking, most live sound operators don't have a pile of microphones to use, let alone 8 to 12 available mixer channels. However, excellent results can be achieved with a few microphones, intelligently positioned and skillfully crafted.

Essential Microphones

You should have a good condenser mic for over the drum set and for cymbals. Condensers are the mic of choice for percussion, and they do the best job of capturing the true sound of each instrument. The fact that condenser microphones respond to transients more accurately than the other types of microphones makes them an obvious choice for percussion instruments, such as tambourine, shaker, cymbals, triangle, claves, or guiro.

Mic Choices

The mic of choice for close-miking toms, snare, and kick is a moving-coil mic, such as a Shure SM 57, Beta 57 A, Sennheiser 421, or Electro-Voice RE20. Though they don't have the transient response of condenser microphones, moving-coil microphones work great for close-miking drums because they can withstand intense amounts of volume before distorting. Also, most moving-coil microphones have a built-in sensitivity in the upper frequency range, which provides an EQ that accentuates the attack of the drum.

Most reasonably priced condenser and moving-coil microphones provide good results. Don't overlook the obscure. As trends come and go, we all start hunting for unique and interesting sounds that imprint a sonic personality. Keep all the mics you can get your hands on. Even a cheap lo-fi mic might be the perfect tool to create an interesting and musical sound.

Positioning the Microphones

Keep the mic out of the drummer's way. Most mic manufacturers make microphones that are designed for getting into tight spots, such as drum sets. If the mic has to be pointing

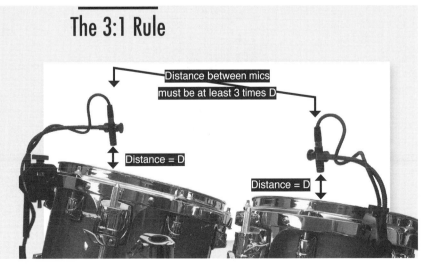

The 3:1 Rule

The distance between microphones in any multi-mic setup should be at least three times the distance from either mic to its intended source. This is especially true in drum mic setups. The fact that there are often several mics used to record the kit provides ample opportunity for destructive phase interactions. Always maximize your efforts to find the perfect mic positions. Small changes can produce huge changes in the overall drum sound.

Distance between mics must be at least 3 times D

Distance = D

Distance = D

straight across the drum due to space restriction, there will be more leakage between drums. It's best to point the mics at the drums, 1 to 2 inches away, and at an angle of 30 to 45 degrees.

From the purist's viewpoint, it's best to use a mic stand for each drum mic, rather than using stands that mount on the drum rims. The less that touches the drum, the better the tone. However, from a practical standpoint a good mount that attaches to the drum provides a cleaner-looking stage and typically provides satisfactory results. When the drums utilize RIMS-type mounts, where the drum hangs freely, attach the microphone mounts to the bracket holding the tom rather than to the drum rim itself.

Phase Interaction between Mics

Whenever you're close miking the kit, be particularly aware of phase interactions between microphones. With up to 10 microphones placed in a small acoustical environment, the leakage between microphones can enhance or rule the sound of the drum set.

Keep the 3:1 rule in mind. When placing microphones, the distance from one microphone to its intended source should be no more than one-third the distance between it and another microphone. Adhering to this suggested ratio between microphones at least provides some assurance that native phase influences will be minimal.

When placing microphones, movements of an inch or two by one microphone make an amazing difference in the overall sound of the drum kit. Experiment with placement and position to get the perfect sound. If you settle for the wrong sound, everything else will be difficult. Phase interactions and leakage increase the likelihood that small changes in mic position will produce large changes in overall sound quality.

Drum Levels

Drum sounds contain transients that can easily overdrive a signal path. Most modern mixers provide plenty of headroom, but there's no good reason to hit the mixer too hard at the mic preamp with the drum levels. When using acoustic drums, the clean transient provides a very important ingredient to the overall drum sound—adjusting the trim for a conservative peak level around −5 on the FSM and −5 to −7 on the VU meters typically provides acceptable headroom and a sufficiently full, strong signal.

Percussion instruments, such as claves, vibra slap, crash cymbal, hi hat, and so on, contain the strongest transients and should be adjusted at the gain trim for a reading between −7 and −9 on the VU meter and around −18 on the full scale meter (FSM).

Isolating the Acoustic Drum Kit

Most live drum sets are isolated through the use of a simple Plexiglas drum shield, such as Clearsonic Panels. Often, the drum volume is less of a reason for using isolation panels than the leakage into other mics on stage. Sometimes the drums are so loud on stage that the vocalists can't hear themselves sing without extreme monitor levels, which cause the sound in the room to be muddy. In addition, the leakage from the drums into the rest of the microphones creates a distant sound with little intimacy and clarity.

For the sound operator, creating an isolation area for the drums helps provide a controllable drum kit sound that has increased presence and punch when compared to the same setting without isolation. It is important that the clear panels interrupt the sound from

One Mic in Front

This setup uses a condenser microphone with a cardioid pickup pattern positioned in front of the set, pointed at the set, approximately 6′ above the floor.

traveling to other mics, and that the direct drum sound is somewhat muted; however, don't overlook the reflection off walls behind the drums. The area behind the drums should utilize absorptive panels to help dampen the reflection into the room. Some isolation packages even provide a ceiling for the isolation area. A tool like this really helps give the control back to the sound operator—it also lets acoustic drummers play strong and loud without creating volume problems in the room.

Miking a Drum Set with One Microphone

Listen to the examples of a complete drum set recorded with one microphone. Audio Examples 20-41 to 20-44 all use the same drum set in the same setting.

In Audio Example 20-41, the drum set was recorded with one mic directly in front of the kit, pointed at the set, and about six feet from the floor.

Audio Example 20-41

Mic in Front

In Audio Example 20-42 the mic is behind the kit, just above the drummer's head, and pointed at the kit.

Audio Example 20-42

Mic over the Drummer's Head

In Audio Example 20-43, the mic is about four feet above the set and is pointed down at the drums. When a mic is placed over the drums and points down at the set, it's called an *overhead*.

Audio Example 20-43

Overhead

Finally, in Audio Example 20-44, we hear the drum set from one mic, positioned about eight feet away and pointed toward the kit.

Audio Example 20-44

Eight Feet Away

Miking a Kit with Two Mics

With two microphones on the set there are two primary options: You can use both mics together in a stereo configuration, or you can use one mic for overall pickup while using the other for a specific instrument.

In Audio Example 20-45, I've set one mic directly over the kit with the second mic in the kick drum. When you use one of the microphones for the overall kit sound, you can place the second mic on the kick drum (or possibly the snare) to get individual control, punch, and definition in the mix. Choosing to close-mike the kick or the snare is purely a musical decision that's dependent on the drum part and the desired effect in the arrangement. This mic setup is more flexible than the single mic technique, but we're still limited to a monaural sound because the kick or snare would almost always be positioned in the center of the mix with the rest of the set.

Audio Example 20-45

One Mic over the Kit, One in the Kick

Audio Example 20-46 uses the two condenser microphones with cardioid polar patterns as a stereo pair. The two mics are placed in a traditional X-Y configuration, directly above the drum set, at a distance of approximately three feet above the cymbals, pointing down at the drums. With this configuration, we can get a sound that has a stereo spread. As we get into the mixing process, we'll see that positioning supportive instruments away from the center of the mix helps us hear the solo parts that are typically positioned in the center of the mix.

Audio Example 20-46

Stereo X-Y

Try the X-Y configuration from different distances and in different rooms. Stereo mic technique is often the best choice for a very natural drum sound, but for contemporary commercial drum sounds, it lacks flexibility.

One Mic Overhead, One in the Kick

1. One cardioid condenser microphone is pointing down at the kit.
2. One cardioid moving-coil microphone is inside the kick drum, aimed at the head, about halfway between the center of the head and the shell.

Miking the Kit with Three Mics

1. One cardioid condenser mic placed 2' above the cymbals, pointing down at the set.
2. One cardioid moving-coil mic pointing at the snare, from a distance of approximately 2".
3. One cardioid moving-coil mic inside the kick, positioned for the best sound.

Another good two-mic technique involves placing one mic on each side of the drummer's head, level with his or her ears, pointing forward toward the drums. Position the microphones with their capsules three to six inches from the drummer's ears to achieve a good stereo image. The drummer's skull will act as a baffle between the two microphones. Audio Example 20-47 demonstrates this technique.

Audio Example 20-47

Head Baffle

Miking a Kit with Three Mics

If you use one mic on the kick, one mic on the snare, and one overhead mic, separate control of the kick and snare is possible. With three microphones, this technique will yield the most commercial and punchy sound. The kick and snare are the two main contributors to the definition of style. Being able to fine-tune their levels, EQ, and effects is an advantage. The drum set in Audio Example 20-48 was miked with one mic inside the kick, one mic two inches above the snare, and one mic about two feet above the cymbals. This configuration produces the most commercially viable results so far, but it doesn't provide a stereo image of the set. The kick, snare, and overhead are almost always positioned together in the center.

Audio Example 20-48

Three Microphones

X-Y Mic Technique

The traditional X-Y technique uses two cardioid condenser microphones positioned together to form a 90° angle. The mics should be overlapping and nearly touching.

This is called a coincident mic technique because the capsules are very close to each other and they share the same horizontal and vertical planes.

Coincident stereo mic techniques such as the X-Y configuration exhibit the least amount of adverse phase interaction when combined in a mono mix. Because the mic capsules are as close together as they can possibly be without touching, they hear the sound source nearly simultaneously; they receive the sound waves in the same phase.

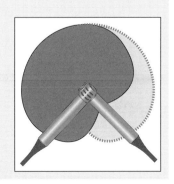

If we use a kick and two overheads, we can get a stereo image of the kit, but we lose individual control of the snare. Another option is to put the single mic on the snare instead of the kick, combining that mic with the two overheads. This can be a usable option, but we sacrifice control of the kick. Audio Example 20-49 demonstrates the sound of a drum set with two microphones overhead, in an X-Y configuration, combined with one mic inside the kick.

Audio Example 20-49

X-Y Overhead, One in the Kick

Miking a Kit with Four Mics

With four microphones on the set, you begin to have good control over the kick and snare sounds, plus you can get a stereo image. Some very acceptable drum sounds can be achieved using a setup with one kick mic, one snare mic, and two overheads. You'll need to experiment with placement of the microphones (especially the overheads), but solid and unique kick and snare drum sounds are possible with this mic technique. The individual microphones plus the overheads used in a stereo configuration can provide an excellent stereo image. The set in Audio Example 20-50 was miked with one kick mic, one snare mic, and two overheads in an X-Y configuration.

Audio Example 20-50

Snare, Kick, and X-Y

Close-Mike Technique

The most common approach to getting good, punchy drum sounds that have unique character is to use the close-mike technique. Each drum will typically have its own mic and channel.

The drum set in Audio Example 20-51 is set up with one kick mic, one snare mic, one mic on each tom, two microphones overhead in an X-Y pattern, and one hi-hat mic.

Kick, Snare, and X-Y

1. Two cardioid condenser microphones in a traditional X-Y configuration above the kit. Experiment with placement and distance above the kit to find the appropriate musical sound.
2. One cardioid moving-coil mic inside the kick.
3. One cardioid moving-coil mic pointed at the top of the snare drum, from a distance of about 2″ above the top head.

Kick, Snare, Toms, and X-Y

1. Two cardioid condenser microphones in a traditional X-Y configuration above the kit. Experiment with placement and distance above the kit to find the appropriate musical sound.
2. One cardioid moving-coil mic inside the kick.
3. One cardioid moving-coil mic pointed at the top of the snare drum from a distance of about 2" above the top head.
4. One cardioid moving-coil mic pointed at the floor tom.
5. One cardioid moving-coil mic aimed between the upper two toms and positioned so that the two drums are balanced and blended.

Audio Example 20-51

Snare, Kick, Toms, and X-Y

Video Example 20-5

Miking the Drum Set with Various Microphone Configurations

Equalizing Drums

Always find the microphone, mic placement, and tuning that sound the best on any drum before beginning the equalization process.

The nature of close-miking a kick drum typically produces a raw sound that's overly abundant in lower midrange frequencies between 200 and 600 Hz, and the sound usually needs EQ to be usable.

When I listen to a raw close-miked drum sound before it's been equalized, I first listen for the frequencies that are clouding the sound of the kick. That frequency range is almost always somewhere between 200 and 600 Hz. Listen to the kick in Audio Example 20-52 as I turn down a one-octave-wide bandwidth centered at 300 Hz.

Finding the Drum Tone

1. To hear the tone of the drum (the head ringing), place the mic near the rim.
2. To hear the attack of the drum place the mic near the center of the head.
3. Move the mic from the rim to the center of the drum until you hear the sound you like, but be sure to keep it out of the drummer's way.
4. When you're sampling a single hit, find the perfect spot for the mic, then be careful not to hit the mic.
5. Be aware that the point of contact where the stick hits the head also affects the sound of the drum. When the stick hits near the center, the sound has more attack. When the stick hits near the rim, the sound has more tone.

Cut 300 Hz

Once the lower mid frequencies are turned down, I'll usually address the low frequencies between 75 and 150 Hz. On the kick drum and maybe the low toms, I might boost a frequency bandwidth in this range. On toms, snare, hi-hat, and overheads, I'll typically cut the frequencies below 100 to 200 Hz. This opens up the sound of the set and lets you isolate the sound of these higher drums.

Next, I'll typically locate an upper frequency to boost that will emphasize the attack of the beater or stick hitting the instrument. Boosting a frequency between 3 and 5 kHz will usually emphasize this attack. Listen to the kick in Audio Example 20-53. A moving-coil mic is pointed halfway between the center of the drum and the shell, from a distance of six inches. At first, this drum has no EQ. First I'll cut at 300 Hz, next I'll boost the low end at about 80 Hz, then I'll boost the attack at about 4 kHz.

Audio Example 20-53

Cut 300 Hz, Boost 80 Hz and 4 kHz

Kick Drum

The kick drum (bass drum) is very important to the impact of the drum sound. Different styles demand different kick sounds. Some sounds, such as jazz and heavy rock kicks, have less dampening and ring longer.

Often in the jazz idiom and some hard rock settings, the kick is not dampened, but the most common kick sound is lightly muffled, with good low-end thump and a clean attack. To achieve this sound, remove the front head and place a blanket or a pillow in the bottom of the drum. The blanket or pillow should be positioned for the desired amount of dampening—the more contact with the head, the more muffling. The weight of the pillow

Typical Kick Drum EQ

The actual amount of cut or boost you use is solely dependent on what it takes to get the sound you want out of the instrument you're miking. First use mic choice and placement to get the best and most musical sound, then use the amount of EQ necessary to create the appropriate sound.

1. A low-frequency boost between 75 and 150 Hz adds a low, powerful thump to the kick drum sound.
2. A mid-frequency cut between 250 and 600 Hz helps clean up the thick, cloudy sound of a close-miked kick.
3. A high-frequency boost between 3 and 5 kHz adds definition, attack, and impact to the kick drum sound.
4. Adjust the bandwidth on each of these bands to fine-tune the overall tone.

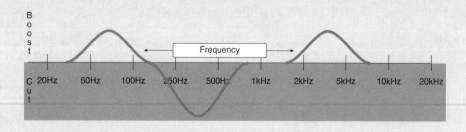

or blanket affects the sound. I've found that a down pillow works great; I'll usually place a brick or a mic stand base on the pillow to hold it in place.

There are also a few very effective kick drum muffling systems available for purchase at your local music store. Evans and Remo have both addressed the need for a solid and punchy kick drum sound.

For a little more tone, leave the front head on the drum, still dampening inside using a pillow or blanket. It has become popular to cut a six- to eight-inch hole in the front head, typically slightly off center. Use this hole to position the mic inside or slightly outside the drum. Move the mic in or out to achieve the balance of tone and attack that best supports your music.

A moving-coil mic, positioned inside the kick about six inches from the head and about halfway between the center of the head and the shell, will usually produce a good sound.

Experiment with mic placement to get the best sound you can before you equalize the sound. On any drum, the attack is strongest at the center of the drum, and the tone is strongest toward the shell. Move the mic to the center of the head if you want more attack. If you need more tone, move the mic toward the shell.

Audio Example 20-54 demonstrates the sound of a kick drum with the mic inside the drum pointing directly at the center of the head, where the beater hits, from a distance of six inches. Notice the attack.

Audio Example 20-54

Kick Attack

Audio Example 20-55 demonstrates the same kick as Audio Example 20-53, with the same mic aimed at the head about two inches in from the shell and about six inches from the head. Notice the tone.

Audio Example 20-55

Kick Tone

Another factor in the sound of the kick is the distance of the mic from the drum head. Audio Example 20-56 demonstrates the kick with the mic three inches from the head and about halfway between the center of the head and the drum shell.

Audio Example 20-56

Kick Three Inches Away

The mic in Audio Example 20-57 is about one foot outside of the drum, still pointed about halfway between the center and the shell.

Audio Example 20-57

Kick 12 Inches Outside

Video Example 20-6

Adjusting the Kick Drum Equalization

As you can tell by these different examples, positioning is critical to the sound of the drum. Not only is the placement of the mic critical, but the tuning of the drum can make all the difference. The tension should be even around the head, and there should be appropriate dampening for the sound you need. It's common to hear a very deep-sounding kick that has a solid thump in the low end and a good attack. In search of this kind of sound, most drummers tend to loosen the head to get a low sound. This can be a mistake. If the head is tuned too low, the pitch of the drum can be unusable and might not even be audible. To get a warm, punchy thump out of a kick, try tightening the head.

Another very important consideration in the kick sound is the drummer's technique. Drummers that stab at the kick with the beater can choke an otherwise great sound into an unappealing stutter-slap.

Snare Drum

Snare drums usually fall into one of two categories: very easy to get good sounds out of or almost impossible to get good sounds out of. Fortunately, there are some tricks we can pull out of the hat to help the more difficult drums sound good. It's important for you to know some quick and easy techniques for getting the snare to work. It's amazing how many decent drummers are lost when it comes to drum sounds.

First, make sure the heads are in good shape. A lot of times the top snare head has been stretched and dented so much that the center of the head is actually loose and sagging, even though the rest of the head is tight. This isn't good. Replacing the head will make a huge difference in the sound.

A good snare sound is dependent on a lot of factors working perfectly together. If you can handle drum-tuning basics, it'll make a big difference in the sound of your live drum recordings, plus you'll have an insight and perspective on drums that will prove to be a valuable asset.

Often, it's a good idea to place a microphone on the top of the snare drum and another microphone underneath the drum. The sound from the top of the head is usually full, with plenty of tone. The sound from the bottom of the drum provides the edgy high frequencies from the rattling snares. It is, however, very important to understand that the microphone on the top is pointing down and the microphone on the bottom is pointing up. Anytime two microphones point at each other, they are 180 degrees out of phase. When these two microphones combine to one track, several frequencies cancel, causing a very thin and weak sound. The solution to this is to reverse the phase of the bottom microphone,

The Snares

If the snares are in bad shape or unevenly tensioned, you will have difficulty achieving a full-bodied and smooth sound. Pay attention to these details and you'll have an easier time recording this important instrument. Keep in mind that often the snare drum sound is fundamental to the stylistic impact of your recording.

1. Be sure the snare strands are even in tension.
2. If the snares are loose or broken, cut them off.
3. If the snares can't be adjusted for constant tension or if there are too many strands missing, replace the entire set of snares.
4. Applying tape to loose snares can minimize extraneous buzzes, but the tape chokes the sound of the drum.

Snares

Two Mics on the Snare Drum

Depending on the snare drum you're recording, the mic on the top head might not produce a sufficient amount of buzz from the snares rattling on the bottom head. When this is the case, simply position a mic aiming up from the bottom at the snares; blend these mics to produce a sound with just the right amount of snare sound.

Anytime two mics point at each other, like the mics described herein, they are 180 degrees out of phase. Notice that when these mics are combined to one track, or both panned to the center position, the resulting sound is thin and unappealing. Simply invert the phase of the bottom microphone, and the combined sound becomes smooth and very usable.

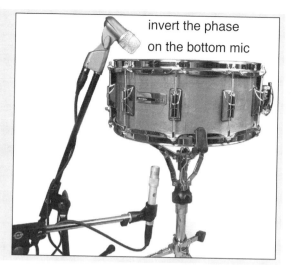

invert the phase on the bottom mic

thereby forcing the two microphones to work together to create one good sound instead of working against each other to create one bad sound. Audio Example 20-58 demonstrates this concept as two microphones work together to capture the complete sound of the snare drum.

Audio Example 20-58
Two Mics on the Snare Drum

There are times when the snare sound has an unappealing and dominant tone. The right solution to this problem is to retune the drum so the obtrusive tone is gone. We can, though, use this opportunity to learn something about equalization. If this happens to you, and for one reason or another you can't get rid of the obnoxious ring, try this technique—it actually applies to shaping many different types of sounds.

On a parametric equalizer, set up a very narrow bandwidth and an extreme boost. Next, sweep the boost until the unwanted tone is obviously as loud as it gets—this procedure is simply locating the problem frequency by sweeping until it reaches the maximum level. After you've located the problem frequency, all you need to do is change the boost to a cut, eliminating the problem ring to whatever degree you desire.

Video Example 20-7
Eliminating an Unwanted Ring in the Snare Tone

Miking Toms

Miking toms is similar in many ways to miking the kick drum and snare drum. It's important that the heads are in good shape, that they're tuned properly, and that the dampening gets the appropriate sound for the track. Tune the top and bottom heads to the same tone, and be sure the tension is even around each head.

If you want more attack in the sound, move the mic toward the center of the drum, but keep it out of the drummer's way. A miked drum sound has more attack when the

Wide Stereo Overheads

- For a wide stereo image, use two cardioid condenser microphones over the drum set spaced 1 – 3' apart.
- The mics should be at 90° angles to each other and should point away from each other.
- If you point the mics toward each other, you'll encounter problems, especially when summing the stereo mix to mono—essentially they become one big mono microphone with phase problems.

microphone is positioned near the center of the drum and more tone when the microphone is positioned near the rim.

Overhead Microphones

Once you've positioned the close microphones for the snare, kick, and toms, use mics over the drums to capture the cymbals and fill in the overall sound of the drums. It's amazing how much separation we can achieve by close-miking the kit. One or two mics over the drums are essential to a blended, natural sound.

Position condenser microphones in a stereo pattern (such as the examples of a two-mic setup). A good pattern to use is the standard X-Y configuration, with the microphones pointing down at the set at a 90-degree angle to each other. This will provide the excellent stereo image necessary for a big drum sound and will work well in mono.

If the drummer's kit is large and covers a wide area, try spreading the X-Y out. Move the microphones away from each other, but be sure they're still pointing away from each other. Also, keep the microphones on the same horizontal plane to minimize adverse phase interactions when listening to the mix in mono.

Overheads on a close-miked kit give definition and position to the cymbals and fill in the overall sound. There isn't much need for the low frequencies because the close microphones give each drum a full, punchy sound. I'll usually roll the lows off below about 150 Hz, and I'll often boost a high frequency between 10 and 15 kHz to give extra shimmer to the cymbals.

We want the overheads to accurately capture the transient information. Because the transient level exceeds the average level by as much as 9 dB, recording levels on the overheads should read between –7 and –9 VU at the peaks to ensure accurately recorded transients. Digital meters should not reach overload (OL).

Pan the overheads hard right and hard left for the most natural sound. The X-Y technique will provide a sound that is evenly spread across the stereo spectrum. The overheads in Audio Example 20-59 are about three feet above the cymbals in an X-Y configuration and are panned hard right and hard left. The lows below 150 Hz are rolled off, and the highs are boosted at 12 kHz.

Audio Example 20-59

X-Y Panned Hard

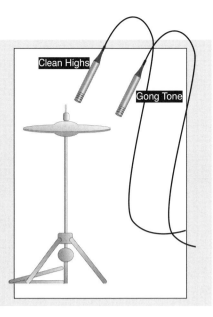

Getting the Desired Hi-Hat Tone

1. Miking the hi-hats at the edge of the cymbals produces a thick, gong-like sound that's not usable for most recording situations.
2. In this position, the air coming from the cymbals closing can cause a loud popping sound as it hits the mic diaphragm.
3. Miking the hi-hats at the bell of the top cymbal produces a good, clean sound with plenty of highs. The sound at the bell of the cymbal contains very little of the gong-like sound that comes from miking the edge of the cymbal.
4. The microphone must be at least 3" from the cymbal to minimize the change in phase interaction between the cymbal and the mic capsule (caused by opening and closing the hi-hat).

We can add different character to the sound of the drums by moving the overheads closer to or farther from the kit. Positioning the mic farther away from the set includes more room sound on the track. This can be good or bad depending on the acoustics of the recording environment.

It isn't typically necessary to add reverb to the overheads in a close-miked configuration. The reverb on the snare and toms is usually sufficient to get a smooth, blended sound.

The Hi-Hat Mic

It's desirable to put a separate mic on the hi-hat so that the definition of the stick hitting the hi-hat can be accentuated. This hi-hat definition typically helps the band stay together in the monitors as much as it helps the sound operator.

Audio Example 20-60
Panning the Hi-Hat

Audio Example 20-61 demonstrates the sound of a hi-hat miked at the outer edge.

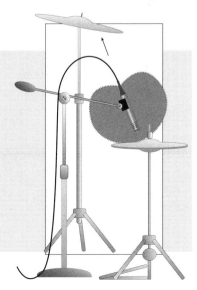

Minimizing Leakage

1. Aim the cardioid mic at the bell of the hi-hat to pick up a good, clean hi-hat sound.
2. Position the microphone so that it aims directly away from the instrument you want to minimize—in this case it's the crash cymbal. This technique won't eliminate an unwanted instrument, but it will decrease leakage and increase separation.

Audio Example 20-62 demonstrates the sound of the hi-hat with the mic pointing down at the bell of the top cymbal.

Audio Example 20-62

Hi-Hat Miked at the Bell

Gating the Drum Tracks

Sometimes we need to isolate the drum tracks to equalize them separately, to pan them, or to add effects to an individual instrument or group of instruments. Gating helps eliminate the constant ringing from the toms from the mix. If the gates are adjusted correctly, the drum channels will remain turned off until the drummer hits that drum, at which time the gate will open up quickly and smoothly turn down as the drum tone decays naturally away.

- Patch the drum track through a gate.
- Adjust the attack time to its fastest setting and the release time to about half a second.
- Adjust the range control so that everything below the threshold will be turned off.
- Finally, adjust the threshold so that the gate only opens when the drum is hit. This will isolate the drum. After the drum is isolated, you can process it alone with minimal effect on the rest of the kit. For example, you can add as much reverberation as you want without leakage adding reverb to the rest of the drums. Listen to the kit in Audio Example 20-63. I'll solo the snare track, then adjust the gate to get rid of the leakage between the snare hits.

Audio Example 20-63

Adjusting the Gate

Aiming the Mics

It's important to get into the habit of aiming microphones away from sounds you want to exclude from a track. Use the cardioid pickup patterns to your advantage. For example, if you're miking a hi-hat and the mic is pointed at the bell of the top cymbal, that's good. Not only should you point the mic at the bell of the hi-hat, but you should point the back of the mic at a cymbal that's close by. Pointing the back of the hi-hat mic at the crash cymbal helps minimize the amount of crash that is recorded by the hi-hat mic. Use the cardioid pickup pattern to reject the unwanted sound while it captures the intended sound.

Click Track

One feature of a professional-sounding mix is a solid rhythmic feel that maintains an even and constant tempo. A sure sign of an amateur band and an amateur recording is a loose rhythmic feel that radically speeds up and slows down.

Most drummers need some assistance to maintain a constant tempo. We call this assistance the *click track*. A click track can simply be a steady metronome pulse, such as that from a drum machine or an electronic metronome. It gives the drummer a rhythmic reference to keep the tempo steady.

A drum machine is a good source for the click because it offers the ability to change the sound. Click sounds with good transients work the best because the transient attack unquestionably defines the placement of the beat.

Using a click track in a live setting requires some practice, but it is worth the effort. Once the drummer is used to starting the click and develops the type of solid rhythmic time required to stay with the click for the duration of the performance, the band will all lock together tighter and more powerfully. A click track is only practical when the band is using in-ear monitors. Most of the time it is best if the entire band hears the click track; however, at first, probably only the drummer should hear it. If the entire band is used to playing with a click, it should be no problem to keep with the click. Be sure to stress to the rhythm section the importance of accurately staying with the click at all times. A lazy band can feel like they're constantly chasing the click.

Effects on Drums

The amount of reverb and ambience that you incorporate in your drum sounds depends on the natural room acoustics, as well as stylistic and musical factors. We seem to seesaw from very wet sounds to very dry sounds in all pop genres. It's up to you to stay in tune with the trends in your musical arena. Be informed to create a competitive and commercial sound, or at least know what you're doing when you break all the rules in pop-dom.

Remember, the more reverb you apply to any sound, the farther away and less intimate it feels.

Compressing Drums

The application of *compression* is a common drum mixing technique. Compression has two primary effects on drum tracks. First, because the compressor is an automatic level control, it evens out the volume of each hit. This can be a very good thing on a commercial rock tune. The compressor keeps the level even so that a weak hit doesn't detract from the groove.

The second benefit of compression is its ability, with proper use, to accentuate the attack of the drum. If the compressor controls are adjusted correctly, we can exaggerate the attack of the drum, giving it a very aggressive and penetrating edge. This technique involves setting the attack time of the compressor slow enough that the attack isn't compressed, but the remaining portion of the sound is.

The following steps detail how to set the compressor to exaggerate the attack of any drum:
- Set the ratio between 3:1 and 10:1.
- Set the release time at about .5 second. This will need to be adjusted according to the length of the snare sound. Just be sure the LEDs showing gain reduction have all gone off before the next major hit of the drum. This doesn't apply to fills, but if the snare is hitting on 2 and 4, the LEDs should be out before each hit.
- At this point, set the attack time to its fastest setting.
- Adjust the threshold for 3 to 9 dB of gain reduction.

- Finally, readjust the attack time. As you slow the attack time of the compressor, it doesn't react in time to compress the transient, but it can react in time to compress the rest of the drum sound.

Panning the Drums

In a live setting, the drums are typically kept pretty close to center position. Any extreme panning creates an imbalance for one side of the room or the other; however, slight separation of the overheads and toms can help open up the center position for the lead vocals, solos, and so on.

Miking Piano

Whether you're miking a nine-foot grand piano, a spinet piano, or an old-time upright piano, it's understood that we're trying to achieve the full, rich sound that only an acoustic piano can provide. A great piano, miked well, has life, transparency, and openness that's hard to beat.

In Audio Example 20-64, listen to a grand piano miked with a couple of good condenser mics for a wide, impressive solo piano sound.

Audio Example 20-64

Stereo Grand Piano

The method of choice for most sound operators when miking grand piano includes two good condenser mics aimed at the strings. One mic is placed over the high strings, and the other is placed over the low strings. When these mics are panned across the stereo spectrum, the piano has a very big sound and provides good support for most vocal or instrumental solos. When they are panned to a single location in the mix, they provide a closer, tighter, and more balanced sound than a single mic, which has been set to cover all the strings.

In order to get a good transition from lows to highs and a good recording of most of the piano range, it's necessary to keep the mics about a foot or more from the strings. If the mics are placed much closer, the mid notes might get lost in the blend. In the context of

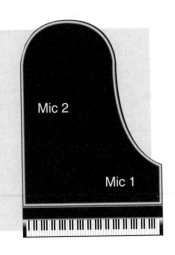

Basic Piano Miking

Two condenser mics are aimed at the strings from above the piano with the lid open on the long or short stick.

- Mic 1 is centered over the treble strings, 6–18" above the strings and 6–18" behind the hammers.
- Mic 2 is centered over the bass strings, 6–18" from the strings and 2–4' from the end of the piano, depending on the size of the piano and the desired sound.

Condenser microphones are the best to use on pianos because of their typically excellent frequency response curve and their accurate transient response.

Piano Miking Variation

Mics 1 and 2 are in a traditional X-Y configuration about 1' inside the open lid.

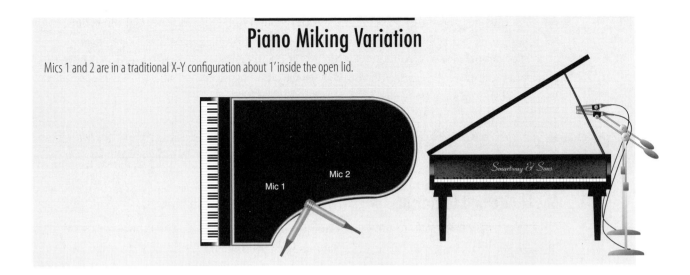

an entire band or music team, the experienced rhythm section musician doesn't frequently play the extreme low notes. In such a case, the sound operator can choose to move the mics slightly closer to the strings and to move the mic covering the low notes up an octave or so to achieve more balanced coverage of the usable range. If there are a lot of grand piano features or solo pieces, it's best to set the mics for an even balance from the lowest to the highest notes.

Always experiment with the exact mic placement for two specific reasons:

* Different musical parts have different musical ranges for the left and/or the right hand. Musical style and consideration dictate the microphone placement.
* The phase interaction between the two microphones is critical. If the distance between the mics changes a few inches, the sound of the piano changes drastically when the channels are panned to the same mix location.

Condenser mics are the best choice for achieving the most accurate and natural piano sound. The piano is technically a percussion instrument because the felt hammers hitting the strings produce a transient attack—condenser mics respond more quickly and accurately to transients than the other mic types.

The intensity of the transient is dependent upon the condition of the felt hammers and the brilliance of the strings. The felt hammers on any piano can be conditioned to produce a sharper attack with a brighter tone or a duller attack with a mellower tone. If you have the felt hammers on your piano conditioned, keep in mind that a brighter sound with more

Piano Miking Variation

Place a stereo pair of mics at the opening of the lid, then use a packing blanket over the top for acoustic isolation and separation.

The packing blanket can do a good job of isolating most grand piano recording techniques, but it can also hinder the acoustic life of the piano sound.

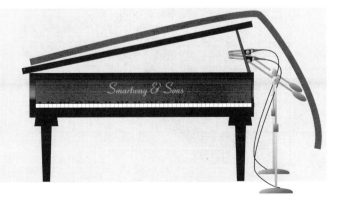

attack stands out in a rhythm section mix better than a dark, warm tone. On the other hand, solo pianists often prefer the mellower tone. This conditioning is called *voicing the piano*.

The grand piano in Audio Example 20-65 is miked with one condenser mic inside the wide-open lid, from a distance of about three feet. It's necessary to keep the mic back a little to get an even balance between the low notes and the high notes.

Audio Example 20-65

One Condenser Mic from Three Feet

Mic Choice and Technique

Each condenser microphone provides a unique sound on a piano—even though two mics might have identical specifications, the sounds they produce might be very different. A grand piano recorded in a world-class studio might use microphones that are very expensive, perfectly matched, or esoterically extravagant. Most live sound systems aren't really capable of reproducing the fine detail and character captured by such mics. However, if you have access to some of these mics, they do produce excellent results. Keep in mind that mics valued anywhere from a thousand dollars to several thousand dollars might be better off stored safely away and saved for special occasions.

The following mics produce excellent results when miking a grand piano. They're also not outlandishly priced, and they're durable enough for live use, although more fragile than the industry workhorse SM 57.

- Shure KSM 141
- AKG 451
- AKG 460
- Shure KSM 32
- Audio-Technica 4041

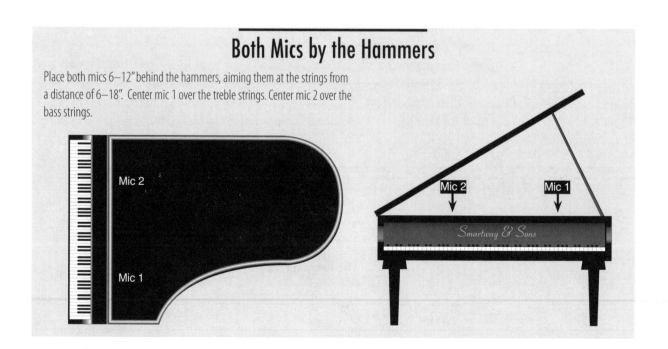

Both Mics by the Hammers

Place both mics 6–12" behind the hammers, aiming them at the strings from a distance of 6–18". Center mic 1 over the treble strings. Center mic 2 over the bass strings.

Mic 2

Mic 1

Mic 2

Mic 1

Smartway & Sons

- Audio-Technica 4047
- AKG 1000
- AKG C 3000
- Shure SM 81
- Shure Beta 87

Virtually any condenser mic is worth listening to on piano. Depending on the piano and the desired sound, there might be several of the new inexpensive imports that sound okay. Be careful when buying cheap microphones—I use the word "cheap" intentionally rather than "inexpensive." In my experience, they tend to fail at the worst time in the heat of a big show. Personally, I think there is a much better value in the purchase of a well-respected microphone from a reputable manufacturer than in purchasing a cheap mic that looks pretty—you get what you pay for.

Sometimes, there just isn't enough funding to buy appropriate microphones. As the sound operator, you might be forced to use whatever is at hand. In this case, you can make anything work. An old Shure SM57 or SM58 will still provide a decent sound, but once you get a good condenser mic, you'll notice the difference in detail it provides.

As a side note, the Shure SM57 and 58 demonstrate my point about purchasing a good mic rather than a cheap mic. SM57s and 58s aren't expensive mics, but there are less expensive alternatives. However, everywhere I go I see SM 57s and 58s that have obviously been well-used and apparently abused, but they still work well after years and years of service. They were—and still are—a great bargain. The updated versions in the Shure Beta series are also an excellent value.

Listen to the piano in Audio Example 20-66 miked with two condenser mics. First I solo the mic for the bass strings, then I solo the mic for the treble strings. Listen as I blend the two mics for a good, even mono sound, then pan the two mics slowly apart for a wider stereo image.

Audio Example 20-66

Two Condenser Mics from about Eight Inches

C-Ducer

Viable Miking Options for Grand Piano

A contact microphone, which fastens to the sound board with adhesive tape, can be an excellent option for miking the grand piano. A mic like this picks up the sound board vibrations—it can provide more isolation than a regular microphone, although it is important that the entire mic is secured correctly and completely. The C-Ducer is commonly used for this application and is capable of providing an excellent piano sound.

Another excellent option is a mic like the Shure Beta 91 (pictured) or the Crown PZM—these microphones capture the direct reflection off their built-in platform. The physical design of this mic makes it fairly easy to fasten to the piano lid. Placing one or two of these inside the closed piano lid can provide an excellent piano sound and superior isolation.

Shure Beta 91

There are several options for mic placement when miking the grand piano. The mics can either both be placed by the hammers, or they can be positioned with one mic over the treble strings by the hammers and the other over the bass strings, about halfway toward the far end of the piano.

In Audio Example 20-67, the piano is miked with two condenser mics a few inches behind the hammers—one aimed toward the high notes and one aimed toward the low notes. The mics are about a foot apart and about eight inches from the strings.

Audio Example 20-67

Two Condenser Mics by the Hammers

To get a wider stereo image or to gain better control of the lows in relation to the highs, move the mics farther apart. Compare Audio Example 20-68 to the previous example. Notice the difference in treble-to-bass balance with the mics farther apart.

Audio Example 20-68

Two Condenser Mics Farther Apart

Set up a coincident stereo X-Y mic configuration with the piano lid up and the mics facing the strings.

Audio Example 20-69 demonstrates the sound of a grand piano with two cardioid condenser mics placed at the edge of the open piano facing in and positioned in an X-Y configuration.

Audio Example 20-69

X-Y Configuration

Stereo or Mono

Even when the piano is miked with two mics, it's not always best to keep the piano stereo in the mix. If the stereo tracks are hard-panned, the sound might be unnatural, with the highs and lows spread far apart in the stereo image. In this instance one side of the audience will hear the right hand, and the other will hear the left. Two mics are simply used to get a good balance between the treble, mid, and bass strings.

Instrument Maintenance

Even if you do everything technically perfect, don't overlook the importance of keeping the piano in tune and properly serviced. Without good conditioning and intonation, there's not much chance of attaining an acceptable piano sound.

It's sometimes tempting to try tuning or voicing a piano yourself. This is a difficult task that is best left to professionals. Be careful! Have the pros service your piano.

Additional Mic Techniques

Listen to the sound changes resulting from these different mic placement combinations. Each Audio Example (20-69 through 20-74) uses the same two condenser mics, panned hard left and hard right, on the same piano through the same console.

Audio Example 20-70 demonstrates a stereo piano mic setup with one mic four inches behind the hammers, centered over the treble strings, and one mic three to four feet behind the hammers, centered over the bass strings. Each mic is about six inches from the strings.

Audio Example 20-70

One Mic over the Treble Strings and One Mic over the Bass Strings

Audio Example 20-71 uses the exact same configuration as the previous example, except the mics are about 18 inches from the strings.

Audio Example 20-71

Mics 18 Inches from the Strings

In Audio Example 20-72, both mics are about six inches behind the hammers and about a foot from the strings. One mic is centered over the treble strings, and the other mic is centered over the bass strings.

Audio Example 20-72

Both Mics Six Inches behind the Hammers

Audio Example 20-73 is the same as the previous example, except the mics are farther apart. The treble mic is aimed at a point about 10 inches in from the highest string, and the low mic is aimed at a point about 10 inches in from the lowest string.

Audio Example 20-73

Mics over the Hammers but Farther Apart

In Audio Example 20-74, there's a stereo pair of mics in an X-Y configuration facing into the piano, about one foot inside the piano with the lid up in its highest position.

Audio Example 20-74

Stereo X-Y Configuration Aiming into the Piano

Audio Example 20-75 demonstrates the same configuration as 20-75, but with the piano lid in its lowest position and a packing blanket covering the opening. This is one of the techniques we might use to help acoustically isolate the piano tracks.

Audio Example 20-75

X-Y with the Lid Lower and a Blanket over the Opening

Video Example 20-8

Miking the Grand Piano

There are several ways to isolate the grand piano sound in a live setting. The previous example demonstrates the sound with a thick packing blanket draped over the lid, which is held open by the short pole (about 12 inches). Crown makes the PZM microphone and Shure makes the Beta 91, which can attach to the underside of the piano lid. They are both direct reflection microphones that are capable of providing an excellent piano sound with the lid fully closed. In addition, C-ducer and Helpenstill offer contact strip microphones that attach to the piano soundboard. Depending on the piano and the installation, these mics are capable of providing a good sound with minimal leakage.

Listen closely to Audio Examples 20-69 through 20-74. These are valuable comparisons that will help your opinions about sound mature to new levels. We can learn as much by noting configuration changes that don't noticeably change the sound quality as we can by noting configuration changes that can make significant sound changes.

Piano Levels

Because the piano is a percussive instrument, care must be taken to consider the transient attack of each note. In addition, pianos voiced for a bright tone exhibit an increased transient when compared to pianos voiced for a dark tone. When metering piano using peak, full-scale meters adjust the trim so the level is near the top of the meter when the pianist plays strongest. This might be difficult to judge accurately because performance adrenaline usually bumps everything up a notch for passionate musicians; however, if, at rehearsal, you ask the pianist to play as loudly as he or she possibly can during the program, you should set the trim so that there is about 5 dB of headroom on the FSM. This level should be satisfactory during the performance, and it will provide an accurate transient reinforcement throughout the duration of the performance.

When using VU meters to set the piano level, keep in mind that the transient exceeds the average level by about 9 dB; therefore, to maintain the integrity of the transient, the loudest part of the performance shouldn't read above about −9 VU. The sound operator should know the amount of available headroom on each channel and adjust transient levels accordingly. Most modern mixers utilize peak meters, which enables the sound operator to avoid guesswork in level adjustments.

When adjusting piano input levels, meters are important, but your ears are more important. Listen to the piano sound. If the piano is loud and the pianist plays loudly, it is possible that the signal from the mic is just too hot coming into the mixer. Most mixers provide an input attenuator that can help compensate for excessively strong signals from the mic. Select the first level of attenuation, readjust the trim for the proper level, and listen for transient clarity. If the sound still isn't acceptably clean and clear, try the next level of attenuation. In many cases, this process will result in a sound that is clear and clean with excellent transient character.

It is also possible that the condenser mic might overdrive its own electronic circuitry before the signal even gets to the board. If the levels look correct but the sound is still slightly distorted-sounding, select the attenuator on the mic body for the first level of attenuation (usually −10 dB). With the attenuation selected, the mixer input trim will need to be raised to regain the lost level. If you hear an improvement but the sound still lacks definition and is slightly distorted, select the next level of attenuation on the mic body. With the next level of attenuation, the input trim will need to be raised again to regain the proper level. There's a good chance that the piano will sound clean and clear with excellent transient clarity at some point in this process.

Equalizing the Piano

As I mentioned before, there's a difference between the sound that works best for solo piano and the sound that works best for piano within a complex orchestration. Solo piano should be full and even in the low end, mids, and highs. Essentially, you need to cover the entire frequency spectrum evenly. Audio Example 20-76 is an example of a good, full solo piano sound.

Solo Piano Sound

Musically, a piano part that works well in a rhythm section is usually percussive. A more aggressive approach to equalization combined with a more percussive musical part usually results in a part that can be heard well in the mix without being in the way of other instruments or voices.

Acoustic piano within the context of a full band orchestration should be somewhat thin in the low end because the kick drum and bass guitar cover the low frequencies quite well. Including an abundance of lows in the piano sound could result in a muddy-sounding mix that's confusing in the low frequencies.

To thin out the lows, try cutting in the range of 60 Hz to about 150 Hz. This is very noticeable to the solo sound, but it won't be noticed in the context of a rhythm section. This will prevent the low frequencies of the piano from conflicting as much with the low frequencies of the bass guitar or kick. Listen to Audio Example 20-77 as I cut 60 Hz, then sweep from 60 Hz up to about 150 Hz.

Cutting the Lows from 60 Hz to 150 Hz

To give the piano more clarity and an aggressive edge, boost slightly between 3 kHz and 5 kHz. Be careful! Dramatic equalization might sound fine on one monitor system and terrible on another. In Audio Example 20-78, I'll start with the EQ flat, then I'll boost the 4-kHz range slowly until I reach a 7-dB boost.

Boosting 4 kHz

Compressing the Piano

The dynamic range of the piano can be very wide, depending on the musical part. Most solo pieces contain some very soft passages and some very loud passages. Sometimes, especially during an emotional and spacious solo piece, it is nice to leave the natural dynamic range intact. This lets the pianist realize the most control and power over the emotional impact of the music. On the other hand, during a rhythmically intense song in a band setting, it's a good idea to compress the piano channel(s) so they can fit into a specific mix position.

To compress the grand piano, try a ratio of about 3:1 with a medium-fast attack time and a medium-slow release time. Adjust the threshold for 3 or 6 dB of gain reduction at the hottest part of the track. This approach will give you natural-sounding compression while letting you control the dynamic range of the piano as you build the mix. The piano in Audio Example 20-79 was uses this technique.

Reverberation

The choice of whether or not to use reverberation in the mix is completely dependent on the acoustical character of the venue. Most large rooms don't require reverberation effects on the piano. Some very controlled and deadened acoustical spaces absorb so much of the natural reverberation that it's appropriate to add a little to regain some of the warmth the reverberation brings.

The solo piano in Audio Example 20-80 uses hall reverberation with a 75-ms predelay and a 2.5-second decay time.

Audio Example 20-80

Hall Reverb, 75-ms Predelay and 2.5-Second Decay Time

On faster songs with busier arrangements, the hall reverb tends to add clutter and can make the mix sound muddy. If you want your mix to sound close, tight, and punchy, a good piano sound with no reverberation works great.

If you want the piano to blend into the mix without sounding like it's at the other end of the hall, try adding a little plate reverb with a short predelay (between 0 and 50 ms) and a short decay time (between .5 second and 1 second). This effect adds an interesting ambience while maintaining a feeling of closeness. The piano in Audio Example 20-81 has a plate reverb with a 35-ms pre-delay and a decay time of .6 second. The track starts dry, then I slowly add the reverberation.

Audio Example 20-81

Plate Reverb with 35-ms Predelay and 0.6-Second Decay Time

Miking Vocals

Vocals are the focal point of almost all songs. If the vocals sound good, the song will probably sound good. If they sound bad, the song will probably sound bad. In addition, the vocals typically contain the most apparent emotional content and impact of the song. Most listeners focus on the vocals first.

Mic Choice

Just as all voices aren't created equal, all mics aren't created equal. Give an excellent vocalist 10 microphones, and he or she will eventually choose one favorite. Give a different vocalist the same 10 microphones, and a different favorite is likely to be chosen. Most singers are fairly content to use the mic on the stand in front of them; however, a conscientious sound operator should always strive to provide the tools that highlight the musician's talents.

The most common live vocal mic of all time is the Shure SM 58. It sounds good, it's rugged, and it lasts practically forever. For some singers, it's the best sounding microphone.

The updated version is the Shure Beta 58—in my opinion it sounds better than the SM 58, but it might not sound better for all singers.

Moving-Coil (Dynamic) Mics

Moving-coil microphones are typically the most rugged and durable of the mic types. They are the most popular type of vocal mic for a few reasons:

- They're the most likely to survive being dropped and mishandled.
- They're typically designed for close-mic applications—they sound best at close range.
- They usually have a built-in presence peak between 4 and 8 kHz.
- They don't require phantom power so they'll work well with any mixer—even with an inexpensive mixer that doesn't provide phantom power.

Common moving-coil microphones used in live sound applications include the following:

- Shure SM57
- Shure SM58
- Shure Beta 57A
- Shure Beta 58A
- Electro-Voice N/D767a
- Sennheiser E845S
- Sennheiser 935
- Beyerdynamic TG-X 60
- Audio-Technica AE6100
- AKG D870
- Beyerdynamic M 69

Ribbon Mics

Ribbon microphones are known for providing a warm, smooth tone with clear presence. Many singers favor them because they complement their vocal tone and style. Ribbon mics are the most fragile of the microphone types.

Historically, the magnets have been physically large in order to provide a strong enough signal, and the ribbons have been long (two inches or so) and very thin—dropping a ribbon mic meant certain breakage.

Modern ribbon microphones take advantage of new technology, which provides greater magnetism in a smaller physical size. Modern ribbon elements are typically between one-half and three-quarters of an inch long and provide ample signal and relatively durable functionality.

Common ribbon microphones used in live sound applications include the following:

- Beyerdynamic M 260
- Beyerdynamic M 500

Condenser Mics

Condenser microphones were historically used primarily in recording studios because their electronic design is capable of providing the most accurate frequency response and transient definition. In the past, the definition they provided surpassed most live sound systems' ability to reproduce. In addition, their acute sensitivity made them impractical for live handheld use.

Modern live sound systems, however, are capable of realizing many of the sonic advantages provided by the condenser design, and new designs have created some of the best-sounding vocal microphones ever made for live sound reinforcement use.

Common condenser microphones used in live sound applications include the following:

- Shure Beta 87
- Shure SM86
- Shure KSM 9
- Neumann KMS 104
- Neumann KMS 105
- Electro-Voice RE510
- Audio-Technica AE3300
- Audio-Technica AE5400

There are obviously many more microphones available other than the ones mentioned in this section. Nearly all manufacturers offer multiple lines of mics, each designed and priced for a specific user demographic, and each manufacturer offers new designs regularly. This is an excellent era for audio equipment consumers because the industry is very competitive. When choosing microphones, it's best to stick with reputable manufacturers and to listen to several options in your price range. Determine the primary goal in your microphone purchase and visit your local audio retailer—know basically what you want and how much you want to spend, and then have an open mind. Use your ears to help make the best selection and make sure that your audio dealer understands that you want something that will withstand regular live sound reinforcement use.

Mic Technique

Good singers can make any sound operator look good because they know how to use excellent mic technique to control the volume and tone of their performance. An inexperienced sound operator will tend to accept whatever mic technique the musician shows up with. An experienced sound operator will help the singer learn how to use the microphone to help craft the best possible performance. He or she understands the importance of the musician's use of the inherent changes in tone and volume associated with changes in mic distance from the source.

Proximity Effect

Any time a singer or narrator moves close to a microphone, the low frequencies get louder in relation to the high frequencies. This can result in a boomy or thick sound, especially if the voice is being recorded through a high-quality condenser mic. The proximity effect describes low frequencies increasing as the mic distance decreases. This effect is the most extreme when using a cardioid pickup pattern.

Experienced singers use the proximity effect to their advantage. During the more intimate and subtle vocal passages, the best vocal sound is usually achieved with the singer very close to the mic. During the loud and stylistically aggressive passages, the singer benefits by pulling back a little from the mic—the sound gets a little thinner, which cuts through the mix better, and because the singer is louder, the volume typically remains appropriately strong for the passage.

The Timid

The overly timid singer must be encouraged to stay close to the microphone, especially during softer vocal passages and while they're speaking into the mic. Most people tend to back off naturally when they sing louder for a few different reasons.

- They can tell in the monitor that they're getting too loud.
- They can tell that the sound is becoming thick and possibly overdriven as they sing strong and loud directly into the mic.
- They're afraid of what's going to happen when they open up and go for it.

It's a good thing to back off the mic naturally for the previous reasons, but some people are just shy—they need to get over it. Timid singers tend to back off the mic if they say something to the audience during or between songs. In reality, during these moments the singer must move close to the mic. The fact that the level is typically set for louder singing makes it likely—almost a certainty, actually—that the singer won't be heard or understood by the audience. Granted, the sound operator should be focused on making sure that the singers are heard, but the singers should be cognizant of the importance of their role in this communication link. In addition, the depth of the sound quality when singers speak closely to the mic helps them command the audience's attention.

The Bold

The bold and overly confident singer typically needs to be encouraged to back off a little when he or she sings loudly. Some inexperienced sound operators instruct singers to stay right on the mic no matter what. This suggestion typically comes from frustrations in dealing with timid singers, but in the long run singers need to learn how to use the mic to their advantage—and that means varying the distance from the mic based on musical and communication requirements.

A singer who sings loudly and boldly into the mic typically causes the mic or the mixer to overdrive, creating distortion to some degree. An astute sound operator will probably see the levels overloading and back the trim off to compensate. Although this is a necessary action, it requires the sound operator to set the trim at a level that's less than optimal for the sound of the times when the singer is at a normal volume. Rather than backing the trim off, the sound operator should ask the singer to simply move a few inches away from the mic during the loud passages—in this way, the best sound quality is possible at all times.

The Outrageous Ham

What can I say? They're the clowns—the extreme personalities. They love attention and they love to show off. They swing the mic by the cable and they scream. They have charisma. They're loved by some, hated by others. They're always on. They're full of energy. They never stop—they don't know when to stop even if they could. They might be clinically ADHD. They're probably insecure, even though they seem to have more confidence than anyone in the room. They are "The Outrageous!"

This category of singers, when skillfully directed, can potentially turn into the best of communicators. In the meantime, it's the sound operator's responsibility to make sure they don't damage themselves or any sound equipment. Traditional church music doesn't usually provide a vehicle for these types of personalities, but non-traditional church music often does. Music provides an emotional connection for its appreciators. Because there are all types of people, there is an audience for all types of music that communicate on all kinds of different levels.

It's the sound operator's duty to protect people and equipment, and especially people from equipment. Outrageous conduct and use of equipment shouldn't be tolerated. Swinging mics by their cables is not only silly, but it's dangerous for the audience, the performers, and the gear. When this type of contact crops up, the sound operator gets to practice diplomacy and authority at the same time.

If extreme and outrageous actions are a part of the stage show—and if someone else is paying for damaged equipment and being responsible for personal injury—the show must go on. If you've committed to the show and if they've committed to take responsibility, then you can decide whether you want to participate. Choose wisely.

The Seasoned Singer

A seasoned and experienced singer knows how to work the microphone for the best possible communication of his or her performance—any inherent outrageous tendencies are in control and channeled to be released in just the right doses to help communicate a point or to create rapport with the audience. The seasoned singer knows what to say, how often to speak, and when to keep quiet—his or her maturity has produced an efficient and effective performer. A seasoned singer makes the sound operator's job much easier. Oh, if they could only all be seasoned singers.…

Holding the Mic

Singers should hold the microphone body. They should never hold the mic around the capsule. The cardioid polar response pattern is shaped via the area just behind the capsule. If this area is sealed off, the mic instantly turns into an omnidirectional mic—this is bad. It dramatically decreases the gain before feedback, it changes the response character of the mic, and it looks amateurish.

Wind Screen

A wind screen, also called a *pop filter*, is used to keep abundant air caused by hard enunciation from creating loud pops as the microphone capsule is overworked. In an outdoor application, the wind screen is also used to shield the capsule from wind.

When a singer enunciates words containing hard consonants, such as "p" and "b," there's a lot of air hitting the mic capsule at once. When the air from these hard consonants, called *plosives*, hits the mic capsule, it can actually bottom out the diaphragm. In other words, this "pop" can be the physical sound of the microphone diaphragm actually hitting the end of its normal travel range. This pop is can be very distracting during the course of a live performance.

Virtually all microphones designed for live vocal applications contain a built-in wind screen, a thin foam layer attached to the wire mesh that surrounds the capsule. This built-

The Wind Screen

There are many different types and shapes of foam wind screens. They work very well when used in the proper context but can adversely affect sound quality. When a wind screen is placed over a mic capsule, the sound arriving at the capsule through the foam wind screen is affected mostly by the type of foam material rather than physical shape.

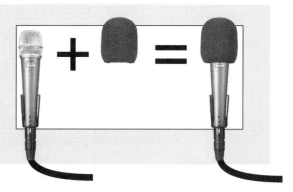

in foam provides protection against most plosive problems; however, external foam wind screens can easily provide additional protection from plosives, as well as wind (in the case of an outdoor performance).

External wind screens are typically made from a dark gray foam rubber, although they are also offered in multiple colors. Colored wind screens help the sound operator keep track of which microphone is which. If performers are sharing microphones during a performance or if mics are constantly being repositioned, colored wind screens are very comforting to the sound operator—they take some of the guesswork out of the ongoing microphone shell game.

Hygiene

External wind screens are also easier to clean than their internal counterparts. When mics are shared, germs are swapped—it's impossible to avoid. Foam windscreens can be washed with antibacterial soap and water—they dry relatively quickly and are soon ready to go back into service. A conscientious sound operator should wash the wind screens regularly. The built-in wind screens can also be washed with little effort. Simply unscrew the screen surrounding the capsule and wash it with soap and water. Once it dries out, screw the screen back on. Always be sure to thoroughly rinse the soap out of the foam.

The Proximity Effect

The proximity effect is functionally simple. When the mic moves closer to the source, the tone gets boomier—the low frequencies increase in relation to the high frequencies. Most singers are aware of the proximity effect, and some instinctively use it to their advantage.

The proximity effect is really a result of a couple things happening as the mic moves closer to the sound source.

- A destructive phase interaction occurs as the sound waves reflect back and forth between the source and the microphone diaphragm. As the mic moves closer to the source, highs begin to decrease to a greater and greater degree.
- As the highs decrease, the low frequencies continue to get stronger and stronger due to the fact that each time the distance decreases by half, the level increases by 3 dB.

Bass Roll-Off

Some microphones exhibit an extreme proximity effect. In such cases, the manufacturers typically include a bass roll-off. The bass roll-off is merely a high-pass filter that rolls off the low frequencies below the specified frequency, usually at rate of between 6 and 12 dB per octave. Rolling off the lows lets you get close to the mic without getting a thick, boomy sound. Some condenser microphones have variable bass roll-offs that will filter the lows below a couple of different user-selectable frequencies.

The low frequencies can also be rolled off on the mixer channel; however, when the roll-off is applied at the microphone, it provides a much more workable signal at the mixer input. Because the boomy low frequencies are filtered before the mixer input, the trim level can be accurately adjusted. Selected level settings will be based on a balanced full-range sound rather than on a thick, boomy tone that contains an overabundance of lows.

Because the vocal tracks don't usually need the frequencies below 100 to 150 Hz anyway, the bass roll-off is an intelligent way to clean up the microphone signal before it even gets to the mixer.

Dynamic Range

Dynamic range is the distance between the softest sounds and the loudest sounds in any given segment of audio. Vocalists almost always use a wide dynamic range during the course of a song. Often they'll sing very tenderly and quietly during one measure, and then emotionally blast you with all the volume and energy they can muster during the next.

Compression can be the single most important contributor to a vocal sound that's consistently audible and understandable in a mix. For this reason, most lead vocals on commercial hit recordings are heavily compressed. Even though vocal channels are often compressed, the sound operator must still attend to the mix, adjusting channels as necessary to build the best mix at all times.

Compressor/Limiter/Gate/Expander

As the compressor's VCA (*voltage controlled amplifier*) turns down the signal that passes the threshold, the entire vocal track occupies a narrower dynamic range. When the vocal is in a narrow dynamic range, the loud sounds are easier to listen to because they aren't out of control, plus the softer sounds can be heard and understood better in the mix.

Vocals are usually compressed using a medium-fast attack time (3 to 5 ms), a medium-long release time (from a half second to a second), and a ratio between 3:1 and 7:1 with about 6 dB of gain reduction at the loudest part of the track.

Listen to the vocal track with rhythm accompaniment in Audio Example 20-82. The vocal isn't compressed. Notice how it sometimes disappears in the mix.

Audio Example 20-82

Vocal without Compression

Audio Example 20-83 demonstrates the same vocal, compressed using a ratio of 4:1 with up to about 6 dB of gain reduction. This time the peak level is the same, but listen for the softer notes. They're easier to hear and understand.

Audio Example 20-83

Vocal with 4:1 Compression

Dynamic Range

These graphs represent vocal energy (the curve) in relation to the rest of the mix (the shaded area). The top graph (not compressed) shows the volume of some lyrics sinking into the mix, probably being covered up.

The bottom graph indicates the same performance and lyrics after being compressed. This time the lyrics peak at the same level, but after they are compressed, notice that the softer lyrics are turned up so they can be heard above the rest of the mix. Once the compressor/limiter has decreased the dynamic range and levels have been adjusted to attain the proper peak level, your vocals should be consistently understandable—more emotional nuance will be heard and felt.

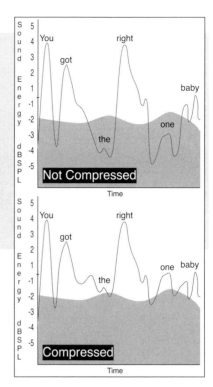

Backing Vocals

Backing vocalists need to adhere to the same suggestions as lead vocalists. Once the backing vocal levels are set, their EQ is adjusted, and their monitors are dialed in, as long as they're hearing a balanced mix, they should be able to adjust much of the blended vocal sound by varying their tone and mic technique.

Choir

Choirs come in myriad sizes and styles. Churches come in a variety of stylistic preferences. Each style and size of choir demands a slightly different approach to miking and processing, but a few considerations are consistent no matter what.

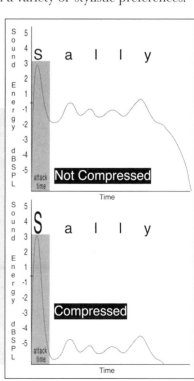

Sibilance

These graphs represent the changes in amplitude over time of the word "Sally." The top graph has an average level of about 0 VU, but the "S" is about 3 dB above the remainder of the word "...ally."

If the compressor's attack time is slow enough that the VCA doesn't begin to act until after the "S" and if the remainder of the word is compressed, exaggeration of the initial sibilant sound results.

The bottom graph represents the result of compressing the word "Sally." Notice the difference in level between the "S" and the remainder of the word. This type of compression technique, when used in moderation (1 to 3 dB), can help increase intelligibility and understandability.

- The mics must close enough to the singers to avoid overpowering the vocal source with ambient leakage.
- A good balance of the choir voices is necessary. Individual voices should blend into overall texture.
- As much gain before feedback as possible is desired.
- The choir needs to hear a primary instrument such as piano or organ as a pitch and timing reference.

Mic Choice and Technique

Condenser microphones are the right choice for a choir.
- They provide a flat frequency response.
- They sound full from a distance.
- Their excellent transient response captures the intricacies of the choir vocal textures, enunciation, and dynamic contrasts.

Moving-coil and ribbon mics are virtually always designed for close-mic applications; therefore, using a moving-coil microphone from a distance of 12 to 24 inches provides a thin sound that is typically edgy and harsh somewhere between 3 and 10 kHz.

The following condenser microphones have become well-respected for use in miking choirs.
- AKG 451
- AKG 460
- Shure KSM 141
- Shure KSM 32
- Audio-Technica 4041
- Nuemann KM84
- CAD CM100
- Shure MX20
- Audio-Technica U853
- Audio-Technica ES933
- Audio-Technica Pro 45

There are plenty of other good choices; in fact, the number of good options seems to increase daily. A good mic choice for a choir includes the following attributes:
- Small diaphragm condenser capsule
- Cardioid polar pattern
- Flat frequency response
- High-pass filter
- Smooth, warm sound
- Sounds great for a choir

Ratio of Mics to Singers

Church choirs vary dramatically in size, from 10 to more than 100. The goal is always centered on achieving an intimate, blended sound with plenty of gain before feedback.

There is a standard rule of thumb—called the 3:1 rule—commonly used to calculate the placement of multiple mics in the same acoustic environment. The 3:1 rule states that when placing microphones, the distance from one microphone to its intended source should be no more than one-third the distance between it and another microphone.

Therefore, because the ideal distance from which to mike a choir is between 12 and 18 inches, microphones should be between 36 and 54 inches apart.

Depending on how the choir is arranged, the 3:1 rule indicates that there should be a mic for every 9 to 15 singers. Some choirs stand shoulder to shoulder, but many choirs have all the singers face toward the center at about a 45-degree angle. Then they scoot closer together so that there can be more people on the choir risers.

Positioning Choir Mics

The typical choir riser has three levels. With the choir angled in and facing the center, there can be a voice every foot or two, so three to five singers could fit within 54 inches. Three rows in this configuration results in between 9 and 15 singers per mic if we adhere to the 3:1 rule.

The 3:1 rule provides a guideline for choosing the optimum number of mics to use on a choir—the 3:1 rule, in actuality, applies more specifically to sound sources with a much smaller point of origin, such as a drum, violin, trumpet, or just a single voice. As we consider miking choir zones of between 9 and 15 people, each zone is the sound source, which extends fully to the beginning of the next zone. There will be varying degrees of destructive phase combinations at each location in the zone.

In actuality, a choir is miked for control and reinforcement. In a recording scenario, we find that the best mic techniques utilize two excellent small-diaphragm condenser mics in an X-Y configuration, located several feet into the hall where the blend is perfect between the choir and the room. Obviously, the recording setup is very impractical in a live scenario.

Typically, the choir mics should be positioned 12 to 18 inches in front of the front row and 12 to 18 inches above the back row, aimed down at the middle row. There are a few important principles involved in this technique.

- As much as is possible, the mic should be equidistant from all singers in its coverage area.

- Aiming the microphone down at the choir helps reduce the influence of the reflections and leakage off the surface behind the choir.

- With the mics positioned above the choir, they are less visually intrusive from the audience's vantage point.

In any permanent installation involving a choir, the mics should be hung from the ceiling. Mics on stands make the stage look cluttered, and there is roughly a 100-percent chance that someone will eventually trip over a stand or cable, sending the mic and stand crashing into the floor, or even worse, into Grandma Betty, who is 80 years old and very, very frail.

Most of the very small mics intended for use on a choir are designed for hanging from the ceiling. They typically come with a hanging mount that aids in the positioning of the mic at the proper angle, as well as helps to maintain the microphone's angle and direction.

Cardioid Polar Pattern

Use condenser mics with cardioid directionality. Feedback is an issue in any live setting, but especially in a live performance—the off-axis discrimination provided by cardioid mics helps reduce feedback. Any omnidirectional mic on stage will dramatically increase feedback potential—several omnidirectional mics on stage is a recipe for disaster. If you want to record your choir in a controlled setting, using mics that aren't in the sound system, try a pair of excellent omnidirectional condenser mics. Positioned properly, they can provide an excellent sound. However, for sound reinforcement, use cardioid mics.

Pitch, Tone, Enunciation, and Blend

The choir must develop a sound that is full and enunciation that is clean. A choir that knows how to sing with support and unified diction will almost always sound good. A choir that doesn't focus its tone and that isn't together in its diction is another story.

With any sound wave combinations, the sound level increases when the sonic ingredients work together; it decreases when they don't. Likewise, when the singers in a choir are precisely together in pitch, enunciation, and tone, the resulting sound can be extremely powerful. However, when several choir members are out of tune, when enunciation is sloppy, and when the singers derive their tone from their throats rather than their diaphragms, the resulting choir sound will be very weak and messy.

It is incredibly important that the choir members learn to blend their voices together to create the choir sound. If there are just a few members with a harsh, edgy, and pointed tone quality, they will be the ones heard over the system. The sound operator should help identify the voices that are sticking out in the mix, working together with the choir director to build the best choir possible. If certain members dominate the house mix, try moving the singers around. Move the distracting voices to the edge of each zone, and move the singers with the best tone, pitch, enunciation, and blend to the center of the zone.

Reinforcing a choir isn't that difficult as long as they do their part, providing excellent sound, blend, pitch, and enunciation.

Monitors

Each church is different. The acoustics are very important when it comes to the choir's ability to hear itself. If the room is very dead, the choir sound will be very dead. When the choir sings with a band accompanying them in a dead room, there's a good chance they won't hear themselves much, and the sound they hear from the band might not be what they need to hear.

On the other hand, in a very live room the choir sound reverberates around the hall, and the choir sounds alive. In a live room, the choir might hear plenty of what's going on vocally and instrumentally, but it will be confusing and not conducive to an excellent performance.

Both of the previous scenarios have an interesting effect on the choir, especially when singing with instrumental accompaniment.

- In the dead room, the choir will probably need a monitor so they can tell what the band is doing and possibly so they can hear where they fit into the picture.
- In a live room, the choir will probably be able to hear everything, but the timing will be very confusing. Depending on where the choir stands in relation to the instrumentalists, they might hear the instruments off the back wall. If they sing in time with what they hear reflected off the back wall, they will always be behind the beat, and the band will feel like they're dragging a hundred people through the mud.

With these factors in mind, it's easy to see that some sort of monitoring is necessary to bring the choir into the blend with the rest of the group.

- In a dead room, the monitors help provide a pitch reference and an accurate musical perspective.
- In a live room, the monitors provide an accurate frame of reference regarding timing and pitch.

Controlling the Band

Any band or instrumentalist that performs with a choir needs to become just a little more selfless. Volume control is fundamental because of the choir mics. The choir microphones will indiscriminately pick up any onstage sound. Therefore, loud acoustic instruments must be controlled, by turning them down, baffling them in some way, or playing them less aggressively.

- Either drums should be enclosed by a Plexiglas shield system, which surrounds the drums with Plexiglas in front and foam in back, or the drummer should use electronic drums.
- Any guitar amplifier should be enclosed or baffled, or the guitarist should run direct through an effects device into the board.
- Keyboards and bass should run direct into the board.
- Acoustic percussionists should be enclosed by a Plexiglas shield, such as that used for the drummer. Acoustic drums and percussion should be in separate Plexiglas enclosures.
- The instrumentalist(s) should be using in-ear monitors.

If all instruments are acoustically isolated and if the instrumentalists are using in-ear monitoring, there is an excellent chance that the choir will be heard and understood, especially if the choir has precise enunciation, great blend, and excellent tone, and is in tune.

Monitor Content

If at all possible, don't put the choir mics in the choir monitors because they'll increase the likelihood of feedback. It's best to keep the choir monitors simple. The choir primarily needs an accurate pitch and timing reference, so a primary keyboard or guitar along with a light rhythmic ingredient should suffice.

Keep in mind that any instrument in the choir monitor could potentially lose some of its closeness and intimacy. The phase combinations resulting from the direct instrument sound and the leakage from the monitors into the choir mics could cause destructive phase interactions that make the instrument sound more distant.

Monitor Position

The ideal number of monitors and their most efficient locations depend on the monitor's coverage area, stage size restrictions, and the surfaces surrounding the choir. Typically, one floor monitor with a 12-inch woofer and a horn should be able to cover a group of 20 to 30 singers, as long as the monitor can be positioned in front of the choir and approximately 10 to 12 feet away.

When choosing floor monitors, make a sketch of your stage area and the exact choir position. Map the space in front of the choir and look for the best monitor location, considering the amount of available space. Next, read the specifications for the floor monitors you have or are considering purchasing. Note the coverage area specs—imagine the coverage area and how it fits into the available stage space. On paper, it's pretty easy to see how many monitors are required to cover the choir area evenly.

- There should be minimal overlap between monitor coverage areas.
- There should be minimal spillover outside the choir.

The monitors should work together to cover the choir zone without overlapping and without extending outside the choir rise area.

Synthesized Sounds

Synthesizers and sound modules connect to the mixer through a direct box. Modern sound modules and keyboards provide a stereo output. If your sound reinforcement system is stereo, use two DIs to connect both the left and right channel outputs to two separate mixer channels.

Panning the Synth

Pan the synth channels hard left and right. Virtually all usable keyboard sounds utilize stereo reverberation and delay effects to make the patches sound wider and more impressive. Panning them out of the center position makes them sound more impressive in the mix, and it helps clear out the center mix position for the vocals.

If your sound reinforcement system is mono, it is usually best to connect one channel from the synth through a DI to one mixer channel. The synth should be panned to the center mix position in order to give entire system coverage area the same balance and mix.

Equalization and Effects

Most of the basic effects and EQ are controlled within the synthesizer. Modern synths allow for ultimate sonic shaping, as well as almost any imaginable effect.

The keyboardists' tonal adjustments provide a full-bandwidth sound; however, most live mixes require less than full bandwidth from the synth most of the time. In the context of a rhythm section arrangement, most frequency bands are well covered without the synth. Including a synth pad, which contains a full complement of tonal colors on top of an already full-sounding rhythm section, results in a muddy sound that loses definition and increases confusion.

However, if the low frequencies, below about 200 Hz, are filtered out and if a primary high-frequency band is boosted by just a few dB, the same pad sound might fit into the context nicely—filling in the texture and adding personality and character rather than muddying the texture and confusing the mix image.

There are times during many services when the rhythm section texture should thin out, often leaving the synth sound alone as support to the pastor or music leader. In these instances, the sound operator should bring the low frequencies back into the sound, and typically any high-frequency boost should be reduced or even cut. High frequencies tend to interfere with vocal intelligibility, so pads that support voice should be mellower in the highs and fuller in the lows.

Acoustic Guitar

The sound of an acoustic guitar is amazingly pure and sonically rich. Like most of the symphonic instruments, year after year, generation after generation, the acoustic guitar sound remains fine just the way it is in its natural environment. Capturing the natural sound of an instrument that is so universally accepted and recognized is an important and fundamental skill in the audio world.

At least 50% of the guitar sound—maybe it's 100%—is because of the musician. A bad player can't make the best guitar sound good; however, a great player can make music on almost any guitar. In your quest to find the ultimate sound, don't forget to give yourself a break if the player isn't cutting it musically.

Strings

If the strings are dead and lifeless on any acoustic guitar, the recorded sound will be dead and lifeless. No amount of equalization of effects will restore the rich tone that a good set of new strings provides.

Keep in mind that heavier-gauge strings provide more low-frequency content. Most acoustic players use a medium-gauge set of strings because of the excellent balance of high and low frequencies they produce, along with reasonable playability. A lot of electric guitarists use light-gauge acoustic strings simply because their fingers are accustomed to playing on guitars with low action and thin strings.

Light-gauge acoustic strings produce adequate highs but weak lows. For the player, it's important to weigh the options. If you need a full strum sound from the acoustic guitar, you might have better luck using medium-gauge strings. However, the guitarist might perform a much more musical and technically acceptable part on lighter-gauge strings simply because he or she doesn't need to struggle to provide adequate finger pressure.

There's one problem with new strings. They squeak more when the guitarist moves around on the neck. The best players usually have enough technique and finesse to play on brand-new strings without much of a problem. For the rest of us, there can be other solutions. The quickest way to get the strings to squeak less is to put something slippery on them. Unfortunately, slippery products usually contain some sort of oil. Your local music store has access to commercially manufactured products designed to make guitar strings more slippery. These products can work very well. I even know people who put the thinnest possible coat of vegetable oil on their strings. Use any of these products sparingly. Oil on strings causes them to lose brilliance and clarity. You might end up with no squeaks at the expense of all that great acoustic guitar sound.

Picks

Another very important factor in the sound of an acoustic guitar is the pick. Playing with a thin pick gives a sound that has clearer high frequencies. The thin pick slapping as it plucks the strings becomes part of the sound. Playing with a thick pick produces a full sound with more bass and fewer highs, plus you don't get as much of the pick sound.

Alternate Tunings

Acoustic guitar has embraced the concept of alternate tunings for a long time. Whereas electric guitarists often do everything they can to avoid open strings, acoustic players rely on the open strings to provide the basis of richness and depth.

There are many alternate tunings that are regularly implemented in acoustic music. Some have become more common than others, but often tunings are selected in the heat of a session, with a player adjusting to the requirements of a specific piece of music.

Some of the most common alternate tunings are:
*   Dropped D tuning – DADGBE
*   Dropped G tuning – DGDGBE
*   Open G tuning – DGDGBD
*   Major D chord – DADF#AD
*   High-strung guitar – EADGBE Bottom four strings tuned up one octave. This tuning requires a change of string gauge on the lower strings. Try, from bottom to top, .034, .024, .013, .010,
    016, .013.

The Impact of Different Wood Configurations on Tone

Each type of wood lends a varied tonal quality to the acoustic guitar.
*   Spruce is the most common wood used for the acoustic guitar top. It provides a bright and full tone.
*   Rosewood is commonly used for the sides and back of the acoustic guitar. It provides a clean and bright tone, with tonal stability and transparency.
*   Mahogany sides and backs typically produce a warmer tone with a smoother midrange than guitars with rosewood.

Whether mahogany, spruce, or rosewood produces a better sound is not the question. They all provide a different musical feel and are very applicable to any number of musical applications.

Electric Acoustic Guitars

Acoustic guitars with pickups can work well in a live performance situation—simply plug directly into the mixer, amplifier, or through a DI into the snake box on stage. You can get a

Electric Acoustic

Connecting the electric acoustic is the same as connecting the regular electric guitar. You can either connect the guitar output directly into the mixer line input or through a DI to the mic input.

There is an amazing difference in the sound quality between direct boxes. Use a high-quality DI and be sure to test several on your guitar. I have found the Radial Engineering, Countryman, and Demeter DIs provide excellent sound quality in this application.

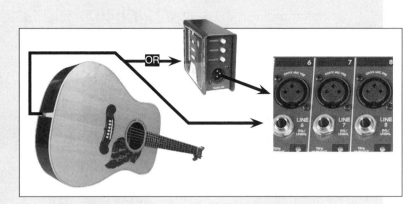

passable sound and eliminate one microphone in the setup. However, though the sound can be okay for live performances, it's hardly ever a great sound for recording. The sound from an electric acoustic pickup typically sounds sterile and small, and it doesn't have the broad, full, interesting sound of the acoustic instrument. To run an electric acoustic guitar directly into a mixer, follow the same procedure as with any electric guitar.

The frequencies that tend to be over-accentuated by most acoustic guitar picks are typically between 1 and 2 kHz. Try using a parametric equalizer; set up a narrow-bandwidth cut and sweep the frequencies between 500 Hz and 3 kHz. You'll probably find one specific point that eliminates much of the brittle, edgy sound. Once you seek and destroy this problem frequency, the electric sound is passable. Most of the time, I find the problem frequency ends up between 1.2 and 1.5 kHz, but it really depends on the guitar, the room, and the sound system.

When the mid-range problem is solved, try adding a little clarity in the highs between 5 and 10 kHz—whatever sounds best in your application. In a recording situation, you might want to warm up the low end by boosting around 100 Hz; however, in a live setting this is a recipe for very boomy feedback.

Video Example 20-9

Equalizing the Electric Acoustic Guitar

Mic Techniques

Typically, the best kind of mic to use on any acoustic guitar is a condenser mic. Condensers capture more of the subtlety of the attack, the sound of the pick on the strings, and the nuance of artistic expression. Moving-coil mics and ribbon mics can produce passable acoustic guitar sounds, especially if that's all you have, but the accepted mic of choice for acoustic guitars is a condenser.

The steel string acoustic is the most common acoustic guitar. These guitars come in many different shapes, sizes, and brands. Each variation has a characteristic sound, but the primary trait of the acoustic guitar is a very clear and full sound. The second most common

Acoustic Guitar Mic Positions

Select the microphone position that provides the sound that best suits the music. Each mic position offers a different tonal balance. When you're setting the mic, try wearing headphones while you move the mic to different locations around the instrument. It's likely that you'll find the perfect sound, and you might be surprised at the mic position when you find it.

I typically prefer the sound I get when I place the mic three to six inches away from the point where the neck joins the body, in front of the guitar—it works great in a live setting but the performer needs to feel comfortable remaining a consistent distance from the mic.

type of acoustic guitar is the nylon string classical guitar. Classical guitars have a warm, full, and mellow sound.

In most cases it's best to keep it simple when miking an acoustic guitar. Selecting one great mic (typically a small-diaphragm condenser mic) and moving it to the one place that provides the sound you're looking for is a highly successful approach. The advantage of a single-mic technique is the assurance of mono compatibility and simplicity.

There are distinct regions on the acoustic guitar that provide predictable tonal character. Aiming the mic at the sound hole provides a boomy, bass-heavy sound.

Video Example 20-10

Changing the Single Mic Position

Compressed Guitar Note with a Slow Attack

The attack time is now longer. Now the attack is not compressed, but the rest of the note is. Notice that now the peak is 9 dB above the rest of the note.

All of the note except the attack has been compressed at a ratio of 5:1. Depending on the musical setting, this technique might result in too much attack. But when used properly, the exaggerated attack results in a guitar sound that is clear, with lots of definition. Fine-tune the adjustment of the attack, threshold, and ratio to get the sound you want.

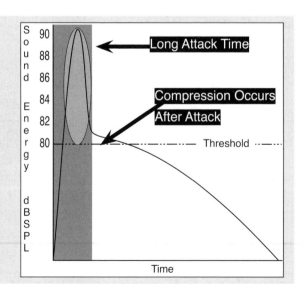

With a condenser mic six to eight inches from the guitar, we can potentially get a sound that has too much bass, especially as we move over the sound hole. We can control the frequency content of the acoustic guitar sound dramatically by changing mic placement. If there are too many lows in your acoustic guitar sound, try moving the mic up the neck and away from the sound hole, moving the mic back away from the guitar to a distance of one or two feet, or turning the low frequencies down.

One way to turn the low frequencies down is by using the bass roll-off switch. Most condenser microphones have a switch to turn the bass frequencies down. These switches may have a number by them to indicate the frequency at which the roll-off starts. The number is typically between 60 and 150. If there's no number, there might be a single line that slopes down to the left. When you use a condenser mic for close-miking, you'll usually need to use the bass roll-off switch to keep a good balance between lows and highs.

Individual acoustic guitars often produce different tonal balance when miked at different spots. In other words, there isn't one microphone placement that works best for every guitar. Experiment with each instrument to find the sounds you like.

Dynamic Processing and the Acoustic Guitar

Acoustic guitars have a wide dynamic range. A compressor can help even out the volume level of the different pitch ranges and strings. Some instruments even have individual notes that are much louder than others. Low notes (on the larger strings) will often produce a lot more energy and volume than higher notes on the smaller strings—it all depends on the instrument.

Try this approach to compressing the acoustic guitar:

- Set the ratio control between 3:1 and 5:1.

- Adjust the attack time. Slower attack times accentuate the sound of the pick. The fastest attack times will de-emphasize the sound of the pick.

- Adjust the release time. Setting this control between one and two seconds usually results in the smoothest sound.

- Adjust the threshold for a gain reduction of 3 to 7 dB on the loudest part of the track.

chapter 21

SOUND CHECK

S ound check is one of the most important parts of the sound reinforcement process. If the gear isn't all working properly and if the musicians can't hear themselves well enough to produce inspired performances, then not even the very best sound operator will be able to build a wonderful-sounding mix. It is very important that the sound operator develops a routine that he or she follows regularly during sound check.

There is a difference between a sound check at a new location, after the system has just been set up, and a sound check at a permanent location, where the same musicians perform regularly. Some of the routine adjustments that are made during a new setup at a new location remain constant at a permanent location.

Adjustments to master EQ and dynamics settings should be performed on the permanently installed system as part of its initial setup—they aren't necessary on a regular basis, although the system should be verified once a year, or when the sound operator notices some problems that are new to the system.

Setup Routine

Obviously, the setup routine is vastly different for a road crew than it is for a permanent location. Assuming that the system is setup and ready to power on, there are some procedures that should be followed and some routines that should be developed. These procedures and routines are intended to protect the equipment while providing a working environment that is efficient, pleasant, repeatable, and manageable during the most intense moments of the program.

Power On/Power Off

The more complex the system, the more involved the powering-up routine by sheer nature of the number of components; however, the procedure is conceptually simple.

- Turn the main FOH level fader all the way down.
- Mute all individual channels.

- Turn on the mix processing racks first (all dynamics, EQ, and effects processors that are connected to the mixer's audio signal flow, such as inserts, auxes, effects sends and return, and so on).
- Turn on all wireless mic systems.
- Turn on the mixer.
- Turn on all system processors (dynamics, EQ, delays, and so on that are inserted between the mixer output and the power amplifier inputs).
- Turn on the power amplifiers.

Depending on how your system is connected to AC, some of the previous steps might all be part of one switched circuit. Although it's best to power up so that every device that outputs an audio pop or surge powers up before the device that would receive the pop, if it's impossible to do so, then at least follow the single most important rule: Turn the power amplifiers on last!

Labeling and Storage

Label all of the main cables, especially when the system is being constantly struck and reset. Keep one case that contains all of the essential mixer connection cables. Spend some time labeling them very clearly. Be very literal about your labels—imagine that an eight-year-old child has to connect everything. This approach will help dramatically when the primary sound operator is ill, detained, or quits—a new person should be able to step right in and quickly attain functionality. In addition, when an audio catastrophe strikes and the sound operator is in that overwhelmed state of confusion—you know, the one where he or she can't think at all, let alone at an eighth-grade level—simple, direct labels will remove any need to think when looking for a cable.

Create logical and segregated storage areas. Keep 20-foot mic cables in one definite area. Keep 30-foot mic cables in an area just next to the 20-foot cables. Label the 20-foot mic cables with a colored heat-shrink tubing (red, for example) and label the 30-foot mic cables with different-colored heat-shrink tubing (blue, for example).

Store ¼-inch TRS line-level cables in an area separate from ¼-inch TS line-level cables. Be sure that all cables are neatly wound and tied with Velcro or string.

Each sound operator must use the space available and devise a storage and labeling strategy that is easy to understand, intuitive, and efficient. There is nothing quite as aggravating as looking for a certain type of cable in a messy storage room while the entire group—or worse, the entire audience—sits and waits. Keep it neat!

Walls and Boxes

It is ideal if there's enough wall space to have all of the various cables neatly organized, where the sound operator can simply stand back and visually locate any possible cable type. Although that would be nice, it's not always practical. One alternative that works well, compacts a lot of cables into one area, and is instantly understandable by the newest of newbies, involves the use of several plastic storage boxes. These can easily stack along a narrow wall, and most of the possible cables and utility devices could easily be stored in six to eight bins.

Use the translucent storage boxes, not necessarily so that the contents can be seen from the outside, but so an 8 ½ x 11-inch sheet of paper can be placed inside facing out at one end of the bin. Use a computer to print the labels so they are large and easy to read from a distance. Fill up the piece of paper with labels such as:

- XLR Cables – 20'
- XLR Cables – 30'
- TRS ¼" – 20'
- Speaker Cables – 100' (Red) 50' (Blue)
- Direct Boxes, Extra Mic Clips, CAT5 Cables

Combine items that don't take up as much space. It's best to combine items that don't look alike—items that are easily spotted next to the other items in the bin.

Build a System and Stick to It

Develop a setup routine. Write it down and practice following it. Look for ways to refine the setup routine until it is efficient, intuitive, and very fast. Build the routine procedure on a computer so that it can be easily updated and readjusted. Even a simple routine is helpful to the sound crew. In addition, a printed routine helps instruct new sound crew members or simply folks who want to help.

Establishing a setup routine also helps ensure that all of the necessary tasks are completed in the proper order. If the sound system is complex and if there are several people setting up, divide the tasks into four or five separate jobs. Give each person a list of assigned duties and help him or her learn how to be most efficient. The fundamental keys to success when including several people in the setup process are organization, intuitive procedures, precise labeling, excellent storage systems, and practice. Eventually, everyone learns the routine, and the entire crew functions like a well-oiled machine.

Be very specific in describing each job. Specify where each piece of equipment should be positioned, exactly what type of cable should be used to make each connection, exactly where each cable connects, and how the cables should be routed. Imagine at each step that you don't know anything and that the next step needs to tell you what to do in a way that is successful and rewarding. Wherever possible, include pictures of the devices and connections in the job description. Remove as much guesswork as possible.

Training the Crew

Walk through the assigned job with each crew member. Permanent sound crew members must learn every job and each routine. Segmenting the entire job into bite-sized chunks makes the setup procedure easier for both experienced and inexperienced crew members.

Once each job is clearly described and the crew is assembled, walk all crew members

---

Setup Routine

The setup routine might include:

- Position amp racks
- Position and connect all FOH speakers
- Position mixer
- Connect the snake to the mixer and route to the stage
- Connect all outboard gear and wireless receivers to the mixer
- Position and connect all floor monitors
- Position and connect in-ear monitor system
- Position and connect all direct boxes
- Position and connect all wired mics
- Position all wireless mics

This simple setup routine could take one sound operator hours to set up, depending on the venue and the system.

Define Jobs for the Team

The sound operator and four helpers could set it up quickly and efficiently, especially once the crew knows the routine. This routine could be easily split into five jobs.

Job 1
- Position amp racks
- Position and connect all FOH speakers

Job 2
- Position mixer
- Connect the snake to the mixer and route to the stage

Job 3
- Connect all outboard gear and wireless receivers to the mixer

Job 4
- Position and connect all floor monitors
- Position and connect in-ear monitor system

Job 5
- Position and connect all direct boxes
- Position and connect all wired mics
- Position all wireless mics

through the entire setup process, explaining each job as you go. The system will make much more sense to crew members if they have an image of how their task interfaces with the other tasks. Include a simple diagram of the entire system—color code each job so that the crew can see where their job fits in the big picture.

Zero the Board

Zeroing the board is a common part of the normal setup procedure for a sound company when they're setting up a new act. This process confirms the status of the board and helps the sound operator feel comfortable that there are no leftover settings that might cause an inadvertent problem during sound check. A board that has been zeroed is the sound operator's blank slate. The permanently installed mixer doesn't need to be zeroed before each use. When the musicians and instrumentation are constant, there is no real need for the sound operator to start from scratch each time. Rather than zeroing the board, the sound operator should verify the status of the board.

Documentation of the Mix

There should be documentation of the basic settings that provide a good mix for the house and in the monitors. This documentation should include a record of all level, EQ, and aux settings, and it should be easy to read and visually accurate. Follow this document from left to right and verify that each control is at the correct setting.

Occasionally, especially when there is an exceptional-sounding mix, a new set of documentation should be developed. It should be clearly marked as to the settings of all mix variables, including:
- Channel assignments

Zero the Board

The zeroed board should have all settings at a preconceived level, such as:

- Master output faders all the way down
- Channel trim at minimum
- All channel attenuators set to unity (no attenuation)
- Channel faders all the way down
- Channel mutes on
- EQ circuit deselected
- EQ boost/cut to flat
- EQ sweep to 12:00
- EQ bandwidth to one octave
- Auxes turned all the way down
- Aux masters turned all the way down
- Effects sends turned all the way down
- Monitor auxes set to PRE
- Effects auxes set to POST
- Channel subgroups unassigned
- Subgroup masters all the way down
- Subgroup masters unassigned
- Outboard gear set to unity gain
- Matrix levels turned off
- Talkback level turned down
- Ancillary playback devices turned down

- Subgroup levels and assignments
- Trim levels
- Pad settings
- EQ settings
- Aux settings
- Fader levels
- Matrix levels
- Master fader level
- All outboard dynamics and effects settings

Mark the Board

There are a few companies that supply stick-on arrows, dots, and labels that can be semi-permanently stuck on a mixer to indicate normal settings. Using arrows and dots to point to normal mixer settings makes the board easy to set up for virtually anyone. All they need to do is set each control to its previously marked position, and the system should be ready to go.

Part of the advantage of a permanent installation is the ability to develop a mix over the course of time that sounds great and works well for the musicians. The sound operator should always be on alert when setting up from a pre-existing mix because something might have changed with an instrument level, or a pad on a DI might have been inadvertently removed, or a guitarist might have changed his or her output level, or any number of things might have changed beyond the sound operator's control.

...and Then There's Digital

Modern digital mixers save a lot of time and energy for the sound operator. Most digital mixers have a default zeroed snapshot. Simply press a button, and the entire console is perfectly zeroed out. In addition, the user can save several snapshots of the mixer status. This is ultimately convenient. Save and name your snapshots according to song, instrumental configuration, vocal complement, or any number of helpful configurations.

Snapshots recall instantly. Also, it's simple to copy from one snapshot into your mix, so once the drum sounds are dialed in and perfect, they should be saved as a snapshot. The next time the same drums are set up, the sound operator can simply open the drum set snapshot, copy the drum channels, and then paste just the drum channels into the new mix.

Digital mixers provide immense capacity for storage of snapshots, EQ settings, channel information, effects settings, dynamic processor settings, and so on. Once the basic snapshot is set up, the mix can be divided into musical sections. The sound operator can easily set up separate mix snapshots for the intro, verses, choruses, bridge, solos, and out choruses. As the group performs a song, the sound operator must simply pay attention to the arrangement and scroll through the snapshots that have been designed for each musical section.

Decide on the Mixer Layout and Subgroupings

Once everything is connected to the mixer, decide which groups of channels need to be subgrouped and which instruments need to go straight to the main mix bus. The mixer layout contributes to the sound operator's efficiency. When the channels are intelligently laid out, the mix is easier to manage, and the sound operator has more time to think about the creative aspects of mixing.

When connecting to the snake, try to connect in logical groups. All drum channels should be in consecutive order; all choir channels should be consecutive and arranged from left to right, from the sound operator's vantage point. All channels on stage that form groups should arrive at the mixer in a logical order.

Modern live sound reinforcement mixers typically provide four or eight subgroups. Individual channels are routed to subgroups by selecting the subgroup in the assignment section near the channel fader. Subgroup masters are used for a few different reasons. They're used to:

- Control the level of an entire group of microphones, such as backing vocals, choir, guitars, or all the drum mics.
- Control the panning of an entire group of mics. Channels grouped to a single subgroup master can all be panned between the left and right pan positions. A stereo group can easily be created by assigning the channels to a consecutive odd/even group masters. In this way, the channel pan controls the function to move the channel between the odd and even channels.
- Control the simultaneous muting of the channels in the subgroup.

- Assign channels to multiple sets of outputs. The same channels can be simultaneously assigned to the main L-R output, while also being assigned to the 1-2 group or any other stereo group. These extra group outputs can be routed to auxiliary zone feeds, broadcast feeds, audio recorders, video recorders, and so on.

Verify Functionality of All Equipment

It's important that the sound operator check all equipment before the rehearsal or performance. There is nothing that will inhibit the creative process more than having to wait while the tech crew seeks the cause of an inoperable device.

If your system is small, follow a mental checklist and work through the system. If your system is large, use a checklist that prompts you to verify everything in your system. Most checklists should include the following:

- Microphone: Check all mics, including lead vocal, backing vocal, wireless handheld, wireless lavaliere or headset, instrument, and recording mics.
- Monitors.
- In-ear monitor system.
- Instrument amplifiers.
- Instruments connected to direct boxes.
- FOH speakers: Verify that all components are functional. This might involve turning on the power amplifiers one at a time. Power amplifiers should all be labeled according to their frequency band (highs, mids, lows, subs) and pan position (left-right).
- Dynamics processors.
- Equalizers.
- Effects processors.
- CD player.
- Audio recorder.

Feedback

The sound operator should ring out the monitor system if it hasn't already been done. In a permanent install, the sound operator should verify that the monitors all sound about the same when they are sent an identical signal. To accomplish this, use a handheld wireless mic and set it to the same level in all onstage monitor mixes. Then, walk around the stage and make sure that the volume remains fairly even as you walk and talk into the mic.

Performer Sound Check

Once the musicians are on stage and ready to go, work through the entire group to set preliminary levels and EQ. The performer's sound check forms the basis of what will eventually be the FOH mix. It also forms the basis for what will be the monitor mix. This all happens at once, and it really should all happen pretty quickly.

Live sound operators need to consider a completely different set of criteria in comparison to studio engineers. The studio engineer can afford to take his or her time to get everything precisely set for the best sound—minutiae are appreciated by those who strive for perfection. The live sound engineer, on the other hand, must keep the flow of the event in mind. Focusing on a small detail during the initial sound check phase will bring momentum to a screeching halt.

The live sound operator must think in the order of ensure functionality, refine, and perfect.

Ensure Functionality

The primary consideration is that everything is functional. Once all of the musicians can hear themselves and they can be heard in the house, it's time to think ahead to the next phase. The best sound operators seamlessly transition from functionality to the refinement phase; however, in a complex setup with 24 channels or more, it can take a while to refine all the settings.

Refine Settings

Once everything is functional, including the FOH system, monitors, instruments, mics, and so on, the group should run through a song so that they can begin to fine-tune their settings, monitor positions, staging, and so forth. This is also the time when the sound operator should begin refining the house mix and effects. In the refinement phase, the sound operator must consider how everything in the mix fits together.

- The drums and bass need to work together tonally.
- The keyboards and guitars must be heard as separate entities that don't conflict and create confusion in the mix.
- The backing vocals need to blend together without being harsh.
- The lead vocal must be in the forefront of the mix, without being piercing and irritating.
- The singers must be able to hear well in the monitors. The majority of the time, singers will sing in tune, so long as they can hear themselves accurately. If a group of singers is really struggling to blend and tune, the sound operator should actively assess their monitors. He or she should walk up into the stage area and listen to what the singers are hearing. Many singers don't even know what to ask for in a mix; however, an experienced sound operator can help provide what they need. The vocal monitors don't need full-on mixes—they need themselves along with at least one rhythm-defining instrument and one pitch-defining instrument.
- All of the FOH zones must be fairly equal in volume and they must sound good. The sound operator needs to move through each zone to verify the audio quality in each.

Perfection

The perfecting process is ongoing. The better the mix sounds, the more the sound operator should focus on small details. The interesting thing about music and live sound is that it is always a moving target. Textures constantly change, natural dynamic level changes require different treatment in the mix, and varying levels of emotion throughout the performance require varying degrees of intimacy in the mix. So, keep on perfecting until the show's over and the lights go black.

Fine-Tune the Monitor Mix

There is usually a difference between what the musicians want to hear in the monitors and what they need. An in-ear system allows musicians to hear exactly what they want—this is by far the most inspiring and effective way to monitor. There is no doubt about what the other musicians are doing, and most in-ear systems can provide a very professional-sounding mix.

The Difference Between What They Want and What They Need

In most situations in which floor monitors provide the link between musicians, each musician wants his or her monitor mix to sound like a well-mixed CD; however, five or six floor monitors playing back the full mix at levels that can be heard by the musicians might overpower the main FOH system. Depending on the demographic of the attendees, this full-blown monitor mix might negate the need for the FOH system as far as overall volume is concerned.

Spend some time explaining to the musicians what type of effect the monitors have on the overall sound. Encourage them to scale back on what they need in the monitors. Sometimes the fact that there is so much in the monitor makes it difficult to hear any particular instrument or voice at any volume.

Invite the Musicians to the Board

One at a time, invite the musicians to come back to the board. Demonstrate the effect the monitors have on the overall sound. While the group plays through a song, slowly move the main output fader level down until it is off to demonstrate how loud the group is without the FOH system turned up. Next, slowly increase the main output level until the mix actually sound good—that is typically very loud if the monitors are out of control. This demonstration is very instructive—it is easy for the musicians to hear what's going on, and afterward they are likely to be much more conservative in their requests for loud monitors.

Start without Monitors

Try beginning a rehearsal or sound check with the floor monitors off. The musicians probably won't play together very well because they'll be hearing a mixture of live stage volume and reflections off the back wall, but they might like part of what they hear. The sounds that come back from the room are some of the most comforting and supportive sounds that the musician gets.

Often, the monitors sound smaller as they get louder because they mask the ambient room tone, which provides all of the depth and warmth that inspires most musicians to give their best performance. Once the house mix sounds great, ask each member what he or she needs the most. As long as the house mix is well-mixed, most musicians are happy with some of themselves and their most closely linked mix ingredients.

- Drummers need bass players.

- Bass players need drummers.

- Everyone needs a little lead vocal.

- Guitar and keys need each other, plus a little bass and drums.

- …and so on.

With the house mix up, the monitor sound can afford to be a little thin. The warmth and depth provided by the room is unnecessary in the monitors. This helps minimize the unattractively thick sound that loud monitors inject into the house.

When you find the perfect monitor mix, mark the board for each setting and note it in the system documentation. It is guaranteed that if you don't precisely document and maintain the monitor mix, the monitor levels will creep back up over the course of several church services, performances, or events.

Communication with the Performers

The inexperienced sound operator spends a lot of time yelling back and forth to the musicians, asking about monitor levels, asking the band to play their instruments for a quick sound check, and so on. The experienced sound operator either takes advantage of the mixer's built-in talkback mic or simply plugs a mic into an extra channel and creates his or her own.

Built-In Talkback Mic

A mixer that includes a built-in talkback microphone typically provides a talkback assignment section in which the sound operator can quickly route the talkback signal to any one (or more) of the auxes or to the main mixer FOH output. The sound operator must know where each aux connects to the monitor system and should be able to speak

The Talkback System

Any serious mixer will provide a means of communicating with the musicians. The system pictured here lets the sound operator speak, through a mic connected to the talkback input, to pairs of aux buses or through the subgroups into the FOH, recording, or other connected destinations.

independently to the musician or group of musicians hearing that aux by simply pushing the corresponding button in the talkback assignment area.

External Talkback Mic

An external talkback mic, although more cumbersome than a built-in one, is often preferable. Many large-format mixers with talkback capability simply provide a dedicated XLR input for use with any handheld microphone. The advantage to using a handheld mic is twofold:

- The sound quality is dramatically superior in comparison to the small, built-in condenser mic that is provided on some consoles. Any time the sound operator doubles as the performance announcer, he or she must simply press the MAINS or GROUPS button in the talkback assignment section, and it's Mr. or Ms. Announcer time.

- There is less leakage when using the handheld talkback mic. The sound operator can press the AUX 4 button, for example, to speak privately to all those who hear aux 4 in the monitor system. It is much more discreet when the sound operator moves in close to the mic and whispers a comment to the band than when he or she speaks at a normal level.

Most of the time, the talkback system is used during sound check or rehearsal—not during a performance—however, there are occasions when stage-to-sound-operator communications are necessary in the middle of a show. This can be a very touchy proposition if the group is using floor monitors because there's a good chance that the audience will hear the conversation. On the other hand, anyone connected to the in-ear system can privately receive communications from the sound operator, music director, pastor, or anyone else who might think it's necessary to add his or her two cents to the mix.

The inexperienced sound operator might think it is less intrusive to communicate with the group by using a loud voice—it's not. It is frustrating as a musician to try to hear the sound operator yelling from the console, because there is always an instrumentalist experimenting with a sound, practicing, or even just telling a joke. It is inefficient from both ends of the hall when the sound operator is too shy to speak into the mic.

Basic Troubleshooting

Always trace a problem with any electronic device from the source, through each possible connection point, to the destination—don't intentionally leave out any stage in the path. For example, imagine that there is no signal coming from the guitar. Trace the signal from the guitar to the speakers one step at a time.

- First, verify that the source is providing signal. If the input meters register signal when the guitarist plays, then the signal is verified as arriving at the mixer input.

- If the input meter shows no signal, trace the signal from the guitar to the mixer. Try connecting the source to a functional input to verify that there is signal coming from the guitar—bypass any effects in the signal path.

- Next, plug the functioning guitar into the effects device(s). Verify that the guitarist's effects are outputting signal. This simplest way to do this is to plug the effects into a small amplifier. If there's signal, it works.

- Patch the functioning output from the effects device into a direct box and verify that it is connected to the mixer channel input. If there is no signal registering on the input meter, verify that the guitar is plugged into the DI input, that all attenuators are bypassed, that the input type selector is not set to Speaker or Amp, and that the XLR output from the DI is indeed connected to the intended mixer channel input.

- If the signal registers on the input meter but the sound can't be heard, patch any device into the channel insert to verify the existence of signal at that point in the signal path.

- If signal registers at the insert, disconnect the insert and verify that the channel is assigned to the master stereo output or to one of the subgroup masters.

- If the signal still doesn't work, make sure that there is nothing patched into the subgroup inserts, or that if there is something patched in, that signal registers on the device input.

- If the signal still can't be heard, verify that the subgroup assignments route the signal to the main stereo output.

- If the signal still can't be heard, verify that the master fader is turned up. If signal registers on the main stereo output meter but signal still can't be heard, verify that the crossovers, equalizers, and system processor are all turned on and functional.

- If signal registers at the system processors, verify that the power amplifiers are powered on and turned up.

There's a good chance that following this type of troubleshooting regimen will result in successful completion of the circuit. There are a couple of side roads possible when troubleshooting this type of signal path. Make your best assumptions at each and follow the signal path religiously. It's pretty much guaranteed that the single step you overlook or avoid in your troubleshooting procedure will be the one that caused the problem.

Keep in mind that there are times when equipment breaks. In addition, any time you're using a digital device that passes audio, every once in a while the device is likely to just stop working. It might even look like it's working, but it's probably not. The standard fix for any digital device is to simply turn it off for about 10 seconds and then turn it back on again. Most of the time, this ridiculously low-tech procedure fixes the problem.

If you've been very thorough and the problem still hasn't been rectified, seek the help of an experienced and qualified technician.

CREATING AN EXCELLENT MIX

Building a mix is like putting a puzzle together—all of the mix ingredients should fit together nicely with notches carved out of one mix ingredient to make room for another. If all mix ingredients occupy the full bandwidth, the resulting mix is a mess. Everything quickly starts to clash, resulting in a mix in which everything is too loud but you can't hear anything in particular—everything jumbles together.

Structuring the Arrangement

In the same way that the sound operator must shape and mold each mix ingredient, the musicians must structure each ingredient of the arrangement so that there is room for everything to be heard and understood. The inexperienced band member will tend to play all the time. If there's music going on, he or she is playing along. This is a recipe for sure textural disaster.

Pay close attention to an excellent recording or watch an exceptional top-rate worship team—you'll quickly notice that the musicians and singers aren't all playing and singing at the same time. They will be listening all the time and contributing when it is appropriate to do so.

As the sound operator, it isn't usually your place to step in and start telling the musicians how to do what they do; however, most music directors will take helpful suggestions when they're appropriate and presented in a non-confrontational manner. If it sounds a little like I'm saying you might need to walk on eggshells around the music director (MD), I suppose I am. Everyone reacts to criticism differently, and the last thing you want to do is offend, alienate, or generally tick off the music director.

Anyway, all ego considerations aside, gently and patiently consult the MD about the possibility of thinning the textures out during some of the verses, or having the acoustic guitar and bass play whole notes on a verse, or even doing a song with a single guitar or piano as accompaniment.

If you don't feel comfortable making suggestions to the MD, use your best judgment and act, or wait until your relationship has grown to the point where you can speak honestly and openly about things that could help the team. Bear in mind that, in a church worship team setting, there is usually time to develop relationships and to patiently implement change. For communications about the creative and structural aspects of any music team, it is important that all parties share a mutual respect at some level. That doesn't necessarily mean everyone has to be musical peers; however, it does mean that everyone must have demonstrated their faithfulness to the team and their obvious best intentions for everyone involved in the team.

In fact, if you don't feel comfortable talking to your MD about issues such as this, give him or her this book with this chapter marked.

How to Structure an Effective Band Arrangement

It is very important that both the MD and the sound operator realize how closely their assigned tasks interrelate.

- If the band hasn't structured their arrangement so that there are interesting textural changes and so that the instruments and voice that have the primary parts can be heard and appreciated, the sound operator's job will be nearly impossible.
- If the sound operator isn't paying attention and setting up a mix in which the pieces fit together nicely into an excellent-sounding mix, the band doesn't stand a chance of communicating the heart and soul of the music they're offering.

If the band and singers are providing a well-thought-out arrangement, with instruments working together rather than conflicting, the mix will almost create itself. The advantage of recording in the studio is that the mix engineer gets to pick and choose which ingredients are included in each section. In a live setting, the arrangement needs to be right to start with.

There are a few common structures that have proven to sound great live and in the studio. The following arrangement formats have been well-used and proven to produce an excellent musical flow, while communicating musical and lyrical passion and power.

Arrangement Format 1

- Intro – Full band
- Verse 1 – Drums, bass, piano
- Chorus 1 – Full band
- Verse 2 – Drums, bass, piano, pad
- Chorus 2 and 3 – Full band
- Bridge – Half time rhythm, guitars and pad playing whole notes and half notes, bass long notes and transitional licks, piano open with sparse licks to highlight between vocal phrases
- Chorus 4 – Drums and voices only
- Chorus 5 and 6 – Full band

Arrangement Format 2

- Intro – Acoustic guitar only
- Verse 1 – Acoustic guitar and vocals
- Verse 2 – Acoustic guitar and pad with vocals
- Chorus 1 – Acoustic guitar, pad, light percussion, and repeating one- or two-note guitar lick (a la U2's The Edge)

- Interlude between Chorus 1 and Verse 2 – Full band
- Verse 3 – Drums thinned out (e.g. groove with snare on 4 instead of 2 and 4), bass long notes, acoustic guitar strum, electric guitar long notes and fills, pad
- Chorus 2 – Full band, electric guitar playing same lick as Chorus 1
- Bridge – Full band, change fell to a more aggressive groove, bigger and louder
- Chorus 3 – Breakdown to simple and driving drums, bass, and acoustic guitar
- Chorus 4 and 5 – Full band

These arrangement formats should provide an excellent point of reference for your team. If the band hasn't played together in this way before, they might be a little uncomfortable at first, but they will quickly realize the power in an excellent arrangement.

Ideally...

Ideally, the arrangement should be structured so that all of the instrumental and vocal channels can be turned up to a strong level and not moved too much during the mix. Band members—if you want your parts to be heard more in the mix, play less. Any musician who plays only when the part is necessary and who knows how to play texturally and tastefully will be heard—simply don't give the sound operator an opportunity or a reason to turn you down.

The real problem in most mix situations that involve unseasoned musicians is that the sound operator is constantly fighting to make sense out of the mix, so ingredients that are unnecessary get turned down to make room for the ingredients that are necessary. This scenario forces the sound operator to choose between guitars and keys and synths, and drums, and percussion, and bass—someone will get to lead the mix, and the rest will be hidden.

When musicians or singers complete a performance or a church service, they're often frustrated when friends and bystanders say that their part couldn't be heard. This isn't necessarily the fault of the sound operator—it might be the fault of the arrangement.

At first, the band must structure an arrangement so that everyone knows what to do and when to do it. This familiarizes everyone on the team with what it takes to present a well-devised and powerfully structured arrangement. The more seasoned the team becomes, the more they will naturally listen to each other and, almost instinctively, play together in a dynamic and musical manner.

There are several benefits to crafting an excellent arrangement and to learning how to naturally play an excellent arrangement, including the following:

- Each instrument can maintain a fuller and more impressive sound because there will be fewer opportunities for conflicting and competing musical ingredients.
- The vocals can be heard more easily because of the controlled textures behind them.
- There will be fewer volume complaints. Even if the overall volume is identical to a more cluttered musical performance, it won't feel quite as irritating and hectic to those who care.
- Instrumentalists and singers will be able hear themselves better so their performances will improve dramatically.
- When musicians know they're being heard in the mix better, they tend to care a lot more about how they play, what they play, and how well their instruments sound.

- Once the team begins to play together in a powerful musical and emotional way, they feel better about participating in the team and try harder, practice longer, and demonstrate an increased trust in their leadership.

Creating Size in a Small Room

Contemporary worship music must be presented as authentic, intimate, and heartfelt. Unlike a secular performing group, a worship team should be unpretentious. If a worship team is in a small room, it's a little overwhelming to build the mix into an arena-like sound. Many inexperienced sound operators scramble to buy a multi-effects processor at the earliest possible opportunity. When they hook it up, they find the big hall setting and crank it up. Lots of reverberation can smooth out some of the rough edges in a performance, but it can also quickly rob the intimacy and believability from the performance.

Any effects that are implemented must be believable, and they should retain a feeling of intimacy. In a small room, such as one that could hold up to about 50 people, try adding a little bit of reverberation, but keep the decay time less than one second. In addition, or as an alternative, try adding a slight delay mixed in just enough to thicken the sound on the long vocal notes. ALWAYS mute any reverberation or delay returns during the speaking portions of the service.

Electric guitarists usually have their own effects, and they should also be tastefully injected; however, effects on electric guitar tend to fill out the guitar sound without distracting from the intimacy.

Essentially, in a small room, keep it simple and fairly clean. In a small room, electronic drums will provide vastly preferable results in comparison to acoustic drums. Acoustic drums are great, but unless your sanctuary can hold 200 or 300 people or more, the volume issues they create will be unbearable for everyone involved.

Controlling the Mix in a Large Room

Visually consider each mix you build. Mixes are multidimensional, so your mix image must include:
- Frequency range – EQ high to low, top to bottom
- Volume – Size
- Depth – Reverberation and delays and the lack thereof
- Pan position – Left to right, LCR, surround

Large rooms that can accommodate more than a few hundred people typically provide so much natural reverberation that the last thing the sound operator needs is a lot of electronically simulated reverberation.

In a large room it becomes especially important that the musicians are working together to provide a well-structured arrangement with conscious combinations of instruments and textures. The fact that there is often an incredible amount of acoustic reverberation demands that the spaces in the music are substantial enough that the listener can easily distinguish the musical ingredients.

Often, a church that has grown large enough to occupy a large facility has an impressive number of people involved in the worship team. This is an excellent way to include people in the church ministry, but it is difficult to manage musically and technically. Most of the time in these settings, a surprising number of the instrumental participants are turned way down in the mix if they're even on, for the simple reasons we just covered.

The Sermon

Modern church services are primarily structured to highlight worship through music and preaching from the Word of God. The music sets the spiritual tone for the service as the music team invites the congregation to join along in the singing and reciting of Scriptural truths. If the presentation of the worship music is heartfelt, and if the team has done a good job of engaging the congregation, the sermon will be much more powerful than it would have been without the musical support.

The music should introduce or re-introduce the congregation to God. Once the congregation has focused their hearts and minds on the things of God, they will be more open and receptive to His message for them, both individually and corporately.

The sermon is incredibly important—it must be heard and understood. It is distracting for the congregation to strain to hear the sermon, and it's equally distracting to have to wince at the excessive volume level every time the preacher yells.

Intelligibility

Intelligibility equates with understandability. It implies the ability to accurately differentiate between words and sounds, whether the source is familiar or not. The pastor's sermon must be intelligible. The congregation must be able to differentiate between similar-sounding words, such as brief, breed, and breach; they must be able to accurately hear the vocal sibilance, which provides so many of the key pronunciation and enunciation cues.

Headset Microphone

Headset microphones are very popular in the modern church. They allow the preacher to preach, unrestricted in any way by a microphone or cable. They are typically worn with wires wrapping behind the ears to hold the mic in place, so the mic remains at a constant distance and angle from the pastor's mouth.

With the headset mic being worn just an inch or two off to the side of the mouth, it provides a much more intimate, warm, and consistent sound than a lavaliere mic worn on a tie or collar. The headset mic typically provides excellent intelligibility and dramatically improved gain before feedback, when compared to the lavaliere mic.

The headset mic often sounds good without equalization; however, depending on the room and the preacher, it might need a high-pass filter at about 80 Hz, a 2- to 4-dB cut between 300 and 600 Hz, and possibly a high-frequency boost of 1 or 2 dB around 4 or 5 kHz.

Lavaliere Microphone

In a large room, using a sound system that has ample gain before feedback, a lavaliere mic is a viable choice. Some pastors are simply not comfortable wearing a headset mic. It is of primary importance that the pastor not be distracted by the forced implementation of technology, so a lavaliere mic might be the best choice.

There is often a feedback issue with a lavaliere mic because of the distance between the mic capsule and the sound source. It is understood that much of the low-frequency depth in the lavaliere sound comes from the pastor's chest tone vibration, but the midrange region is often difficult to control. There is usually an overabundance of the frequency band between about 250 Hz and 800 Hz.

An advantage that the headset mic has over the lavaliere is its stationary position in relation to the pastor's mouth—no matter which way the pastor turns while preaching, the sound quality will remain constant. A lavaliere clipped to a tie, lapel, or collar is stationary in relation to the preacher's chest, but not in relation to the preacher's mouth. Depending on the exact location of the lavaliere, when the preacher turns his or her head to one side, the volume might increase, and then when the head is turned to the other side, the volume might nearly disappear—it's a simple matter of proximity to the sound source.

Lectern Microphone

The lectern, or podium, microphone mounts at the end of a long thin tube to bring the small condenser capsule up to the presenter. Small diaphragm condenser capsules offer the potential of excellent sound quality when used from the correct distance. They are very popular in conservative denominations, where the preacher stands at a podium to deliver his or her message.

The biggest problem for many preachers is that the lectern mic is stationary. It doesn't move at all, other than a slight bendability of the tube holding the mic capsule.

Often, the preacher stands a foot or so from the lectern mic, which reduces the potential gain before feedback to near the same level as a lavaliere mic. If the pastor has a loud and projecting voice and if his or her mouth is within 6 to 12 inches from the mic capsule, lectern mics can function very well; however, if the preacher has a light and timid sound and stands a foot or so away from the mic, there will be a feedback problem.

The lectern mic might not need EQ if the preacher is aggressive, or it might begin to feed back around 300 to 600 Hz or possibly around 5 or 6 kHz. Once the frequencies that tend to feed back are cut by a few dB, the lectern mic will sound good in the right application.

Handheld Microphone

Some pastors prefer to hold the mic. They get used to holding a mic, and they often count on it for security. Most preachers want to keep their hands free for communicating, expressing emotion, holding a Bible, and so on. There are several advantages to the handheld mic, including the following:

- The handheld mic provides the best sound quality when used properly.
- The fact that the handheld mic is held within a couple inches of the preacher's mouth makes it the least likely to feed back.

- It is the fullest-sounding when held close to the mouth.
- It provides the most impressive and dominating sound.
- It can be pulled away from the preacher's mouth when he or she wants to speak loudly.

No matter how many sonic advantages a handheld mic might offer, the hands-free feature of the lavaliere and headset mics makes them too attractive for most pastors to turn down. The headset mic most closely approximates the sound quality of the handheld mic.

Volume

No matter which mic type suits the pastor's needs and desires, the level must be sufficient so that everyone in the sanctuary can hear clearly and succinctly. If the congregation constantly has to strain to hear the preacher, they will become fatigued and disenchanted quickly. As long as the pastor provides divinely inspired insight and the sound operator provides excellent audio quality at an appropriate volume, the congregation will receive the message—and just maybe someone will get their life radically changed.

Compression

Compression is a valuable tool in the communication of most sermons. As with vocal tracks or any other instrument or voice, it helps keep the source in a constant dynamic range. Most preachers get a little excited from time to time—when they're excited, they raise their voices or they might even yell. If their level varies dramatically, there's an excellent chance that there will be portions of the sermon that will be too loud and portions that will be inaudible to many.

The compressor can easily control the loudest portions of the sermon so that the level of the entire sermon can be raised, exposing the quiet passages and helping the congregation hear the message the pastor has prepared.

There is one danger in compressing the preacher's mic: feedback. If the compressor reduces the gain by 6 dB, for example, and then the sound operator makes the gain up, it's the same as turning the whole channel level up by 6 dB. Unless there was more than 6 dB of gain remaining before feedback, this compression will cause a feedback problem.

Most high-quality systems provide ample gain before feedback in order to allow microphones on stage, including the pastor's mic, to utilize compression.

To set up the pastor's compression, following this procedure:
- Set the ratio between 2:1 and 7:1.
- Set the attack time to medium-fast.
- Set the release time to medium.
- Adjust the threshold for up to 6 dB of gain reduction at the loudest portion of the sermon.

Riding the Level

Most often, the pastor's mic level should be roughly set before the service, fine-tuned during the first few minutes, and then left alone throughout the sermon. Typically, once the level is set, it is counterproductive to try to ride the fader to help bring out the quiet parts and tone down the loud parts—just as you're bringing the level up because you think the pastor is bringing it down, he or she blasts out an emphatic 120-dB yell to make a point, then you quickly turn the fader down and nobody can hear a thing.

If the pastor's mic is being compressed so that outbursts are under control, then the quiet portions are likely to be audible and present. Competently adjusted compression should provide a level and dynamic range that is functionally appropriate and consistently audible without many fader adjustments during the presentation. However, there are a couple reasons to adjust the fader during a service.

- If your church has powerful contemporary worship or a large team with brass and strings, by the time the music service is completed, the congregation's ears will have become accustomed to the volume. Part of the ear's defense mechanism causes it to become less sensitive in the presence of loud sounds. It takes a while to regain its original sensitivity. Therefore, it might take five minutes or so for the congregation's ears to regain their normal sensitivity. Practically speaking, this means that the level that sounds just right for the sermon, immediately following worship, will probably sound too loud five minutes later. In this case, the sound operator will be experiencing the same phenomenon and should naturally respond by realizing that it sounds a little loud and turning the level down to the newly perceived perfect level. Once this level is set, it should be acceptable for the duration of the sermon.

- Many modern churches meet in rooms that were never designed for church or any other meeting where vocal intelligibility and minimal background noise were issues. Air conditioners, heaters, and generator noises are all aurally intrusive—they get in the way of speech sounds. Typically, the pastor's mic level is perfect until the blower turns on, and then half the room can't hear. The sound operator must be aware of how the entry of a noise into the sanctuary affects each zone in the system coverage area—those closest to the source will experience the intrusion most dramatically, and each time the distance from the source doubles, its comparative level will decrease by about 6 dB, depending on frequency content and acoustical considerations. The sound operator must remain astute so that when the blower noise comes and goes, the level is adjusted accordingly.

- Sometimes, as the sermon progresses, even though nothing has changed at the board, the level might seem too loud or too soft. This might be due to other variables, such as fading battery level, increase or decrease in surrounding noises, or a shift in the pastor's microphone position. The sound operator should constantly assess the level and, even though the norm states that the sound operator shouldn't typically ride the pastor's level, he or she should assess, decide on, and adjust the level according to previous experience and informed judgment.

- The sound operator should always be aware of his or her physical health. During the cold and flu season, an affected sound operator might experience inconsistency in his or her hearing. Ears are affected by these viruses, and their response characteristic can change radically from minute to minute. Sound operators who are sick should seek a replacement until they are well. If, in the case of emergency, the ill sound operator is asked to help, he or she should be upfront with those in charge about the potential hearing inconsistencies, asking for assistance from a staff member or a trusted congregant regarding sound quality and volume.

- Beware anytime the sound operator has traveled on a commercial airliner within 24 hours of an event. Flying and the air pressure changes involved can easily block the passenger's hearing. Often, the low frequencies are primarily affected, so the sound operator might hear a very thin-sounding mix even though it is actually very thick and boomy.

Body Packs On or Off

Preferences vary in regard to the pastor's microphone or body pack On/Off switch. Some sound operators prefer to instruct the presenter to leave the switch in the On position, so that any time the mic needs to be on, the sound operator can turn it on. This process strives to eliminate one variable in the signal path so that if the mic needs to be on, the sound operator and pastor won't go back and forth, switching switches in the heat of the moment until they land on the correct combination and achieve success.

The other school of thought puts the pastors and other presenters in complete control of the On/Off switch. In this arrangement the corresponding mixer channel is left up, on, and unmuted all the time. When the pastor needs the mic on, he or she turns it on, and when it needs to be off, he or she turns it off.

I've used both of the previous procedures. Either method is satisfactory as long as everyone knows the system, although in church settings I've grown to prefer leaving the control in the hands of the pastor or presenter, especially in a weekly event in which the same presenter uses the same system. Once they understand that the sound will always be there when the switch in the On position and that it will only be off if they switch it off, most speakers appreciate this method. This is most convenient when the pastor needs to confer privately in the middle of a presentation, clear his or her throat, or just mute a cough or sneeze. If you use this system, as the sound operator, be certain that you cover the pastor in case he or she forgets to mute the mic in an embarrassing moment. Always be alert to the status of the mic. At the end of the service, be certain that the mic is off and definitely not being sent to the service recording or television feed—and, of course, watch out for the occasional bathroom break.

When working with a new presenter or a large rotating pool of presenters, it is a better idea to put the control in the hands of the sound operator, instructing the speaker to leave the body pack in the On position all the time—many body packs can be set to bypass the local controls. This way, the presenter won't ever need to flip the On/Off switch, eliminating the variable on one end of the communication pathway. If the mic is not on whenever the presenter speaks, it will be known by all that either the sound operator still has the channel muted or the battery in the body pack is dead. An attentive sound operator won't miss many cues, so this is the preferred method of operation in most large-scale professional applications, where the event involves various speakers who are unfamiliar with the system and body pack controls.

Whatever system you choose, be sure that it is clearly communicated to all parties and that it is agreed to be the best system.

Watch the Wireless Mic Status

Most modern wireless systems provide helpful displays on the receiver and body pack that register battery life, On/Off status, internal levels, and so on. The sound operator can easily see whether the mic is turned on and whether the battery is getting close to giving out. These tools are very convenient in the middle of any event.

At the beginning of any event, be absolutely certain that the batteries contain ample voltage levels to outlast the event. Be careful about trusting the level meters or even a voltage meter. After a battery has been used for a while, the power switched off, and the device stored for a few days, the battery might develop an artificial charge. This charge might read "Full" at the beginning of the event, but within minutes might be dangerously close to dying. Therefore, if a battery has already been used a few times but the level indicator shows "Full," it is likely a false reading and a new battery should be installed.

In a church setting, the sound operator can track the battery life from service to service, gaining a fairly accurate insight about how much to expect from the batteries in each system. It's not uncommon to use the same battery in a body pack for several services, but it is never a good idea to test the limits—batteries aren't that expensive, and failure in the middle of a presentation is very distracting. Most body packs will operate for around eight hours on one 9-volt battery.

Building the Mix

Building the mix is a multifaceted task. While creating an overall musical and appropriately powerful sound, the sound operator must understand the musical intent as well as the musicians and their tendencies. In addition, the mix must be built with a target volume in mind. Obviously, there is a big difference between acceptable rock concert and church service volumes. It doesn't help to build a mix during rehearsal that is perfect for a concert but way too loud or too quiet for the event because the relative acoustical interactions from the monitors, surrounding surfaces, and onstage instruments change with the FOH volume.

Keep in mind that the mix changes when the seats are filled with people—in fact, it varies depending on how many people fill the seats and proportionally how many seats are full. There isn't a hard-and-fast rule that quantifies the difference between the level in an empty venue and one filled to capacity. The sound operator must trust his or her ears and rely on SPL meters and other audio assessment tools to help guide him or her through the mixing process.

The Decibel Meter and the Mix

Much of the basic mix is built during rehearsal, while adjusting monitor levels and fine-tuning instrument and vocal sounds on the mixer. Throughout this process there is a continual give and take as the musicians adjust their sounds and monitor levels. Be sure that there is either freedom to set the volume at the sound operator's discretion or an agreed-upon maximum SPL. Whether the target volume is 88, 95, or 105 dB, that's what the sound operator should aim to achieve during rehearsal. Any variation in level should be a few dB quieter during rehearsal than the target for the event.

The sound operator must verify the SPL throughout the venue, especially when there are specifically designed zone-coverage systems. He or she should walk the room during the rehearsal, using the portable SPL meter to measure the dB at several locations. If there are discrepancies in volume, they should be established as intentional or adjusted to provide a consistent listening experience throughout the room. In a church setting, it's a good idea to intentionally construct a zone that is a few dB quieter than the rest of the room. This area helps people with hearing-sensitivity issues to enjoy the service as much as the rest of the congregation. If the volume is exactly consistent throughout the room, the majority of the congregation is subject to the hypersensitivity of a few people.

The Difference Between the Rehearsal and the Event

Once the audience is in the room, it's likely that the mix level will decrease—it's always easier to turn it up and keep the mix integrity than it is to turn the mix down. When the mix level increases in the FOH system, the mix is cleaned up to a degree because the relative FOH levels increase in relation to the monitor and onstage levels. Also, an increase in the FOH level feels good for the musicians—they sense the increase in power as the acoustic interaction increases. This increased fullness usually results in more inspired performances. On the other hand, if the mix is set up too hot in rehearsal and then the level needs to be decreased by several dB at the start of the event, the musicians feel the decrease in warmth and power—this is very uncomfortable for most musicians, and it tends to decrease the quality of their performances.

Most musicians are intent on getting their sounds right and their monitor levels perfectly set during rehearsal. Although their performances might be accurate, they aren't usually passionate or inspired. Once the event begins, the singers and instrumentalists usually become adrenalized, while at the same time focusing more intently on precision and presentation. This phenomenon typically results in an increase in energy, which often equates to an increase in volume. The sound operator should be aware of the musicians' tendencies—if they constantly perform substantially louder than they rehearse, the sound operator should simply be on alert and adjust the mix accordingly.

First Things First

Once the sound check is under control—the instruments and vocals are functional, and as the monitor mix is built—it's time to start building the FOH mix. Each sound operator develops a routine that helps him or her to be confident that a solid mix is being built, but there needs to be an intelligent starting point for each sound operator to develop his or her system. The following procedure takes advantage of the experiences that I, and others like me, have had. Building a mix, although an artistic endeavor, is results-oriented and typically time-restricted.

It's important that the sound operator moves quickly and makes decisions without hesitation. A typical stage contains between 3 and 30 musicians with other things to do before the performance than wait for the sound operator to endlessly tweak their sound. With all things in proper perspective, the team—including the sound operator—must realize that adjustments will be made throughout the event. As the audience enters the room, the sound changes; if the humidity changes, the sound changes; during each song, the mix adjusts for the specific voices and instruments highlighted; and so on.

The fundamental routine should follow a consistent order. The following routine is effective and can serve as an excellent starting point for customizing a routine that matches your system and team. Keep the SPL meter handy. Check it throughout the process to verify that you're staying within the target decibel range.

- Be sure that basic vocal levels and EQ settings have been established and that all vocal mics are live.

- Build the drum sound starting with the kick drum. Set each mic sound, leaving the mics live as you build the sound. In order, establish the kick, snare, toms, overheads, and hi-hat microphone EQ and levels.

- Build the bass sound.

- Listen to the bass guitar and kick drum together and adjust the tone of each to work with the other.

- Listen to the bass and drums play together. Don't linger; just keep moving through the process.

- Build the piano and keyboard sounds.

- Build the electric guitar sound.

- Build the acoustic guitar sound.

- Have the rhythm section play together and adjust each ingredient so that they all fit together to create one sound. Keep in mind that there needs to be space and volume headroom left over to add the vocals.

- Build the lead vocal sound.

- Build the backing vocal sound.

- Have the rhythm section and vocalists run through a song section in which everyone plays and sings. Adjust the levels and EQ settings for a smooth and clean mix.

- Build the choir sound as the choir sings a cappella—mute their mics and turn their channels down.

- Build any other instrumental ingredients, such as brass and strings—mute their mics and turn their channels down.

- Have the entire team perform a song that includes everyone.

- Unmute the choir mics and slowly raise the fader levels until the choir can be heard. Adjust the equalization on the choir mics so that they blend nicely into the mix.

- Unmute the brass mics and position them in the mix.

- Unmute the string section mics and position them in the mix.

- Assess the volume using the SPL meter. Adjust accordingly.

- Be sure the vocals are clear, audible, and intelligible.

This process shouldn't take more than a few minutes, The sound operator must be proficient enough at setting levels and equalization that there is very little randomness to his or her selections and adjustments.

Subtractive Versus Additive Mixing

It is very important that the sound operator understands and implements the concept of subtractive mixing, avoiding an additive mixing style. Very much like our previous study of equalization, the sound operator should listen for mix ingredients that are in the way of other mix ingredients, turning them down to reveal the hidden instrument or voice.

The additive mixing style is instinctive. It is the typical knee-jerk reaction on which most inexperienced sound operators rely. It operates under the premise that if the guitar can't be heard, then turn the guitar up; if the bass can't be heard, then turn the bass up; if the vocals can't be heard, then turn the vocals up; if the drums can't be heard, then turn the drums up. This probably brings us back to, if the guitar can't be heard, then turn the guitar up. It is easy to see how this approach to mixing could spiral out of control. Before you know it, the volume is so loud that even the most ardent volume appreciator is wincing.

The subtractive mixing style is much more effective and intelligent than the additive style. By turning a mix ingredient up, you always run the risk of obscuring another mix ingredient, whereas turning a mix ingredient down will never obscure another ingredient—it will only create space for other ingredients to be heard.

If you can't hear an instrument, listen intently for what other mix ingredient could be in the way and turn it down. Once the instrument or instruments that are in the way are turned down, you still might need to boost the previously obscured instrument; however, because other ingredients were diminished, the mix should still be under control.

Mixing Techniques for Rhythm Section and Voice

Audiences and musicians aren't usually short on opinions when it comes to music they love. Become intimately familiar with the styles you're mixing. In addition, maintain an open mind during the process. Only when you are very familiar with the expectations of those who love the style of music you're mixing can you legitimately and authentically push the creative boundaries—you should always strive to add your personal edge to the music you're mixing. As the sound operator, you can help the musicians realize their vision for the music. You must be familiar enough with the tools at hand to accommodate the musical vision through the transparent implementation of technology.

Take note through this chapter of specific techniques that will help you develop better mixes. Many techniques involve adjusting frequency ranges, pan settings, and effects that primarily increase the power of the mix without dramatically affecting its sound. Other techniques radically affect the musical, emotional, and sonic impact of the mix.

Mixing music involves several factors other than the basic sound of the music. Although the artistic and musical aspects of mixing are very important, many of the techniques described herein are designed to maintain the musical sound while increasing the musical impact and power.

It is important that the sound operator is a music lover. In particular, it's important to listen to great recordings of the type of music being mixed. Critically listening to recorded music sets the standard for what we want to achieve in a live setting. The sound operator should strive to build a mix that sounds like a recording while the musicians perform with the energy and life that is characteristic of a great live show.

Equalization: The Cumulative Effect

During a performance, the sound operator must make many decisions concerning the use of equalization, levels, panning, effects, and so on. When it comes to equalization, the big picture is what matters. The sonic character of an individual track is almost always secondary to considerations about how the tracks all work together to form a complete mix.

Previously, I noted that when building a mix, a sound operator should look for channels to turn down before turning channels up. Likewise, when adjusting equalization, the sound operator should look for frequency bands to turn down before turning frequency bands up.

In addition, as the sound operator shapes each sound, he or she must be aware of how those sounds fit together with the other sounds in the mix.

Combining EQ

As you build each sound, always keep in mind that boosts and cuts at specific frequencies accumulate. Boosting the same frequency on several channels at once causes the entire mix level to increase artificially. Even though you might like the individual sound of each instrument, boosting one frequency band constantly causes that band to accumulate, adversely affecting the overall mix level.

For example, if you boost 100 Hz on the kick drum by 6 dB, then boost 100 Hz on the bass guitar by 3 dB, then boost 100 Hz on the synth pad by 4 dB, you could increase the mix level by up to 13 dB when all three instruments play aggressive full musical parts.

This concept also applies to cutting similar frequencies on multiple cuts. Multiple cuts at the same frequency create a mix that is unpredictable within an acoustical space. The hole in the broadband spectrum caused by multiple cuts at the same frequency could result in a mix that sounds fine at a particular location in the room but completely unacceptable at others.

Allocating EQ across the Audible Spectrum

It's the mix engineer's job to build a mix that is strong, full, and balanced. Rather than boosting the same frequency on several tracks, allocate varying bands to each instrument. Instead of boosting 100 Hz on the kick, bass, and pad, try boosting different low-frequency bands on each to spread the low-frequency energy across a broader range. This way, there won't be such dramatic impact on the overall mix level, and the mix will sound fuller and better balanced throughout the room.

Learning from the Recording Engineer's Bag of Tricks

Most acoustical spaces are imperfect—most sound systems are imperfect. Notice, however, that a professionally mixed CD sounds good on most live systems and in most acoustical spaces. The techniques used in the development of a mix for commercial release lead to a mix that is well balanced from a level and equalization standpoint. An excellent live sound operator uses most of the same techniques as the studio engineer to create a mix that is clean, consistent, and powerful.

A recording engineer has to keep in mind the fact that the mix he or she creates will be heard on anything from an inexpensive portable radio to an esoterically hip and sonically stunning hi-fi audio system. In comparison, the live sound operator must consider acoustical summing and canceling due to imperfect acoustic design.

The same mix could sound radically different at a handful of different locations in the room. If the sound operator is only basing the mix on what it sounds like at the mix location, and not intellectually reasoning through intelligent mix decisions, a large portion of the audience might needlessly experience a very bad and possibly annoying mix.

It is a disadvantage for the sound operator that he or she can't simply trust in the mix as it's heard at the mixer, but unless the venue is acoustically perfect and the sound system

Combining EQ Curves

When the same frequency is boosted on several full-range instruments, the mix quickly elevates to an unnecessarily hot level. In addition, the similarly equalized tracks conflict with each other, losing their individual character and personality.

Instruments such as the bass guitar and kick drum, when boosted in the same low-frequency range, quickly raise the overall mix level. Since both instruments contain ample content at 100 Hz, when they're both boosted by 6 dB at 100 Hz the entire mix level could be increased by 12 dB and the sound of the low frequencies throughout the room are likely to be muddy and confusing.

Kick Drum EQ Curve

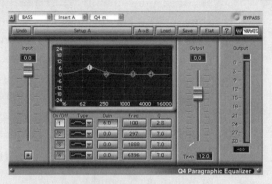

Bass Guitar EQ Curve

is precisely tuned, this non-linear room response is just a fact of life. In a fixed location, it is easier to establish a mix position that is a good representation of the entire venue; however, any time that a system is consistently struck and set up, slight changes in mix positions and speaker angles could create a varied mix perspective. Building a mix that contains an even balance of frequencies provides a mix that sounds better throughout the room.

Low-mid and low frequencies are the least consistent throughout the acoustical environment. Because low-frequency waveforms are long and contain substantial amplitude,

Allocating EQ

Distribute equalization boosts and cuts across the audible spectrum. For example, if you need to boost the low frequencies on the kick drum, bass guitar, and synth pad, choose different frequencies to boost on each track. This procedure distributes the lows across a broader bandwidth, helping to minimize amplitude accumulation.

In this illustration the kick drum, bass guitar, and synth EQ curves reveal a 4-dB boost at separate and unique low-frequency bands.

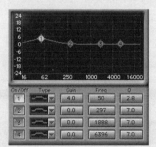

Bass Boosted 4 dB at 50 Hz

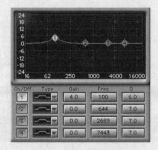

Kick Boosted 4 dB at 100 Hz

Synth Pad Boosted 4 dB at 150 Hz

they are the most capable of forming standing waves. Previously, we saw that an acoustical standing wave provides the opportunity for summing (doubling of amplitude) and canceling (extreme decrease in amplitude). Standing wave patterns interacting between surfaces typically result in continual variation throughout the room dimension (for example, between the front and back walls) between summing and canceling. Therefore, while listening to a single low frequency, it is likely that its acoustic level will be dominant at one location and nearly non-existent a few feet away.

It isn't uncommon to mix a live show and, as the sound operator, think pretty highly of the mix—it had full lows and an excellent balance between mids and highs, and so on—just to have a few people walk up afterward and say that they couldn't hear the bass guitar at all. Although this might be attributed to people who love bass guitar–heavy mixes, it is more likely that they were simply setting in a seat that was acoustically canceling in a low-frequency band. The opposite is also possible, where the bass sounds great at the mixer but some audience members are being blown away by massive lows.

Fitting the Puzzle Together: Consciously Combining EQ

An experienced sound operator allocates equalization boosts and cuts across a broad frequency spectrum so that the mix ingredients fit neatly together like puzzle pieces. For example, if you need to fill out the kick drum and bass guitar tracks, try boosting the kick slightly at 60 Hz, and then boosting the bass guitar at 120 Hz. With this done, create a dip in each of the two tracks to create a space for the boost in the other. In other words, boost the kick track at 60 Hz, and then cut the kick track at 120 Hz. Next, boost the bass track at 120 Hz, and then cut it slightly at 60 Hz.

Using equalization in this way will produce a mix that sounds fuller on each track while the mix level is relatively unaffected. Without the accumulation of equalization boosts, the entire mix is more powerful and will end up sounding much louder when completed.

Sweeping a Peak to Find a Problem

Sometimes you'll run across a sound source that possesses interesting tone problems. Annoying rings from a snare drum or tom can adversely affect the overall mix impact. An irritating high-frequency edge on certain guitar sounds can ruin the best of mixes.

If you find a problem that needs to be eliminated, try sweeping a very narrow peak across the suspected problem area.

- Use a parametric EQ with the narrowest bandwidth selected.
- Create a severe boost.
- Sweep the boost until the problem frequency is most noticeable.
- Cut the selected frequency to eliminate or minimize the problem.

This technique works wonderfully well in a controlled environment, such as a recording studio. It also works well in a live sound application, but special care must be taken due to feedback sensitivities as the boost is swept across the band.

Video Example 22-1

Equalization Techniques

Create Complementary EQ Curves

The previous two illustrations demonstrated two valuable concepts: EQ accumulation and EQ allocation. In this illustration we create equalization curves for instruments with similar frequency content that fit together like puzzle pieces—if 120 Hz is boosted on the kick drum, cut 120 Hz on the bass guitar. Likewise, if you boost 60 Hz on the bass guitar, cut 60 Hz on the kick drum.

If you simply use this technique on these two instruments alone, the resultant power and punch in your mixes will increase dramatically. Whereas boosting both the kick and bass guitar at 120 Hz by 6 dB could result in a 12-dB increase in mix level, using the technique described in this illustration could create a more impressive sound while affecting the mix level very little.

The slider positions on the graphic equalizers below provide an excellent illustration of this concept of creating complementing EQ curves.

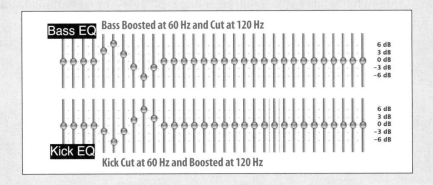

Basic Procedures for Building the Mix

Each instrument and voice demands individualized treatment while building the mix. The sound operator must assess every mix ingredient and determine how it can best support the impact, power, and musicality of the performance. The following considerations will help you build a high-quality mix; however, it is completely up to each individual to develop a sense of what high-quality music is. Listening to excellent music is a crucial ingredient in the process of learning to create a great-sounding mix.

Electronic Drums

Modern electronic drum kits contain many classic synthesized drum sounds along with high-quality sampled sounds from several genres and eras. Although drum machines and sound modules typically provide excellent and realistic sounds, the challenge is to present them in a musical way that adds to the overall mix power. Not many ingredients will degrade the quality of a performance more than bad drum sounds that are awkwardly triggered.

Since the Roland V-Drums came out several years ago, electronic drums have provided decent sounds and acceptable triggering. The drum pads feel pretty much like drums, and the cymbals pads feel pretty much like cymbals. The triggers work well, and the sound module is flexible enough to conform to many different playing styles.

Although modern electronic drums provide several outputs for connection to the mixer, many sound operators prefer to receive grouped sends. Typical electronic drum output assignments should at least send the kick and snare drums to separate channels. It is theoretically ideal if the drum module sends each instrument to a separate mixer channel; however, this isn't usually practical. There usually aren't enough inputs available and, even when there are, it is usually more reliable to have the drummer adjust the levels for each patch than it is to expect the sound operator to adjust 8 to 12 drum and percussion channels for each mix. The following output configurations provide some control for the sound operator while taking advantage of the unit's ability to store premixes of certain groups of instruments.

If there are plenty of available channels, the electronic drums should be panned in stereo for all grouped instruments within the sound module saved as part of the patch data. Although this takes up a lot of channels, it provides the widest image and takes better advantage of built-in effects.

If the drum sound module provides enough outputs, send the effects output to the mixer separate from the instrument sounds. These effects channels are also ideally connected as a stereo pair and hard-panned left/right in the FOH mix.

Stock Versus Custom Patches

Most of the stock drum set patches that I've heard sound like they're designed to show off the "wow" factor of the sound module's capabilities rather than to be functionally brilliant in a live or recorded sound environment. It's important that the drummer learn to use the sound module and that he or she works with the sound operator to create custom patches that are musically relevant and sonically impressive.

Perform as much processing, panning, and equalization as possible within the sound module. This essentially automates the drum mix for most songs. The more the sound operator and drummer work together to achieve the very best sounds, the better it will be for the both the performers and the audience. The sound operator should use the talkback system to guide the drummer through fine-tuning the sound of the kit during rehearsal and, in the case of an emergency need, during the performance.

Phones and the FOH

Because the electronic drummer is typically wearing headphones, there will probably be instances in which the sound operator needs one sound and the drummer needs another. The best approach, in this event, is to build a kit that is just right for the FOH system and then to work with the drummer's monitor system—usually, changing the overall equalization will solve the problem.

The Aviom headphone monitoring system provides global tone controls that let the drummer vary the low- to high-frequency relationship—this is often sufficient to solve the problem. If additional processing is needed, such as outboard equalization or compression, it should be inserted between the mixer feed and the monitor channel inputs.

If the drum sound module sends are split into more than four mixer channels, the sound operator should set up a mix of all the drums and percussion except the kick and snare drums in an available mono or stereo pair of auxes—this send should then be routed to the headphone system inputs. Once the relative balance is adjusted, this grouped send should suffice, especially if the drummer and sound operator are able to pre-set the patches and fine-tune over the course of time.

Equalization and Effects

Electronic drums don't usually require radical equalization. Most of the sounds are pretty well tested by the time they're burned into the sound module's memory. Slight adjustments are often necessary to help the sounds blend with the mix or simply to add a little sizzle to some of the transients.

Effects are almost always too hot in the stock sound module patches. The effects make them fun to play in the music store, but they are typically inappropriate for the live mix. If the built-in effects are used, they should be separated from the rest of the sounds and routed either to available channels or effects returns so they can be mixed at the proper levels for the FOH mix.

Kick and Snare Drums

The kick and snare should be panned to center, each with its own channel on the mixer. Giving these instruments separate mixer channels from the rest of the drums and percussion makes a big difference in the punch and impact of the mix. The snare might need reverb to create the impression of a large hall, but that same reverberation intensity should not be applied to the rest of the kit.

On this song, I want a solid and punchy kick without an overly exaggerated slap. I'll boost the lows at 80 Hz, cut the mids at 250 Hz, and leave the highs flat. These are only suggestions based on the sound of the instruments. Though the techniques are common

Standard Electronic Drum Output Configurations

The more control the sound operator has over the electronic drums, the better. The typical patches that come standard with electronic drum modules are designed to sell kits in the music store—they're not designed to sound great through a live system. It is usually best to turn the internal effects off in the drum sound module, letting the sound operator adjust effects to fit the mix and the room. When running short on available channels, at least separate the kick from the overall mix and send it to its own channel. Realistically, the kick and snare should be on separate channels from everything else. Ideally, the kick, snare, toms, cymbals, and percussion should have their own mixer channels.

Instrument	Suggested Output	Comment
Kick	Output 1	
Snare	Output 2	
Overheads and Percussion	Output 3 (or stereo 3 & 4)	Contains toms, cymbals, and auxiliary percussion

Instrument	Suggested Output	Comment
Kick	Output 1	
Snare	Output 2	
Toms	Output 3	
Cymbals and Percussion	Output 4	This is the most common configuration used to amplify electronic drums.

Instrument	Suggested Output	Comment
Kick	Output 1	
Snare	Output 2	
Toms	Output 3	
Cymbals	Output 4	
Percussion	Output 5	Separating the percussion parts gives the sound operator much more control over the drum mix.

Instrument	Suggested Output	Comment
Kick	Output 1	
Snare	Output 2	
Toms	Output 3	
Cymbals	Output 4	
Percussion	Output 5	
Groove	Output 6 or stereo 6 and 7	Many modern songs utilize groove that play along with the drummer and the band. This can be easily routed to the main left/right outputs, separate from the individual outs.

on these types of instruments, the actual frequencies you alter, if any, are subject to the instruments you're working with and your personal musical taste.

I want the snare to be full and clean. I like the sound of this snare pretty well, but I'll boost 5 kHz by 2 or 3 dB just to help the clarity of the sound once the song starts coming together. I'll add a little bit of a plate reverb with about a one-second reverberation time to the snare.

I've got the cymbals, toms, and percussion all coming from two outputs of the drum module, and on the mixer I've panned these two channels hard right and hard left—individual instrument panning must be done within the module. I'll bring up the drums starting with the kick and snare, rough in their levels and EQ, and then work in the rest of the drums. The panning of the cymbals, toms, and percussion is adjusted to achieve a balanced feel from left to right.

Audio Example 22-1

Building the Drums

Pan the Cymbals for a Natural Sound

It isn't natural to hear one cymbal completely from one side and another cymbal completely from the other, so avoid this kind of panning.

It is natural to imagine the drum set occupying a particular zone in the mix. The drum set is usually in the center of a band, but it still has a little spread across the center zone. I've found that panning the cymbals between about 10:00 and 2:00 in the mix provides enough of a spread to hear separation, but not so much that the drums sound too wide.

Electronic drums aren't always placed in the center of the band. Often, they're placed to one side or the other for staging purposes. Even though they might physically be positioned to the left or right, they should still be positioned around the mix center position so that the entire house receives an equivalent balance.

Acoustic Drums

An excellent drummer will usually provide a very usable drum part, even on a substandard kit. On the other hand, even with the best of drums, a mediocre drummer usually provides a mediocre drum track. There are many tools available to help the sound of a bad drummer, but virtuosity can't yet be simulated by even the most outrageous piece of outboard gear. If you're hiring drummers, pay the price to hire the very best players. If you're working with existing talent or volunteers, provide the very best-sounding monitor mix. If the drummer is hearing an uninspired mix, he or she will probably provide an uninspired performance.

Pay close attention to the drums. Be sure that they are in good condition and that the heads are not dented or stretched. Take time miking the kit, and build the very best possible sound for the room.

Establish an environment in which the drummer can really play the drums. If a drummer has to be too volume-conscious, he or she won't be able to provide the solid rhythmic foundation required by most modern music. Either baffle the kit or set up a Plexiglas enclosure to contain the drum level. Be sure to enclose the front and back of the kit. If the back of the kit isn't enclosed by Plexiglas, it should at least be treated with absorbent material to minimize reflection into the room. Some Plexiglas drum shields also provide a ceiling for ultimate containment. Providing an enclosed environment for the drums is one of the best ways to provide a great mix; it is also a surefire way to help the drummer provide the very best performance. There is nothing worse for a drummer than trying to play solidly without really being able to play into the drums.

Kick Drum

Establish a kick drum sound that is solid but not overwhelming in the low frequencies. Most of the kick drum sound usually comes from the high-end attack through the FOH system. A good drummer and bass guitarist will usually play the same rhythmic parts with the kick drum and bass guitar. Because the bass guitar defines the pitch through the low frequencies and since the room is probably active in the low end, there isn't a lot of room for the kick drum to contain much depth. The sound operator should choose one or the other—either the bass guitar or the kick drum should dominate the low frequencies, but not both.

Sweet Spot

If the kick is selected to drive the low end, the sound operator should find the natural sweet spot for the drum. A good kick-drum sound typically has a warm low end that is controlled and blended with the mix. Try this procedure to locate the sweet spot in the kick-drum sound.

- Set up a boost in the low-frequency range of a parametric equalizer.
- Adjust the bandwidth to about a half octave.
- Sweep the boost in the low-frequency range, between about 60 and 150 Hz, while the drummer plays the kick. Continue until you locate the precise boost that accentuates the drum's warm tone.

There is almost always one spot where the low-frequency boost makes the kick sound come alive.

Mids

As you've seen previously, a simple cut in the low midrange band (between 250 and 600 Hz) helps isolate both high and low frequencies. When this cut is combined with a boost in the low-frequency band (between 60 and 150 Hz), the kick sound quickly takes on a personality and character that is appealing and supportive to the mix.

Highs

The genre determines whether the kick drum should have a high-frequency attack boost. The kick sound is very influential to the overall feel of the mix. Certain songs demand that the kick blends in with the mix; others call for an aggressive kick sound with exaggerated highs and lows.

To accentuate the kick drum attack, boost the highs between 3 and 5 kHz. Use a bandwidth between one octave and a half octave. A narrow-bandwidth boost runs the risk of sounding unrealistically exaggerated on certain types of monitor systems. A wide-bandwidth boost tends to increase leakage from cymbals and other drums.

If the bass guitar has been chosen to dominate the low-frequency range, the attack from the kick drum becomes even more important. Listen to the entire group playing together—adjust the high-frequency equalization so that the kick is clean and tight. Even if there is very little low-frequency content in the actual kick drum sound, it will sound very large when combined with the bass guitar and the rest of the band.

Complement Bass Guitar

Avoid using the same EQ on the kick drum and bass guitar. As stated previously, create EQ characteristics on the kick and snare that work together, resulting in increased definition and power for each.

Snare Drum

Mixing the snare drum is sometimes very easy—all you need to do is turn up the snare, and everything comes together. Usually the snare drum requires the perfect equalization to blend into the mix, supporting the groove while providing a solid backbone for the mix.

To blend the snare with the track, boost the frequencies between about 250 and 600 Hz and cut the highs slightly above about 5 kHz. To highlight the snare, boost the frequencies between 3 and 7 kHz and cut slightly between 300 and 500 Hz. In addition, eliminate unwanted leakage in the low frequencies by using a high-pass filter to reduce the frequencies below 150 Hz.

The snare sound is heavily influenced by the acoustic environment, so the sound operator's mix decisions will be influenced by the amount of shielding or baffling around the kit.

Toms

Set the toms so that they have a clean attack and warm, smooth lows. They can tend to be boomy in many rooms, so they might not need full-range reinforcement. Providing a high-frequency boost between 3 and 6 kHz, along with a high-pass filter that rolls off below 100 to 150 Hz, could provide a sufficiently clean tom sound. Often, the mids need to be reduced between 250 and 600 Hz to clean up the tom sound.

Drum Set Overheads (Live)

Blend in the overhead mics just enough to fill in the sound of the kit and cymbals. Use a high-pass filter to cut the lows below about 150 Hz. The punch and lows of the set come from the close mics.

Overheads are primarily for transient definition on the cymbals and other percussion included in the drum kit. They also serve to blend the sound of the drum kit.

If there are two overheads, boost a different high frequency on each mic. I like to boost 10 or 12 kHz on one and about 13 to 15 kHz on the other.

Overheads are often compressed in an effort to blend the kit together and to maintain an aggressive and intimate sound. Depending on the genre, the overheads might be extremely compressed. Aggressive rock and pop recordings make frequent use of this technique. The amount of compression that should be applied to any drum microphone is completely dependent on the FOH system and the amount of available gain before feedback. The compressor turning up and down in response to signal level can easily create a scenario in which the overheads feed back when the drums stop playing.

Video Example 22-2

Building the Drum Set Sound

Bass Guitar

Once the drums are roughed in, you can move on and add the bass guitar. Fine-tuning the drum sounds will have to wait until the mix is further along and you can hear how the sounds are combining.

You'll need to assess the bass sound in the room. If it needs more lows, don't boost the same frequency that you boosted on the kick drum. Try boosting 150 Hz on the bass if you boosted 80 Hz on the kick, or vice versa. If you boosted 80 Hz on the kick, you're best off to cut 80 Hz on the bass, and if you boosted 150 Hz on the bass, you're best off to cut 150 Hz on the kick. This approach results in a more controlled low end.

If you need to boost highs in the bass for clarity, find a frequency that works, but don't use a frequency that's predominant in any of the percussion instruments. Also, the bass should be panned to the center mix position. Low frequencies contain a lot of energy and can easily control the master mix level. If the bass is panned to one side, the entire mix level will be artificially hot on one side. This would be senseless because bass frequencies are omnidirectional; even if you had the bass panned to one side, the listener might not be able to tell anyway.

In Audio Example 22-2, I'll add the bass to the drums and adjust its EQ to fit with the drum sounds.

Audio Example 22-2

Adding the Bass

One of the biggest concerns in a mix is how the low frequencies fit together. If they accumulate, your mix will be muddy and boomy—in other words, it won't sound good. If you've been able to fit the pieces together well, your mix will sound full but very controlled and clean.

When combining the bass and drums, remember that the kick drum and bass guitar rarely have reverb, though the snare and toms often do. Also the hi-hat, shaker, overheads, and tambourine hardly ever need reverb, while drum machine cymbals and some percussion instruments, such as congas and some very sparse percussion parts, can benefit from the appropriate reverb sound.

Mixing Guitars

Guitars, whether electric or acoustic, are a very important part of defining the musical style. Typically, acoustic guitar provides both tonality and fundamental rhythmic momentum. Electric guitar usually adds fills and guitar performance effects—it might also be responsible for the primary rhythmic and tonal content. If the guitar sounds are not well-suited to the genre, the music won't sound authentic and the mix quality will be in jeopardy.

Acoustic Guitar

The acoustic guitar provides the foundation for many songs, whether they're fast or slow. Its tone must be full and clean. Acoustic guitars can either be miked or run direct into the mixer via the built-in pickup.

By far, the best acoustic guitar sounds are achieved by using a condenser microphone positioned near the front of the sound hole where the neck and body join. This is an impractical technique for most live settings because:

- The additional open mic can decrease the overall FOH system gain before feedback. The open mic can easily feed back.
- The resonance of the guitar body can quickly circulate into a howling feedback.
- During performances, the aggressive guitarist is likely to hit the microphone either with his or her hand or with the guitar itself. This results in a loud and distracting pop in the system.

Even considering these drawbacks, if the guitarist prefers to use a microphone and if he or she is willing to avoid hitting the mic, this technique can provide the best acoustic guitar sound.

The vast majority of time, the acoustic guitar pickup is used to feed the guitar signal to the mixer. Most electric acoustic pickups sound bad. They are usually tinny and brittle-sounding. They usually have an irritating edge around 2 kHz, and their sustain is abbreviated and choppy. However, they are very convenient and they don't feed back very easily.

Running Direct

The direct box is crucial to the direct acoustic sound. Its ability to respond to the complex transient audio that is provided by the acoustic guitar establishes its ability to sound good. Use a great direct box and a great guitar with a well-respected electric pickup.

I have a Gibson Artist Series acoustic guitar that I like a lot. It uses a Fishman bridge pickup that sounds very nice running direct into the mixer, and the guitar sounds incredible miked. I also have a Taylor acoustic with a bridge pickup and a built-in condenser mic inside the body. The ability to blend the bridge pickup with the internal mic can produce an excellent live sound; however, if the volume is up much, the internal condenser mic is very capable of producing feedback.

For direct boxes I've found that the Radial Engineering, Countryman, and Demeter devices work very well with most guitars.

Dynamic Control

Any time the acoustic guitar is the primary rhythmic and tonal instrument, the proper use of compression and limiting helps blend it into the desired mix space.

Set the ratio control between 3:1 and 8:1, and then adjust the threshold control for a maximum gain reduction of about 6 dB. Set the attack time to its fastest position to blend the track into the mix, or adjust for a slower attack time to accentuate the attack of the pick plucking the string. Adjust the release time to a fast setting to control more transient information or to long release times for more global gain control.

EQ

To help the acoustic guitar stand out in the mix, create a slight boost between 3 and 7 kHz. Because it contains most of the clarity and presence of the acoustic guitar sounds, a boost in this frequency range has the effect of turning up the entire track. Most of the low-frequency content on the acoustic guitar track is nonessential to the mix, so it is a good idea to use a high-pass filter to eliminate the content below 100 or 200 Hz.

The irritating upper midrange peak that is present in most direct electric acoustic guitar sounds is typically somewhere between 1.2 and 2.5 kHz. To find the exact location:

- Set up a narrow-bandwidth boost on a parametric equalizer.
- Sweep the frequency selector slowly between 1 and 3 kHz until the unwanted portion of the sound jumps up in level.
- Once the frequency is located, adjust the boost so that it's an extreme cut.

This procedure should easily identify the problem frequency and just as easily diminish it.

Effects

Acoustic guitar is a traditionally pure instrument, so effects such as chorus, flanger, and phase shifter should be used only when artistically and musically called for. Depending on the arrangement and orchestration, an effective technique utilizes a delay line between 11 and about 23 ms. Pan the original track to one side and the delay to the other to widen the acoustic guitar sound without creating an obvious effect. This technique is only effective when the FOH system is running stereo. When assigning the acoustic guitar to a mono monitor system, don't send the delayed signal because it could cause the guitar sound to be thin and hollow.

These stereo delay techniques are common in the recording world but they are also very effective in live audio mixing. They produce an accurate balance throughout the venue while providing a rich full tone and an uncluttered center image.

Electric Guitar

Because electric guitar parts are driven by the instrument sound, the guitarist determines the sonic character as he or she develops the tones for a specific song. Keep in mind that an electric guitar, like any instrument, is going to sound different because of the player. A bad player playing a great guitar won't provide as good of a sound or musical ingredient as a good player playing a bad guitar. Music comes from the soul and is a result of the touch, performance nuance, and heart of the musician.

Given a clean and simple electric track that includes dynamic processing and the perfect distortion sound, the size and scope of the guitar image can be easily shaped and molded in the FOH mix. Most modern guitar processors provide a vast number of available high-quality effects. These effects are easily written as part of the stored patch; however, when the reverberation and delay effects are part of the guitar feed to the mixer, they cannot be controlled by the sound operator during the performance.

Most of the time the guitarist can adequately adjust the effects over the course of a few performances, or, at the sound operator's discretion, the guitarist can simply eliminate delay and reverberation effects from the patches so they can be added back into the sound at the mixer. This is impractical if there aren't enough effects devices at the mixer, but in large systems this approach could be preferable.

Even if the guitarist gives control of the delay effects and reverberation to the sound operator, certain fundamental sound-shaping attributes should always be controlled by the guitarist—they are compression and distortion. These attributes are fundamental to the guitar tone and musical feel, and are best left to the guitarist and his or her arsenal of boxes and gadgets.

Creating a Huge Sound

In the rock genre, guitar sounds are often supernaturally huge. There are several factors involved in creating a huge guitar sound.

- **Natural distortion**. If the distortion sound is small and buzzy, it is very difficult to achieve a massive guitar sound. If the distortion sound is warm and smooth, it is much easier to get an aggressive tone that sounds huge while still being smooth. A buzzy distortion sound gets completely lost in most mixes.

- **Width**. Whether in stereo or surround, the use of short delays (similar to what I previously covered with the acoustic guitar) to widen the sound of the guitar across the panorama is very effective.

- **Tone**. A very broad tone with extended high and low frequencies is the fundamental basis for the development of a huge sound. A thin and piercing tone can project well in a mix, but it is difficult to transform into a huge sound; however, thin and pointed sounds work very well as part of an overall mix as long as they are crafted and positioned in the mix in a way that they are supportive to the musical vision.

- **Depth**. Long delays and reverberation can help a guitar to sound and feel like it's being listened to in a large room, but they don't guarantee that the actual guitar sound will be huge. Keep in mind that a single slapback delay defines the size of the room that the guitar is being placed in. Because sound travels roughly at the rate of one foot per millisecond, combining the guitar track with a 400-ms delay of the guitar will create the image that the guitar is being heard in a space that is 400 feet long. Effective delay lengths are often in time with the quarter or eighth note. Most modern delays provide some way to tap in the tempo with a button or foot pedal.

- **Dynamics control**. Compressing a guitar track helps keep the track precisely in a consistent mix space. A fundamental consideration of most electric guitar sounds is the compression setting. It is very important to the intimacy of the guitar sound, and it highly influences the way the player performs the part.

- **Low tuning**. There are no rules stating that the electric guitar must be tuned to standard tuning. Try tuning the lower string or two down a whole step or more. Next, build a musical part that takes advantage of the low tuning for a really larger-than-life sound.
- **Add guitarists**. There is also no rule that says you can't add more guitarists. Simply adding another guitarist to double the electric part exactly will help create a very large sound. Pan the guitars apart in the mix and bask in the wash of guitar power. Try doubling the electric part with a low-tuned guitar. This technique can easily provide a very big sound, but it is very important that the electric guitarists play with precision and that any time they're not doubling each other, the parts are well-constructed and accurately performed.

Video Example 22-3

Building the Electric Guitar Sound

Mixing Keyboards

Modern electronic keyboards provide ample control to shape nearly any sound imaginable. Spend some time with the keyboardist, listening to and adjusting sounds so that they are impressive and perfectly matched to the music. Because there is so much control within the keyboard or sound module, much of the sound shaping should be done within the device and stored within the patch. It is ideal to shape each keyboard sound so that it can be run through the mixer without EQ or effects. This method takes a little time to master, but it results in a FOH mix that seamlessly flows from song to song and texture to texture.

Piano

Mixing piano is often more about defining the mix space than it is about radically changing the instrument sound. Piano should generally sound like piano, but it needs to either occupy a wide physical section of the mixing panorama or it needs to be positioned in a specific region with another important mix ingredient, such as acoustic guitar, panned slightly across the panorama for the sake of the mix balance.

Panning the piano can help open up the center image, but it shouldn't be panned so far that there is an imbalance on one side or the other. Because the piano is typically located to one side or the other on stage, pan the FOH position to the opposite side from the physical location. In this way the acoustic output of the piano and the reinforced sound offset each other. This helps provide a wide and impressive image while opening a spot in the center image for the lead vocal and other solos.

Synth and Pad Effects

Roll off the high frequencies above 3 kHz on these synth pads to create a warm and smooth tone. The apparent volume of many synth effects and pads is very dependent on a boost or cut in the high-frequency band, so rather than turning the entire synth track up or down, it is often advantageous to simply adjust the highs.

The desired function of synth sounds is often to simply fill up the mix and to help everything blend into a cohesive musical entity. Be cautious, however, that the synth sound does not artificially boost the mix level by emitting low-frequency energy that is essentially inaudible.

Standard synthesizer sounds are frequently designed to sound great on the showroom floor—they exist to help sell units. However, the broad frequency content that sounds great when the synth is alone in a room is typically destructive to a dense musical production.

It is advisable to blend the synth sound with the mix and then to use a sweepable high-pass filter to get rid of unnecessary low-frequency information. Simply trim the low-frequency range until you actually hear the sound thin out when the mix is running, and then back the filter off slightly. Eliminating unnecessary low-frequency information helps remove confusion in the mix. If you are continually fighting to hear low-frequency instruments, it is likely that you have multiple ingredients competing for the low-frequency band. To create a clean and powerful mix, allocate frequency bands to specific mix ingredients. To blend a mix together, create increased overlapping in the frequency range of the mix tracks.

Mixing the Lead Vocal

Typically, the lead vocal track requires special attention to maintain visibility and impact. The dynamic range exhibited by most vocal tracks creates the need for either constant level adjustment or automatic control through the use of compressors or limiters. The primary focal point of the mix is almost always the lead vocal. Because of this, it has to maintain a constant space in the mix. The style of the music generally determines exactly how loud the lead vocal should be in relation to the rest of the band; however, once that's been determined, the relationship must remain constant.

In a heavy R&B or rock song, the lead vocal is often buried into the mix a little. The result of this kind of balance is rhythmic drive and punchy drums. In addition, the bass and primary harmony instruments are accentuated.

In Audio Example 22-3, I've mixed the vocal back a bit. When the volume is turned up on this kind of mix, the rhythm section is very strong and punchy.

Audio Example 22-3
The Vocal Back Mix

In Audio Example 22-4, I boosted the vocals in the mix. When the vocals are forward in the mix, it becomes very important to avoid vocal passages that are overly loud in relation to the rest of the track. Notice in this example that the vocal is uncomfortably loud.

Audio Example 22-4
The Vocals Louder

In a country or commercial pop song, the lead vocal is usually predominant in the mix, allowing the lyric and emotion of the vocal performance to be easily heard and felt by the listener.

The sound operator must always be attentive to the lead vocal level. Even if the compressor is set perfectly, there are typically several times during a song where the level might be too loud or too soft. It's up to the sound operator to make sure that the lead vocal can be heard and understood.

Compression

The lead vocal in a live sound setting should be compressed. The fact that the voice has a wide dynamic range, coupled with the fact the sound system is probably capable of reproducing at very loud volumes, dictates that some automatic level adjustment should be applied. I suggest a fairly high compression ratio, between 7:1 and 10:1, with a fast attack time and a medium release time. Adjust the threshold for gain reduction on the loudest notes only; most of the track should show no gain reduction. Your purpose here is to simply even out the volume of the track without extreme compression. In a studio mixdown, the vocal might be heavily compressed with a ratio between 2:1 and 3:1. This approach sounds good in a recording but leads to feedback problems in a live setting.

Generally speaking, the more musically dense your production, the more appropriate extreme compression is on the vocal tracks. Conversely, the more open your production, the less necessity for extreme compression.

Expander

Expanders are often used on the lead vocal track to help minimize leakage between vocal passages. The main concern when utilizing an expander is that the threshold will be set incorrectly and some parts of the lead vocal won't trigger the expander to open; however, a conscious effort to get the settings adjusted properly during a rehearsal should provide satisfactory results.

Be sure that the expander attack time is set as fast as possible and that the release time is between a quarter and a half second. Also, the range should be set to turn the mic down between 6 and infinity dB when the lead vocalist isn't singing. Carefully adjust the threshold so that the expander is triggered to open only when the singer is singing. In a particularly loud band, it will be difficult to get the threshold set correctly because of the extreme leakage. In this type of setting, the singer needs to sing loudly and stay close to the mic.

Audio Example 22-5

Expander on the Lead Vocal Track

Vocal EQ

The basic vocal sound should be full, smooth, and easy to listen to.
- Try not to create a sound that is edgy and harsh.
- There isn't much need for frequencies below 100 Hz because they are well covered by the rhythm-section instruments—it's usually best to use a high-pass filter to roll off the lows below 100 or 150 Hz.

- If there's a lack of clarity, try boosting slightly between 4 and 5 kHz.

There is a wide variety of vocal types and sounds, so the sound operator needs to assess each vocal

Simple Delay

On the vocal in Audio Example 22-6, I'll add a single slapback delay in time with the eighth-note triplet to help solidify the shuffle feel. Listen to the mix with the vocals. After a few measures, I'll add the delay. Notice how much more interesting the sound becomes. This delay is panned center with the vocal.

Audio Example 22-6

Simple Slapback on the Lead Vocal

As your equipment list grows, try setting two or three aux buses up as sends to two or three different delays. This way, they'll all be available at once, and you can pick and choose what to send to which delay and in what proportion. This technique requires restraint and musical taste to keep from overusing delay, but it's a convenient way to set up. When I set up a mix, I often set each delay to a different subdivision of the tempo; I usually use quarter-note and eighth-note delays and sometimes a sixteenth-note or triplet delay.

Video Example 22-4

Building the Lead Vocal Sound

Add Backing Vocals

Background vocals often include the same kinds of effects as the lead vocal, but to a differing degree. Usually there's more effect on the backing vocals than the lead vocal. If the lead vocal has less reverb and delay, it'll sound closer and more intimate than the backing vocals, giving it a more prominent space in mix.

These choices are purely musical. For the performance to come across as authentic and believable, the sound operator must do some stylistic homework. Listen to some highly regarded recordings in the same style as your music.

Backing vocals don't usually need to be thick in the low end, so I'll roll the lows off between 100 and 150 Hz. Audio Example 22-7 demonstrates the backing vocals on our song, soloed. After the first couple of measures, I'll cut the lows below 120 Hz.

Audio Example 22-7

The Backing Vocals Cut at 150 Hz

Listen to the backing vocals in Audio Example 22-8, along with the rhythm section and the lead vocal.

Audio Example 22-8

Add the Backing Vocals

Backing vocals should support the lead vocal without covering it. If the parts are well written, background vocals should nearly mix themselves. Well-written parts fill the holes between the lead vocal lines without distracting from the emotion and message of the lyrics; they also support the lead vocal on the key phrases of the verse or chorus while offering a musical and textural contrast. There's an art to writing good backing vocal parts that are easy to sing, make musical sense, and aren't corny.

Practice and diligence pay off quite well when it comes to these very important parts. If the drums, bass, lead vocal, and backing vocals are strong and mixed well, most of your work is done.

Compressing the Group

If you are mixing a group of backing vocals, try inserting a stereo compressor on the entire group. Develop a good blend of the backing vocals, pan them appropriately across the stereo panorama, and equalize each part for the best possible sound. When the group sounds good without compression, patch the stereo sub-mix of the vocals through a high-quality stereo compressor.

A vocal group that is compressed together typically sounds much more blended and unified. Deciding whether to compress the vocal group is a creative decision that affects the musical feel. There are many applications in which the backing vocals should maintain a greater degree of transparency and individuality, so compression might not be appropriate. However, when you want to develop a backing vocal sound that is very tight, consistent, blended, and supportive, compression is very effective. Compressing the backing vocal mics could decrease the amount of gain before feedback, so be careful not to compress too heavily. A typical backing vocal compressor would be set in the following way:

- Set the ratio between 4:1 and 7:1.
- Select a medium-fast attack and release.
- Adjust the threshold for a maximum gain reduction of around 6 dB.
- Make up the lost gain with the output control.

Panning

Let's start with a very simple arrangement of a rhythm section with a single melody line. When all instruments are panned to the center, it's possible for each part to be heard, but the overall sound isn't very natural or spatially interesting. Many sound operators prefer to setup the FOH system in a mono configuration because it's really the only way to guarantee that the entire house gets the same mix. Audio Example 22-9 demonstrates the mono reference point.

Audio Example 22-9

Mono Mix

Using simple panning, you can obviously build a more interesting audio picture. Here's the same piece of music but, as the music plays, I'll pan the guitar and keyboards apart in the left-to-right spectrum. This is the beginning—but definitely not the culmination—of a stereo image. Notice how much easier it is to hear everything when these two instruments claim a different space in the soundscape.

Even though in a live sound application it isn't workable to pan the guitar and keys that far apart in the mix, it's instructive to hear the difference it makes when they're panned apart. Most of the time it is safest to build a fundamentally mono mix with stereo reverberation and delay effects. The stereo effects returns create a dimensional width that doesn't confuse the instrument placement or compete with the instrumental or vocal sounds.

Reverberation

Listen to the reverberation on the snare drum in Audio Example 22-11; I've selected a long decay time. Listen intently to the effect it creates. Listen to the changes that occur as the reverb fades away. Try to imagine a real room that would sound like this. How big would it be? What kind of surfaces would be around?

Audio Example 22-11 was an example of stereo reverberation. Stereo reverb is designed to have slightly different combinations of reflections and tonal character from left to right. A stereo reverb is very helpful in opening up the stereo spectrum and leaving more space for the music. Audio Example 22-12 demonstrates the same reverb sound as the previous example, but this time it's in mono. At the end of the Example, I'll open back up to stereo. Mono reverberation can be very useful when placing an instrument or voice within one area of a large stereo image, but generally stereo reverberation is preferable.

Combining Wet and Dry

If you're designing textures with multiple instruments performing the same musical part, it isn't usually necessary to add reverb to all pieces of the texture. In fact, the overall image can change substantially depending on what part of the texture you send to the reverb. If you have similar timbres in the different sounds you're layering, try adding reverb only to certain sounds.

As a practical example, it's common on the chorus of many songs to hear the lead vocal singing along with the background vocals. When this scenario occurs with vocals or other textures, you need to determine which part of the texture you want to be in front of the other parts. With the vocal chorus section, try adding reverb to the background vocals, but leaving the lead vocal totally dry. The fact that the background vocals have a similar timbre to the lead vocal will give the impression that there is really reverberation on all of the vocals, including the lead part, because the other parts will activate the reflections. But the overall image will be that the lead vocal seems more present—further forward or closer to the listener—in the mix. This technique is commonly used in the recording industry.

Delay

With the original instrument sound panned to one side of the stereo spectrum and a short delay—below about 35 ms—panned to the opposite side of the spectrum, the originally mono image spreads across the panorama, leaving more room to hear the rest of the instruments. This technique widens the stereo image, but almost more importantly, it makes room in the mix to hear other instruments and their images in the stereo soundfield. As a quick review, listen to the guitar in Audio Example 22-13. It starts mono in the center, and then I pan it left and turn up a 17-ms delay on the right.

Audio Example 22-13

Creating a Stereo Guitar Sound

Mixing the Event: Focus, Focus, Focus!

The competent sound operator is constantly focusing on how to make the mix better. While riding vocal levels, he or she is assessing and reassessing all the mix ingredients. Mixing is not a passive activity. Even when the mix is going well, the sound operator should be constantly on the move, looking for mix ingredients that need help.

Blending the Drums and Bass

First, take a look at blending the drums and percussion with the bass guitar. The drum sound is so fundamental to the sonic signature of any production that I could never overstate the importance of mixing and blending them in a musically effective and powerful way.

Build the drum sound from the bottom up. Start with the kick and add the snare, overheads, and toms.

The exact kick sound you build is very genre-dependent. Some genres depend on the kick sound for the tonal foundation of the mix, in which case there should be substantial low-frequency content and not many conflicting low midrange frequencies. In these types of mixes the bass guitar typically contains more lows, mids, and highs, but not as many deep lows, diminishing or eliminating any tonal conflict with the kick sound.

Listen to Audio Example 22-14. Pay close attention as I adjust the kick-drum tone from a deep and rich sound to very pointed midrange sound.

Audio Example 22-14

Kick Tone

Now listen as I build the drum-set sound and integrate a compatible bass-guitar sound.

Audio Example 22-15

Building the Drum and Bass Mix

Video Example 22-5

Building the Drum and Bass Mix

Blending Electric Guitars

Sometimes the best guitar sounds are clean and simple. Other times several guitar sounds combine to make one massive sound. Listen to the two different types of guitar sounds in the following Audio Examples. Each is very appropriate for the type of music it serves.

When blending multiple electric guitars to form one sound, evaluate each part for its strengths and weaknesses, and then highlight the best aspects of each. If one part has a smooth high-frequency content that's easy to listen to and fairly unspectacular lows and mids, try a high-pass filter to filter out everything other than the highs. This then would obviously need to be combined with the lows and mids from other tracks. Using this concept, you can create unique and powerful sounds that add to your production.

In the following Audio Examples, I'll demonstrate and specify each sound and how I equalize, pan, and blend the sounds together.

First is a guitar track from a band I produced called Glimpse. We spent quite a bit of time miking cabinets and combining different sounds. We had several amps available (Marshall, Mesa, Matchless, Fender, Line 6, and so on, and eventually, throughout the project, we used them all. We also combined the live amps with some of the Line 6 direct outputs of several modeled amplifiers.

Audio Example 22-16
Several Amps Combined to One Sound

Second is a guitar track from a solo artist I produced, Faith Ecklund. This is still a great guitar sound, but we simply used a single guitar track, processed and blended for a big sound.

Audio Example 22-17
One Guitar Track Blended into the Mix

Blending Acoustic Guitars

When faced with production decisions, the temptation is usually to build a bigger sound by adding more and more instruments. Often, the more tracks you add, the smaller your mix sounds. In fact, many tracks fighting for the listener's attention typically result in diffusion, rather than a concentration of focus.

A simpler production and mixing approach often creates a more impressive sound than a complex and busy approach.

Blending the Mix

It's important that your mix blend together so that the rhythmic component is tight and punchy while the harmonic component is clean and precise. Avoid the temptation to set up a wishy-washy mix—one in which there are no real lead instruments or voices.

A good mix has power—it commands the listener's attention throughout the arrangement. In a great mix, the backing vocals are nearly the same volume as the lead vocals, yet they are supporting the lead.

Much of the sound operator's success in the creation of a great-sounding mix is due to the musical arrangement. Great musical parts are easy to mix. Musical ingredients that fight against each other create confusion in the mix—confusion that is difficult to get rid of.

Maintain a very high standard throughout the mixing process. Take what you've learned in this book and apply it to your own unique situation. Practice and learn! Before you know it, you'll be confident and proficient—people will seek you out to help with their mixes. Have fun and enjoy the ride! God bless.

Video Example 22-6

Building the Backing Vocal Sound

DVD-ROM AND ONLINE MEDIA LIST

SELECTED REFERENCES

General Texts

Colloms, Martin. *High Performance Loudspeakers*. New York: John Wiley & Sons, 1978.

Duncan, Ben. *The Live Sound Manual*. London: Backbeat Books, 2002.

Eargle, John. *Sound Recording*. New York: Van Nostrand Reinhold Company, 1980.

Eiche, Jon F. *Guide to Sound Systems for Worship*. California: Hal Leonard Publishing Corporation, 1990.

Everest, F. Alton. *Sound Studio Construction on a Budget*. New York: McGraw-Hill, 1997.

Huber, David Miles and Runstein, Robert E. *Modern Recording Techniques: Fourth Edition*. Massachusetts: Focal Press, 1995.

Stark, Scott Hunter. *Live Sound Reinforcement*. California: Mix Books, 2003

White, Glenn D. *The Audio Dictionary*. Washington: University of Washington Press, 1987.

Yakabuski, Jim. *Professional Sound Reinforcement Techniques*. California: Mix Books, 2001.

Websites & Blogs

Bogatin, Dr. Eric. "What Is Inductance?" *Printed Circuit Design & Fab*. July 1, 2007. http://www.pcdandf.com/cms/magazine/95/3614

Bohn, Dennis. "Setting Sound System Level Controls," *RaneNote*. http://www.rane.com/note135.html

DellaSala, Gene. "Audio Cables—Fact & Fiction Revealed." An interview with Gene DellaSala from *Audioholics.com*. November 3, 2003. http://forum.ecoustics.com/bbs/messages/5/6528.html

Engdahl, Tomi, "Ground Lift," ePanorama,net, http://www.epanorama.net/documents/groundloop/groundlift.html

Engdahl, Tomi. "Ground Loop Problems and How to Get Rid of Them." ePanorama,net. http://www.epanorama.net/documents/groundloop/index.html

Furchgott, Roy. "A Spat Among Audiophiles Over High-End Speaker Wire." *The New York Times*. Technology | Circuits, December 23, 1999. http://www.nytimes.com/library/tech/99/12/circuits/articles/23down.html

Sampson, Michael J.: NASA Official. "Wire and Cabling." *Workmanship Standards*. May 26, 2009. http://nepp.nasa.gov/index.cfm/14257

Sampson, Michael J.: NASA Official, "Wire and Cabling." *Workmanship Problems Pictorial Reference*. December 2, 2008. http://nepp.nasa.gov/index.cfm/5575

United States Department of Commerce—John T. Connor, Sectretary, U.S. National Bureau of Standards. "Table 5. Wire Table, Standard Annealed Copper." *Copper*

Wire Tables. National Bureau of Standards Handbook 100. Issued: February 21, 1966. http://library.bldrdoc.gov/docs/nbshb100. pdf

"Common Wire Gauges Based on Diameter in Inches." Last revised April 7, 2008. http://www.sizes.com/materls/wire.htm

"What is Impedance, Anyway?" BlueJeansCable.com. http://bluejeanscable.com/articles/impedance.htm

Recommended Organizations

The Audio Engineering Society. http://www.aes.org

The National Association of Broadcasters. http://www.nab.org

The Recording Academy. http://www.grammy.com

The Society of Broadcast Engineers. http://www.sbe.org

The Society of Professional Audio Recording Services. http://www.spars.com

Recommended Publications (Print and Online)

Billboard Magazine

Church Production Magazine

Christian Musician

EQ Magazine

Front of House Magazine (FOH)

Journal of the Audio Engineering Society

Live Sound International

Mix Magazine

Mixmag

Pro Sound News

Professional Sound Magazine

Sound on Sound

Tape Op Magazine

Voice•Council

Worship Musician

Education

There is an excellent list of college-level audio educational programs at http://www.aes.org/education/directory/

INDEX